FreeCAD 0.19 Learn by doing

Tutorial Books

FreeCAD 0.19 Learn by doing

For resource files contact us at
freecadtuts@gmail.com

Table of Contents

Introduction

Welcome to *FreeCAD 0.19 Learn by doing* book. This book is written to assist students, designers, and engineering professionals in designing 3D models. It covers the essential features and functionalities of FreeCAD using relevant tutorials and exercises.

Topics covered in this Book

- Chapter 1, "Getting Started with FreeCAD," gives an introduction to FreeCAD. The user interface and terminology are discussed in this chapter.
- Chapter 2, "Sketch Techniques," explores the sketching commands in FreeCAD. You will learn to create parametric sketches.
- Chapter 3, "Pad and Revolve features," teaches you to create basic 3D geometry using the Pad and Revolve commands.
- Chapter 4, "Dress-up Features," covers the features which can be created without using sketches.
- Chapter 5, "Patterned Geometry," explores the commands to create patterned and mirrored geometry.
- Chapter 6, "Sweep Features," covers the commands to create swept and helical features.
- Chapter 7, "Loft Features," covers the Loft command and its core features.
- Chapter 8, "Modifying Parts," explores the commands and techniques to modify the part geometry.
- Chapter 9, "Assemblies," helps you to create assemblies using the bottom-up and top-down design approaches.
- Chapter 10, "Drawings," covers how to create 2D drawings from 3D parts and assemblies.

Chapter 1: Getting Started with FreeCAD 0.19

Introduction to FreeCAD

FreeCAD is an open source CAD application. It is a parametric and feature-based system that allows you to create 3D parts, assemblies, and 2D drawings. The design process in FreeCAD is shown below.

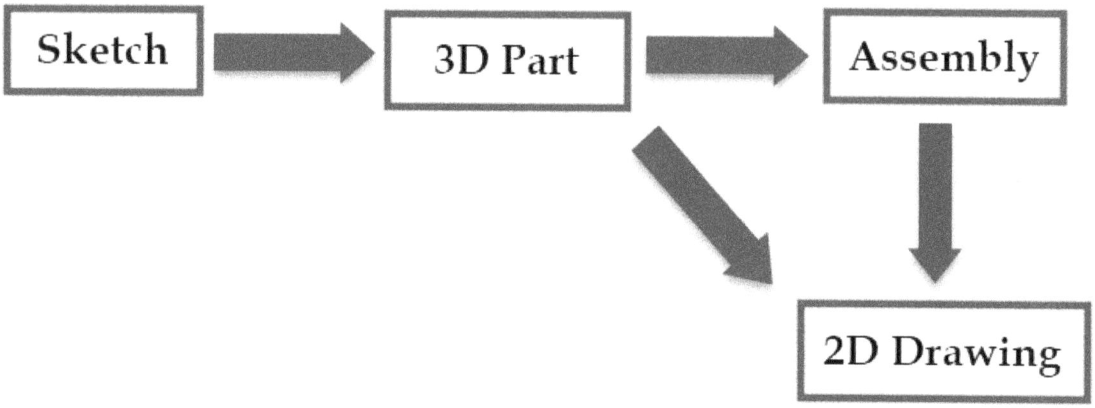

In FreeCAD, everything is controlled by parameters or constraints. For example, if you want to change the position of the hole shown in the figure, you need to change the constraint that controls its position.

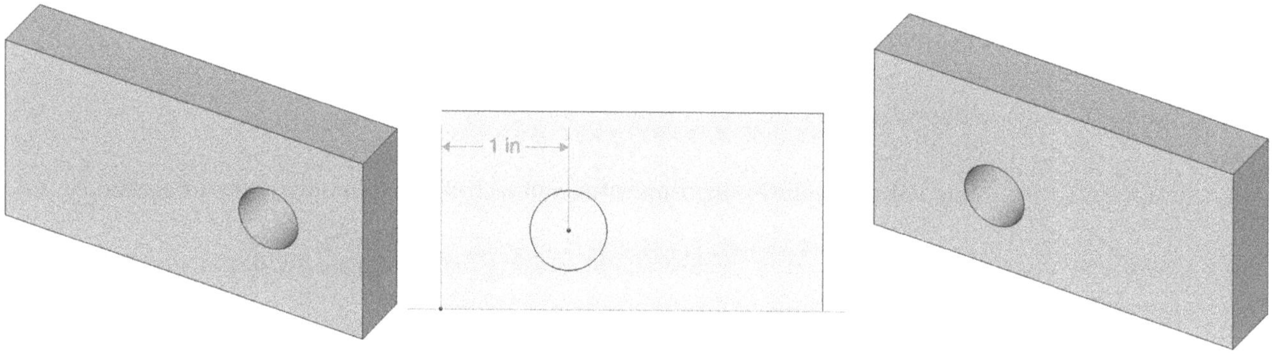

The parameters and constraints that you set up allow you to have control over the design intent. The design intent describes the way your 3D model will behave when you apply constraints to it. For example, if you want to position the hole at the center of the block, one way is to add constraints between the hole and the adjacent edges. However, when you change the size of the block, the hole will not be at the center.

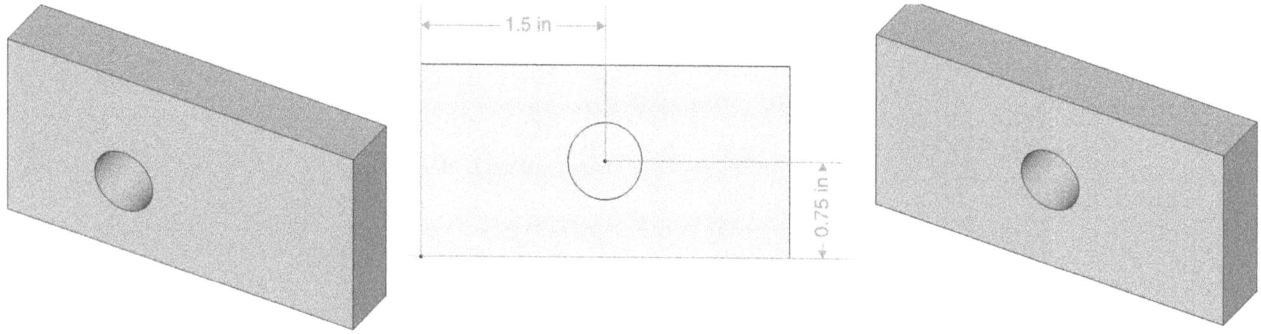

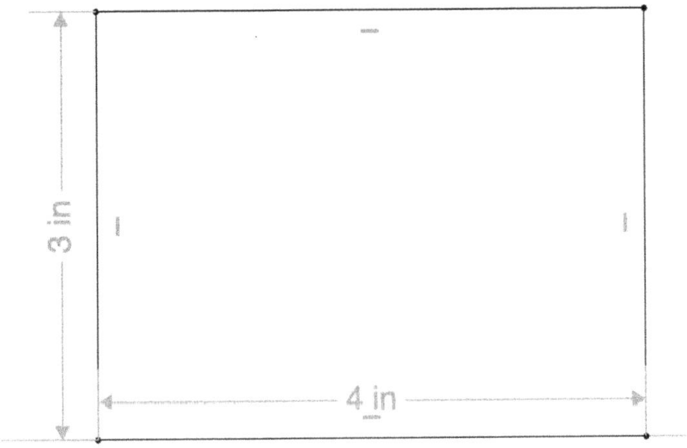

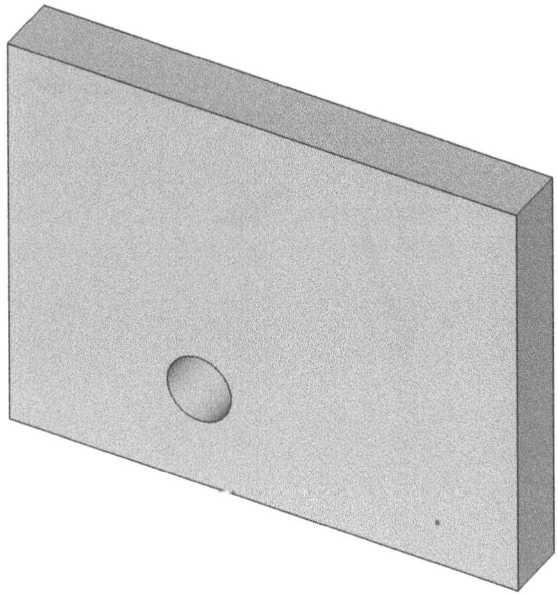

You can make the hole to be at the center, even if the size of the block changes. To do this, right click on the sketch used to create the hole and select **Edit sketch**. Next, delete the constraints and create a diagonal construction line. Apply the **Symmetrical** constraint between the hole centerpoint and the endpoints of the construction line. Next, click **Close** on the **Combo View** panel.

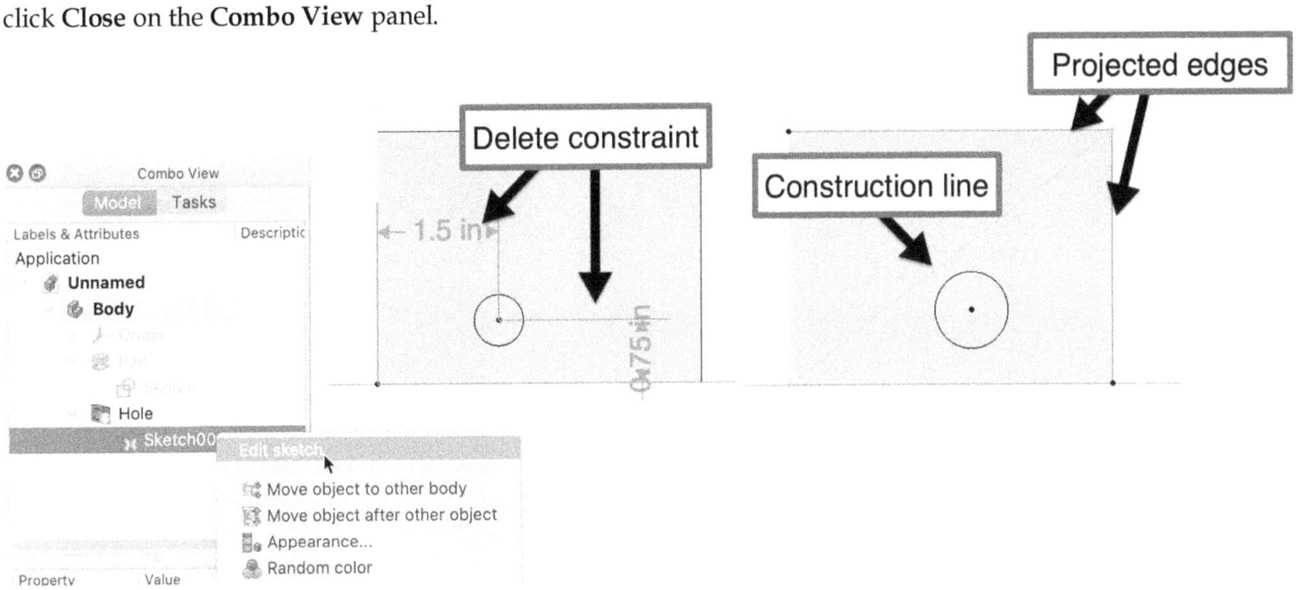

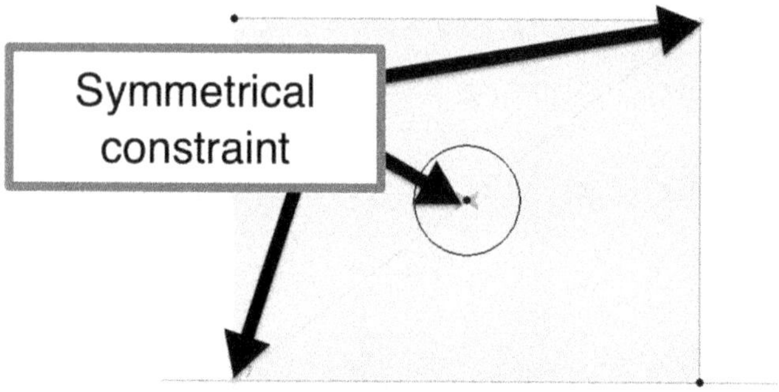

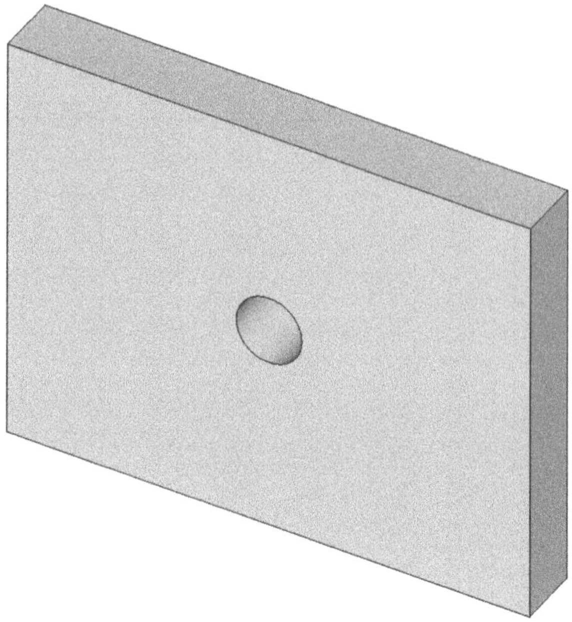

Now, even if you change the size of the block, the hole will always remain at the center.

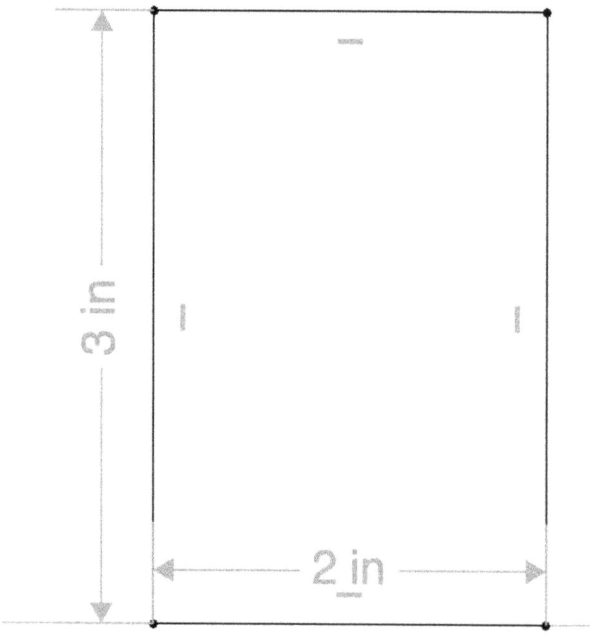

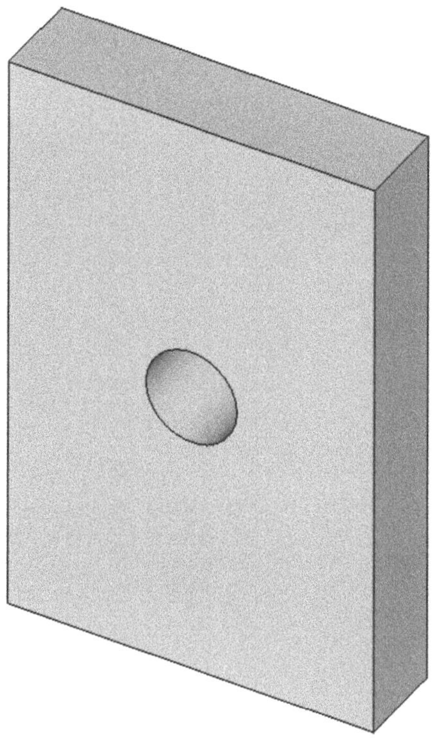

Starting FreeCAD

To start **FreeCAD**, click the **FreeCAD 0.19** icon on your Desktop. On the menu bar, click **File > New**; a new design file will appear on the screen. You can change the working units of the file. To do this, click the **Edit > Preferences** on the menu bar. Next, click the **Units** tab and set the **Unit System** and click **OK**.

User Interface

The following image shows the **FreeCAD** application window.

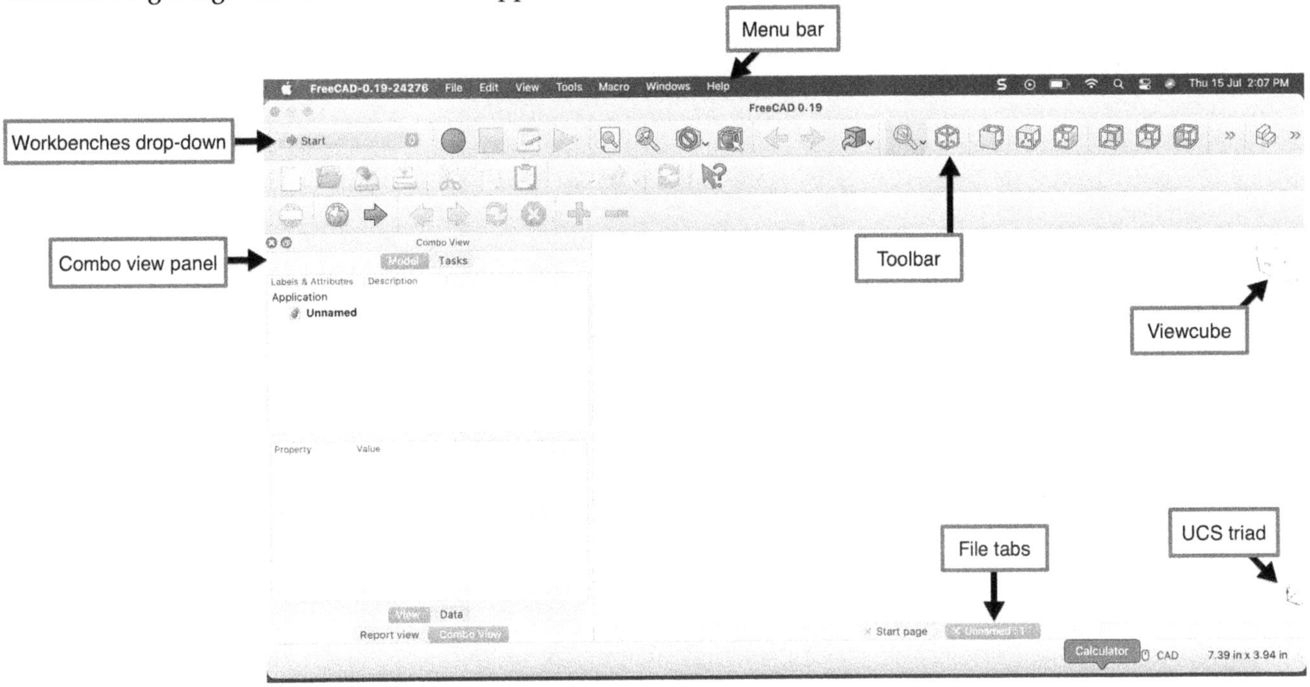

Graphics window

The Graphics window is the blank space located below the toolbars area. You can draw sketches and create 3D geometry in the Graphics window.

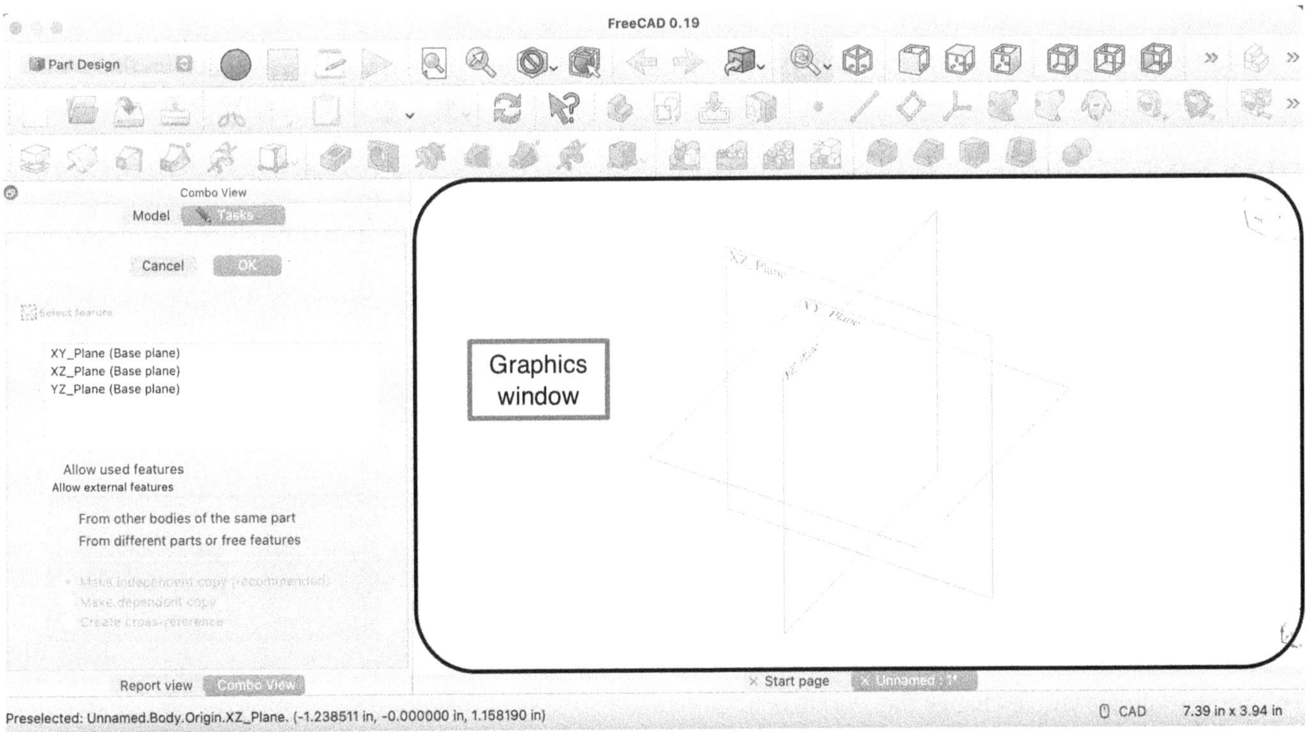

File Menu

The **File** menu appears when you click on the **File** option located at the top left corner of the window. The **File** menu has a list of open menus. You can see a list of recently opened documents under the **Recent Files** sub-menu.

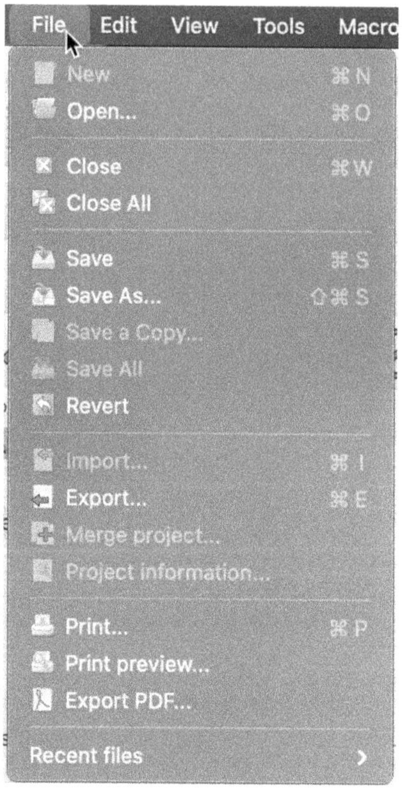

Toolbars

A toolbar is a set of tools, which help you to perform various operations. Various toolbars available in the FreeCAD are:

File Toolbar

This toolbar contains tools such as **New**, **Open**, **Save,** and so on.

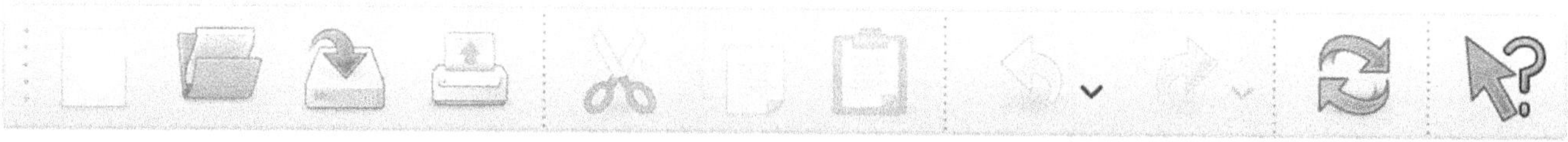

View toolbar

This toolbar has the tools to modify the display and orientation of the model.

Workbench Toolbar

This toolbar allows you to change the workbench.

Macro

This toolbar has tools to create and execute macros.

Structure

It has tools to create or open part files.

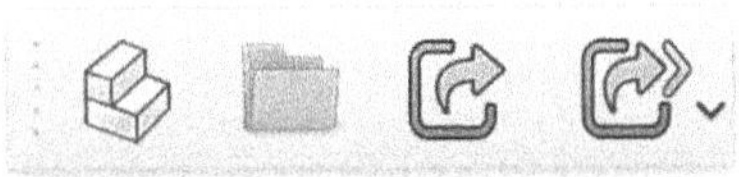

Navigation

It has tools to open a website in FreeCAD.

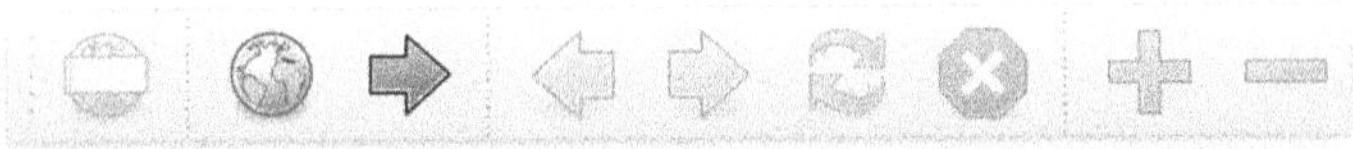

Sketcher Toolbars

Sketcher

This toolbar has tools to start or exit a sketch.

Sketcher geometries

This toolbar has tools to create sketch elements.

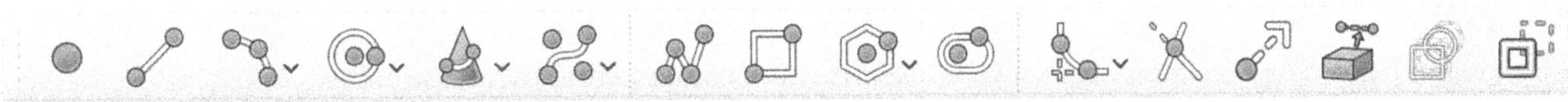

Sketcher constraints

This toolbar has tools to apply constraints between sketch elements.

Sketcher B-spline tools

This toolbar has tools to create and edit B-splines.

Sketcher Virtual Space

This toolbar helps you to hide or show constraints.

Sketch Tools

This toolbar has various selection tools and options that aid you in creating sketch elements very fast.

Part Design Toolbars

Part Design Helper

This toolbar has tools to create new bodies, datum points, axis, datum planes, and clones. You can also create or leave a sketch.

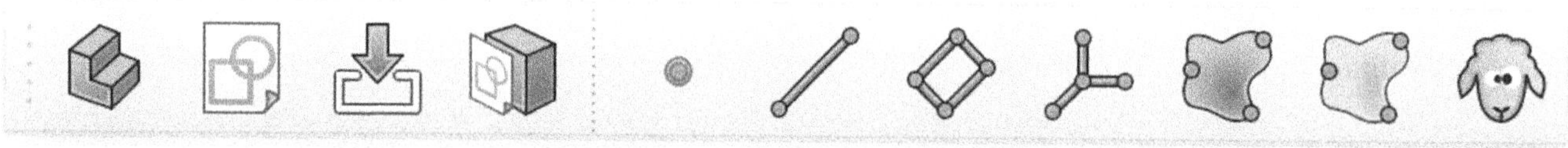

Part Design Modeling

This toolbar has commands to create solid features based on the sketch geometry.

A2plus Toolbars

A2p_Part

This toolbar has tools to create components or insert existing components into an assembly.

A2p_Constraint

This toolbar has tools to apply constraints between components.

A2p_Solver

This toolbar has tools to change the direction of the constraints, solve constraints, and so on.

A2p_view

This toolbar has tools to show the labels of degrees of freedom, toggle transparency of the assembly, and so on.

A2p_misc

This toolbar has tools to repair the tree view, export the parts list, and so on.

TechDraw Toolbars

TechDraw Views

This toolbar has tools to generate standard views of a 3D geometry.

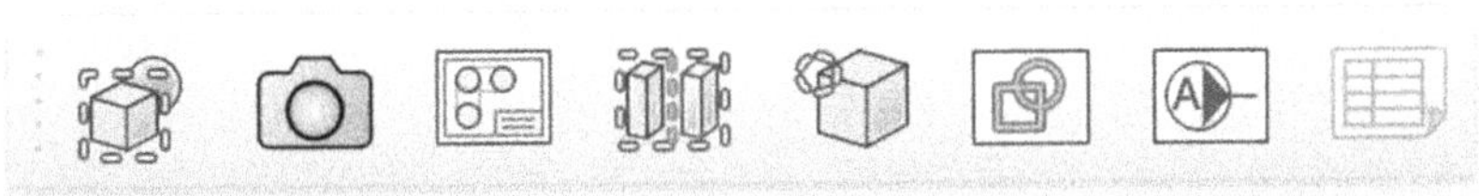

TechDraw Clips

It has tools to add or remove clips to the drawing sheet.

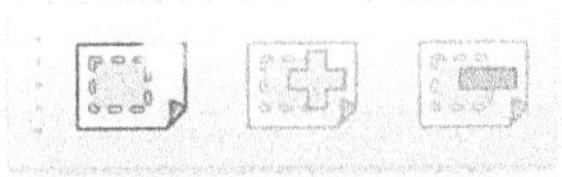

TechDraw Pages

The tools on this toolbar help you to add a new page and refresh the drawing page.

TechDraw Dimensions

The tools on this toolbar help you to add dimensions to the drawing views.

TechDraw File Access

This toolbar helps you to export the drawing page to the SVG format.

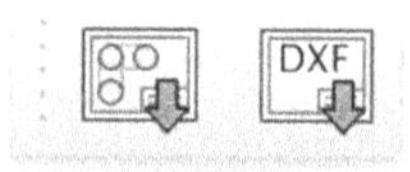

TechDraw Decoration
The tools on this toolbar help you to change the hatch patterns and insert images.

Some toolbars are not visible by default. To display a particular toolbar, click **View > Toolbars** on the menu bar. Next, select the toolbar from the sub-menu.

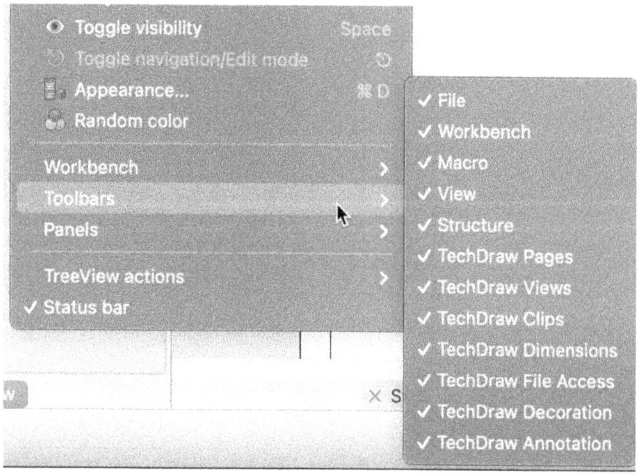

Menu
The menu is located on the top. It has various options (menu titles). When you click on a menu title, a drop-down appears. Select an option from this drop-down.

Status bar
The status bar is available below the graphics window. It shows the current status.

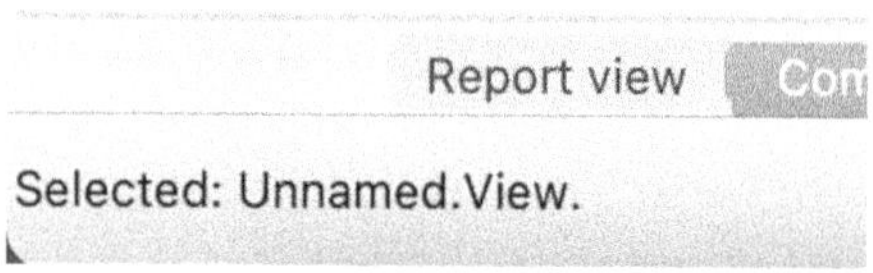

Combo View panel
The **Combo View** panel is located on the left side. It has two tabs: **Model** and **Task**.

Model tab

It contains the list of operations carried while constructing a part.

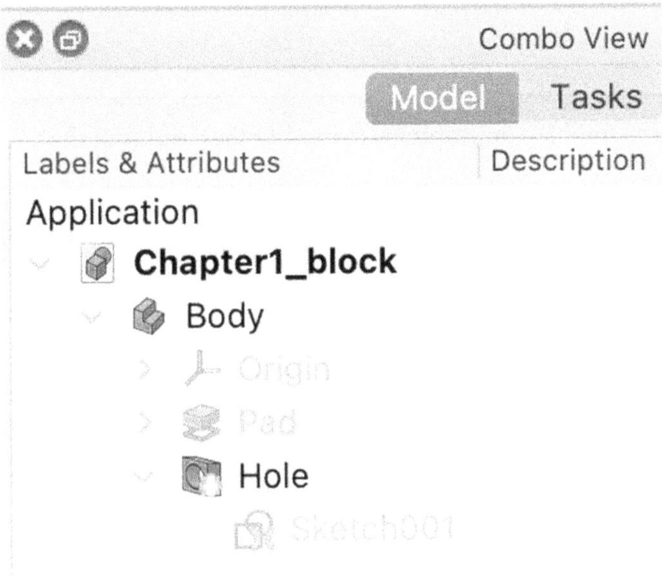

Tasks tab

When you execute any command in FreeCAD, the options related to it appear in the **Tasks** tab. The **Tasks** tab has various options. The following figure shows various components of this tab.

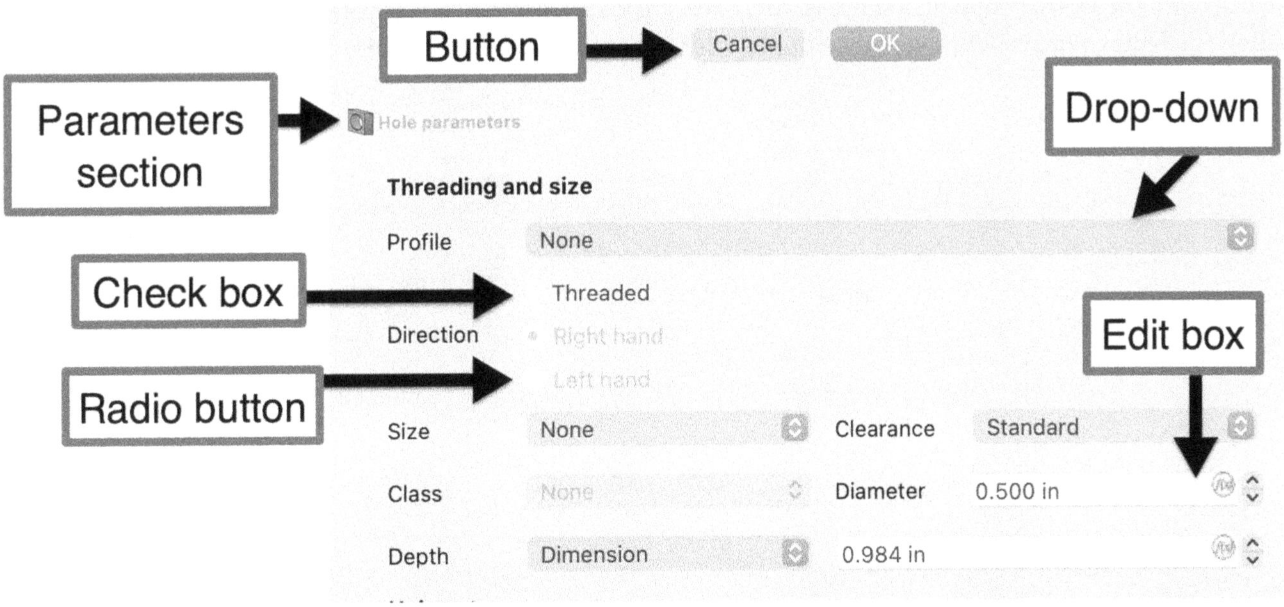

Mouse Functions

Various functions of the mouse buttons are:

Left Mouse button (MB1)

When you double-click the left mouse button (MB1) on an object, the parameters section related to the object appears in the Tasks tab. Now, you can edit the parameters of the objects.

Middle Mouse button (MB2)

Click this button to pan the view.

Right Mouse button (MB3)

Click this button to open the shortcut menu. The shortcut menu has some options to modify the display of the model.

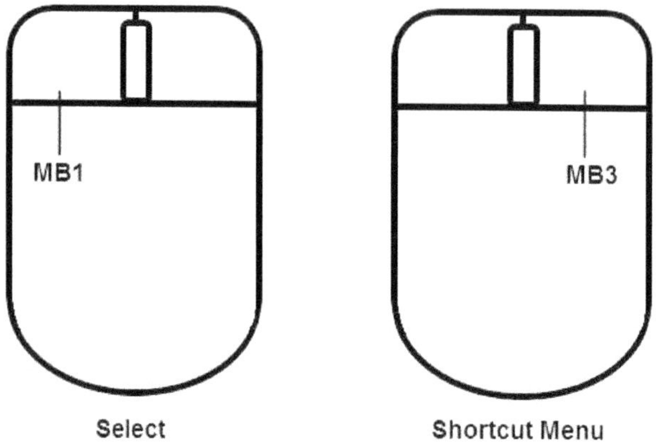

Edit Background

To change the background color of the window, click **Edit > Preferences** on the menu bar. On the **Preferences** dialog, click the **Display** option on the left side. Next, click the **Colors** tab and select **Background color > Simple color**. Click the color swatch next to the **Simple color** option. Select the white color from the **Select Color** dialog and click **OK**.

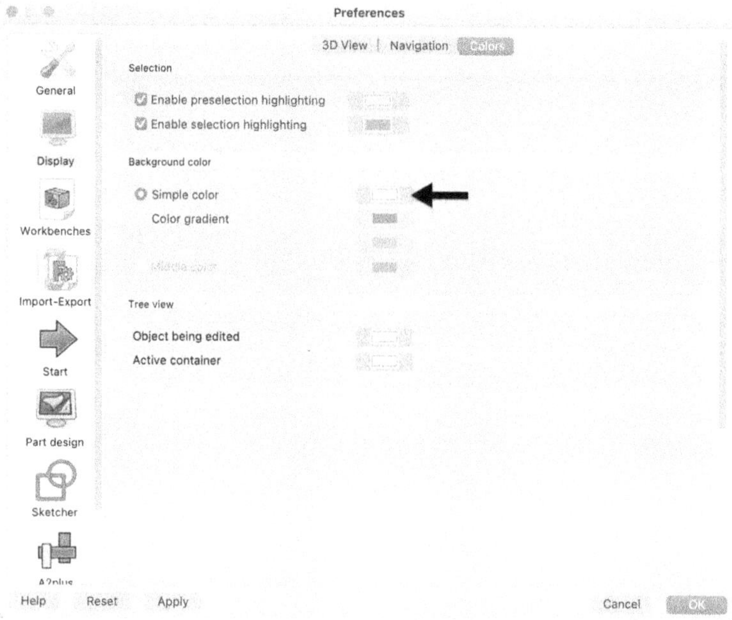

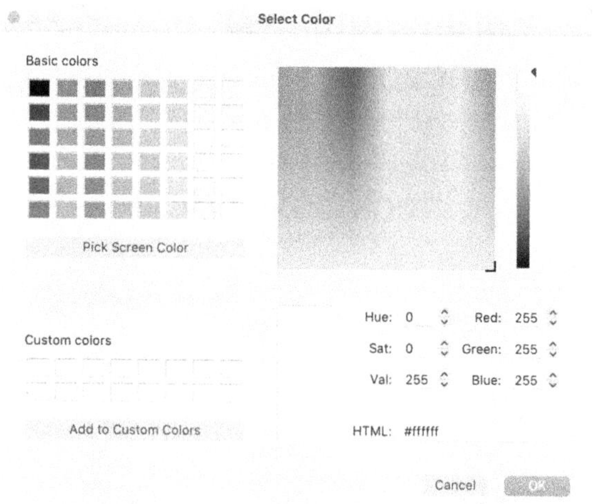

Click the **Sketcher** option on the left side of the Preferences dialog. Next, click the **Colors** tab and change the **Default edge color**, **Default vertex color**, **Edit edge color**, and **Cursor crosshair color** to black. Click **OK** to apply the changes.

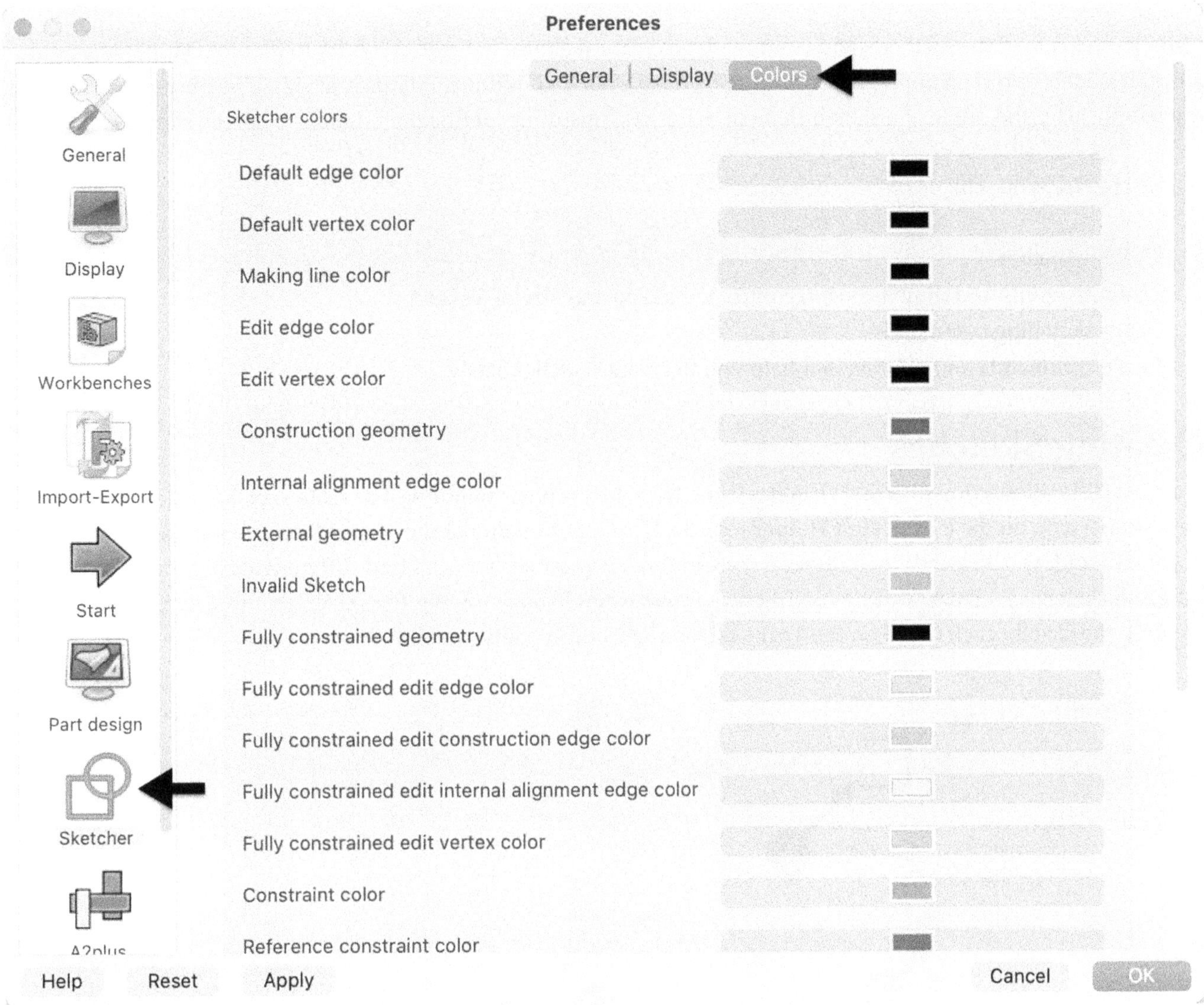

FreeCAD Help

FreeCAD offers you the help system that goes beyond basic command definition. You can access FreeCAD help by using any of the following methods:

- Press the **F1** key.
- Click **Help > Help** on the menu bar.

Questions

1. Explain how to hide or display toolbars.
2. Give one example of where you would establish a constraint between a part's features.
3. List any two procedures to access FreeCAD Help.
4. How to change the background color of the graphics window?
5. How is FreeCAD a parametric modeling application?

Chapter 2: Sketch Techniques

This chapter covers the methods and commands to create sketches in FreeCAD 0.19. In FreeCAD 0.19, you create a rough sketch, and then apply dimensions and constraints that define its shape and size. The dimensions define the length, size, and angle of a sketch element, whereas constraints define the relations between sketch elements.

In this chapter, you will:

- Create sketches
- Use constraints and dimensions to control the shape and size of a sketch
- Learn sketching commands
- Learn commands and options that help you to create sketches easily

Creating Sketches

FreeCAD 0.19 allows you to create sketches directly in the graphics window. To create sketches, click the **Create a new sketch** icon on the **Part design Helper** toolbar. Next, click on any of the planes displayed at the center of the graphics window. Click the **OK** button on the **Combo View** panel. You can find different sketch commands on different sketcher toolbars. You can use these commands and start drawing the sketch on the selected plane. After creating the sketch, click **Close** on the **Tasks** tab on the **Combo View** panel to finish the sketch.

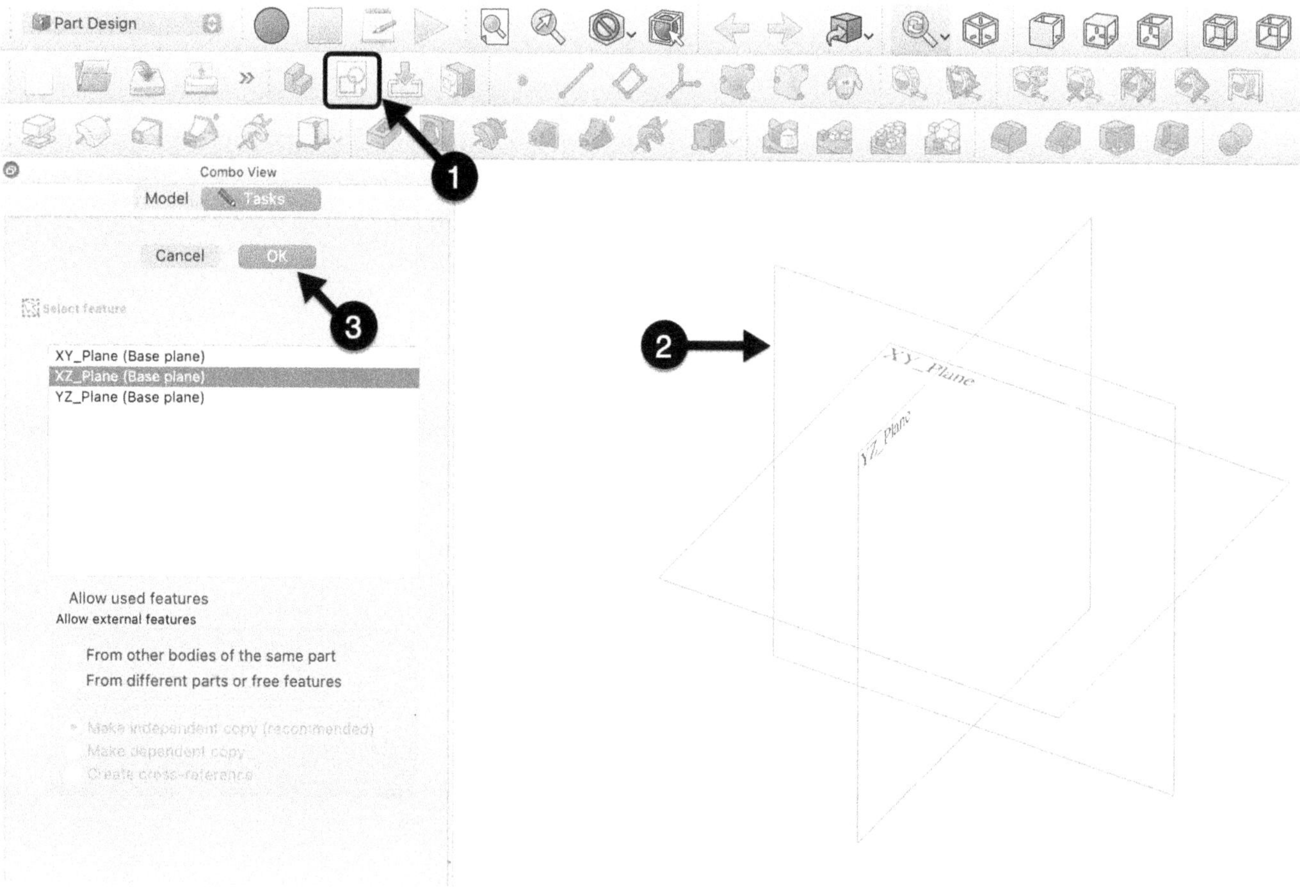

Sketch Commands

FreeCAD 0.19 provides you with a set of commands to create sketches. These commands are available on **Sketch** tab of the menu bar (or) the **Sketcher geometries** toolbar.

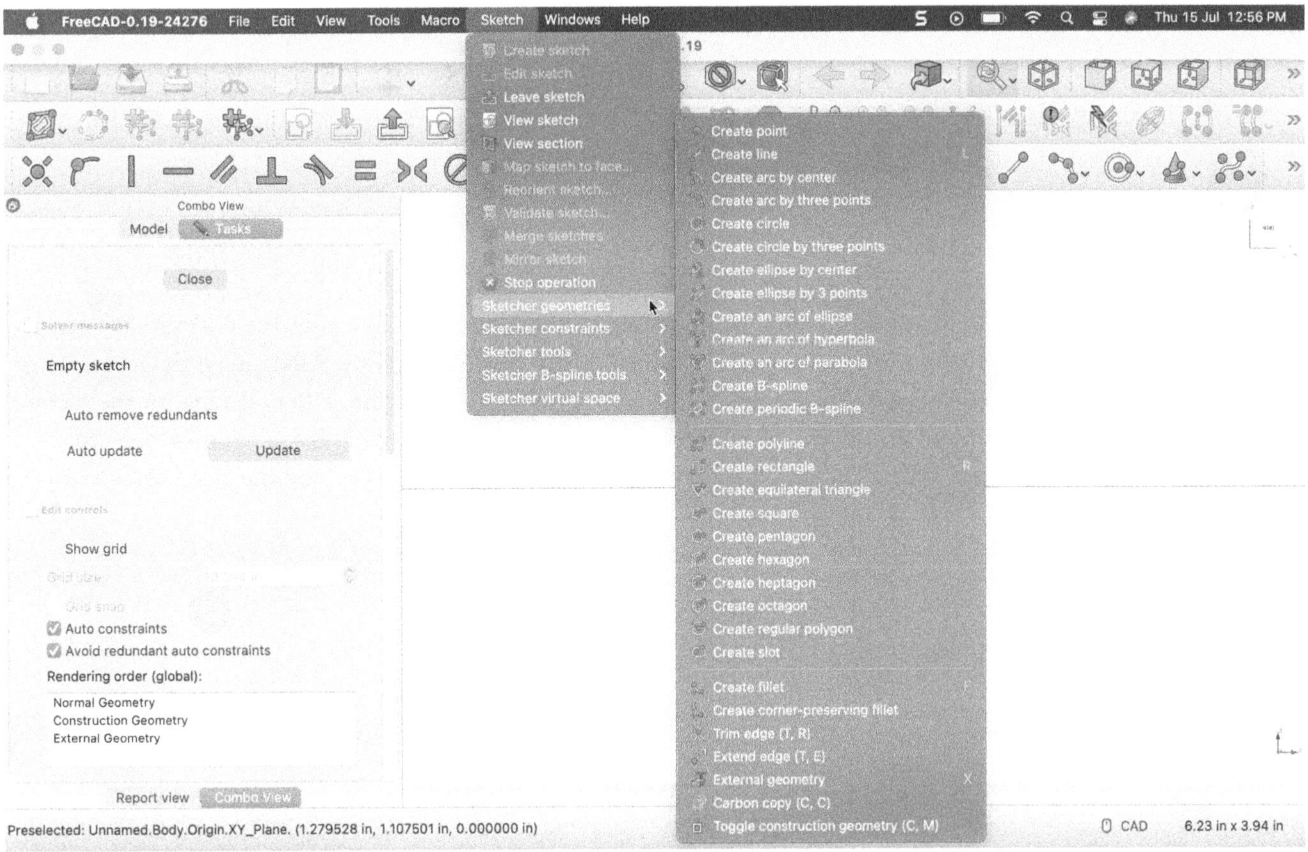

The Line command

This is the most commonly used command while creating a sketch. To activate this command, click the **Line** 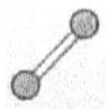 icon on the **Sketcher geometries** toolbar (or) click **Sketch > Sketcher geometries > Create Line** on the menu bar. To create a line, click on the graphics window, move the pointer and click again. To create a horizontal line, specify the start point of the line and move the pointer horizontally; the Horizontal constraint glyph appears below the line. Click to create a horizontal line with the Horizontal constraint applied to it. You will learn about constraints later in this chapter. Likewise, you can create a vertical line by moving the pointer vertically and clicking.

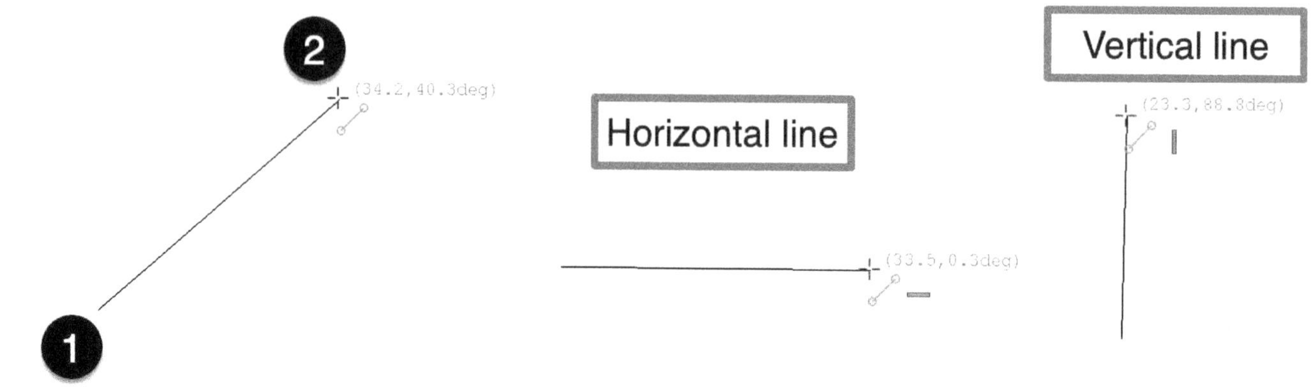

The Polyline command

The Polyline command is used to create continuous lines. To activate this command, click the **Polyline** icon on the **Sketcher geometries** toolbar (or) click **Sketch > Sketcher geometries > Create Polyline** on the menu bar. To create a polyline, click on the graphics window, move the pointer and click again. After clicking for the second time, you can see that an endpoint is added, and another line segment is started. This is a convenient way to create a chain of lines. Continue to click to add more line segments. You can right-click on the graphics window to end the line chain. Press ESC if you want to deactivate the **Line** command.

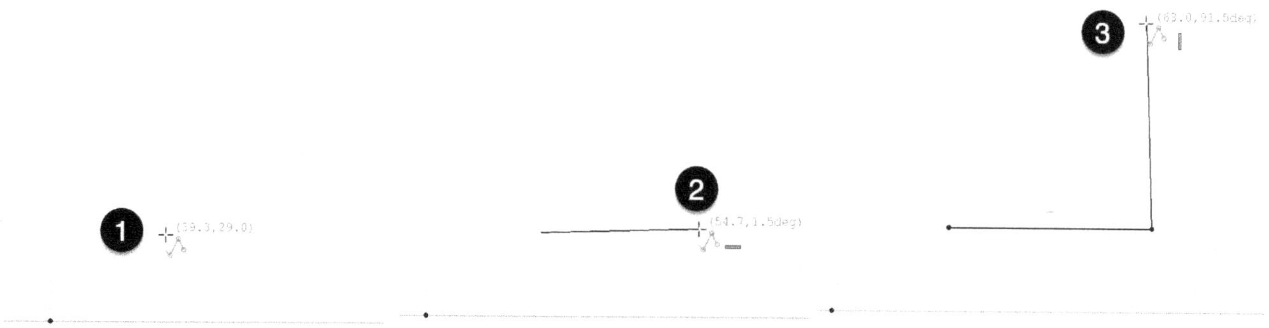

Creating Arcs

FreeCAD 0.19 allows you to create arcs using two commands: **Center and end points** and **End points and rim point.**

The Center and end points command

This command creates an arc by defining its center, start, and end. Activate the **Center and end points** command (Click **Arc** drop-down > **Center and end points** on the **Sketcher geometries** toolbar). Click to define the center point. Next, move the pointer and click to define the start point of the arc. Move the pointer and notice that an arc is drawn from the start point. Once the arc appears the way you want, click to define its endpoint.

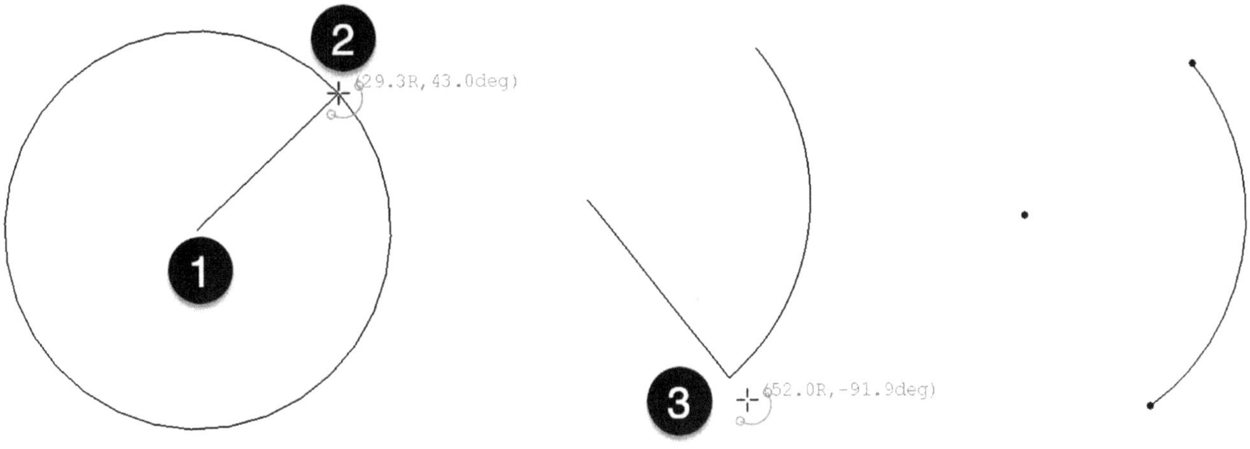

The End points and rim point command

This command creates an arc by defining its start, end, and radius. Click **Arc** drop-down > **End points and rim point** on the **Sketcher geometries** toolbar (or) click **Sketch > Sketch geometries > Create arc by three points**. In the graphics window, click to define the start point of the arc. Next, move the pointer and click again to define the endpoint of the arc. After defining the start and end of the arc, you need to define the size and position of the arc. To do this, move the pointer and click to define the radius and position of the arc.

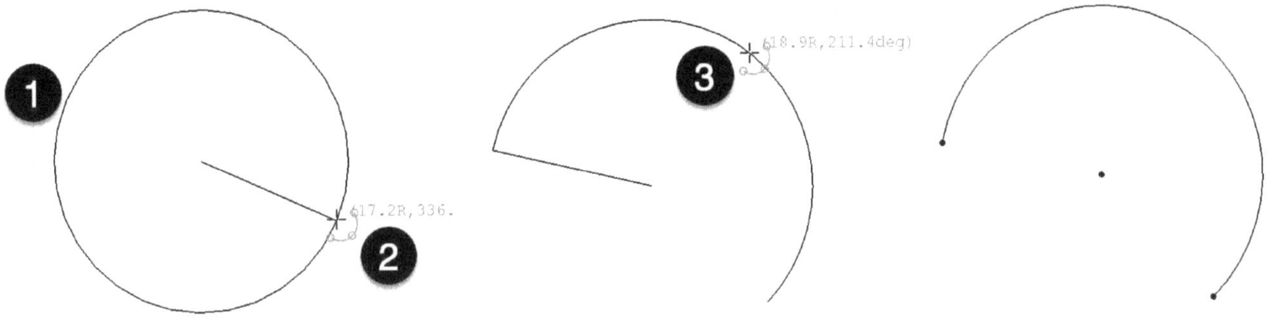

Creating Circles

FreeCAD 0.19 allows you to create circles using two commands: **Center and rim point** and **3 rim point**.

Center and rim point

This is the most common way to draw a circle. Click **Circle** drop-down > **Center and rim point** on the **Sketcher geometries** toolbar (or) click **Sketch > Sketcher geometries > Create circle** on the menu bar. Click to define the center point of the circle. Drag the pointer, and then click again to define the diameter of the circle.

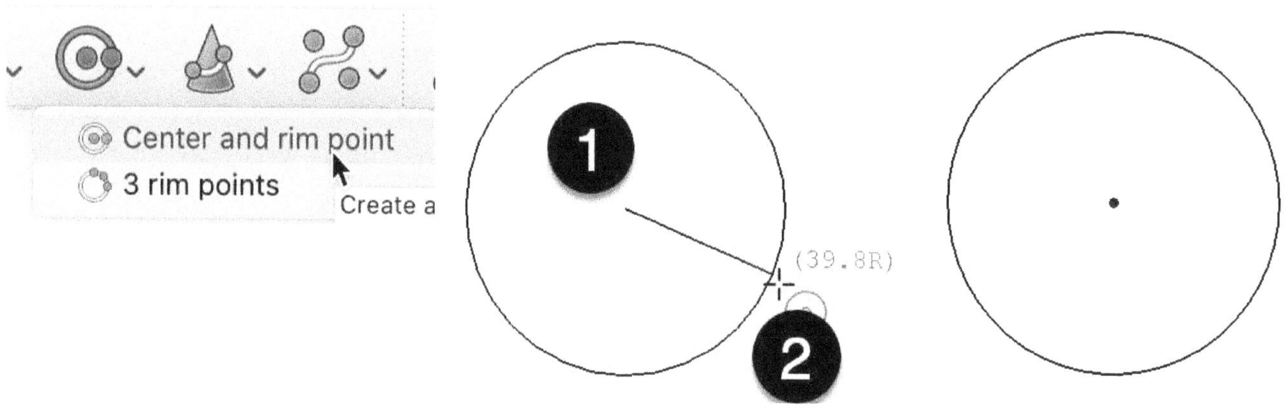

3 rim points

You can create this type of circle by specifying three points or by selecting three existing points. Click **Circle** drop-down > **3 rim points** on the **Sketcher geometries** toolbar (or) click **Sketch > Sketcher geometries > Create circle by three points** on the menu bar. Next, specify the three points of the circle, as shown.

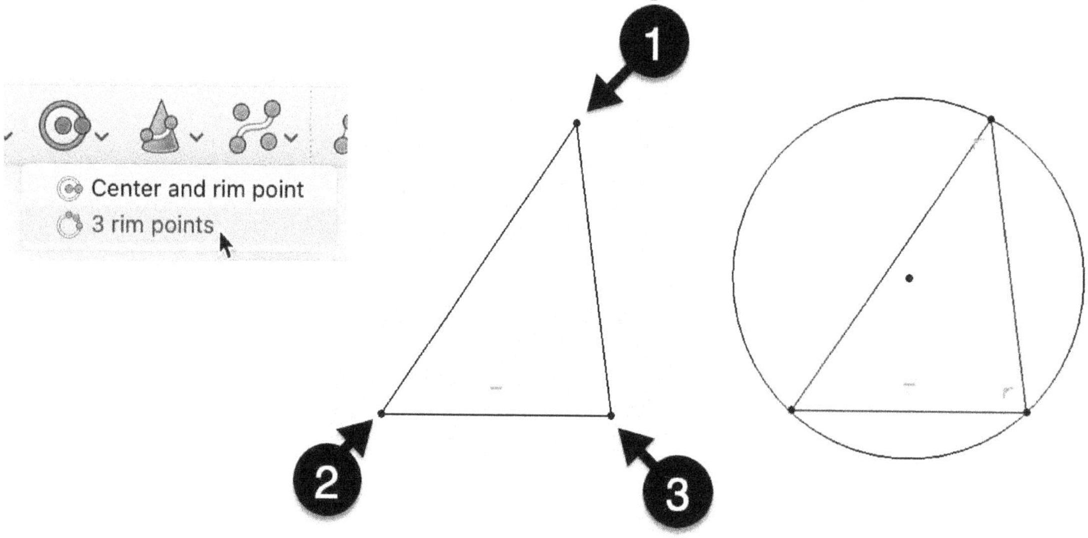

Create Rectangle

This command creates a rectangle by defining its diagonal corners. Click the **Create Rectangle** icon on the **Sketcher geometries** toolbar (or) click **Sketch > Sketcher geometries > Create circle by three points** on the menu bar. Click on the graphics window to define the first corner of the rectangle. Move the pointer and click to define the second corner.

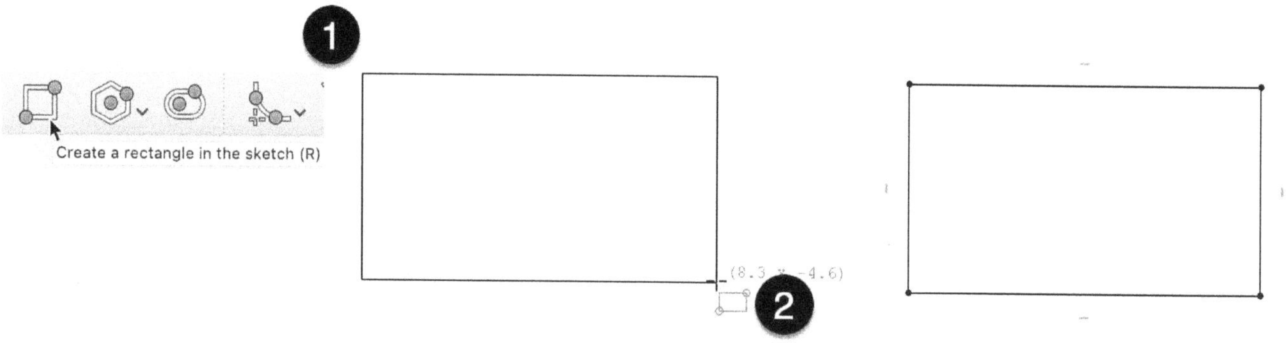

Create Slot

This command creates a straight slot by defining its start and width. Click the **Create Slot** icon on the **Sketcher geometries** toolbar (or) click **Sketch > Sketcher geometries > Create slot** on the menu bar. Specify the start of the slot. Next, move the pointer outward and click to specify the slot width and length.

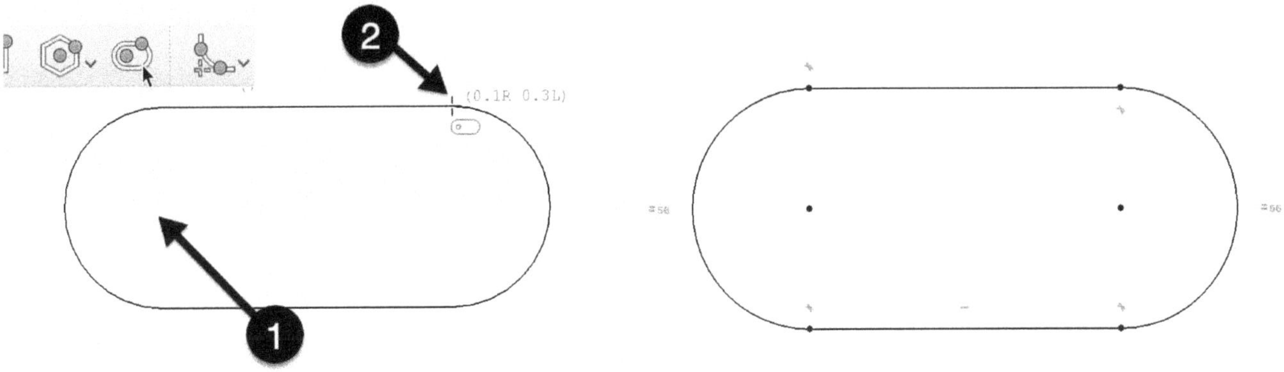

Creating Polygons

In FreeCAD 0.19, you can create a polygon using seven different commands: **Triangle, Square, Pentagon, Hexagon, Heptagon, Octagon,** and **Regular Polygon**.

Triangle

Click the **Polygon** drop-down > **Triangle** on the **Sketcher geometries** toolbar (or) click **Sketch > Sketcher geometries > Create equilateral triangle** on the menu bar. Specify the center point of the triangle. Move the pointer outward and click to specify the vertex point of the triangle.

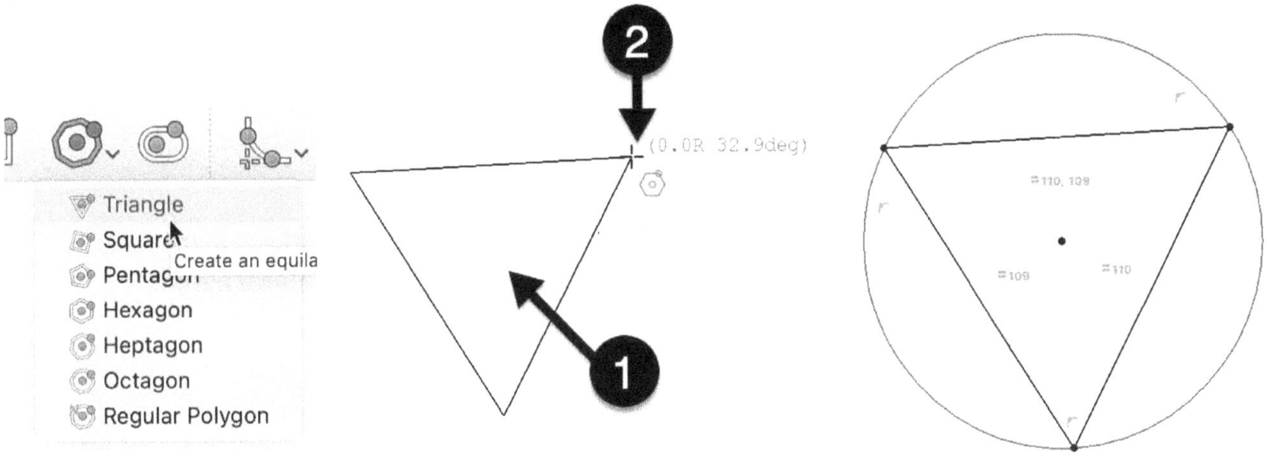

Square

Click the **Polygon** drop-down > **Square** on the **Sketcher geometries** toolbar (or) click **Sketch > Sketcher geometries > Create square** on the menu bar. Specify the center point of the square. Move the pointer outward and click to specify the vertex point of the square.

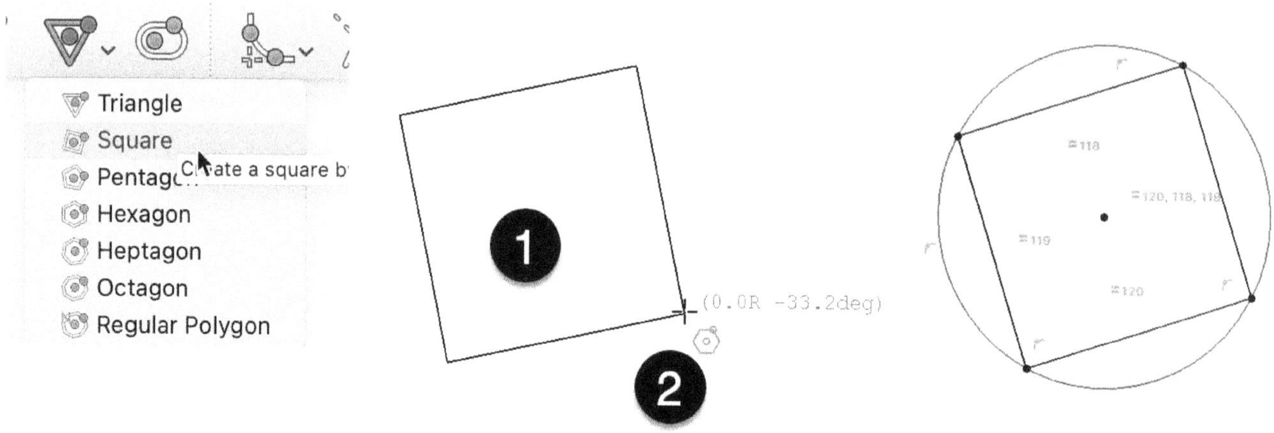

Pentagon

Click the **Polygon** drop-down > **Pentagon** on the **Sketcher geometries** toolbar (or) click **Sketch > Sketcher geometries > Create pentagon** on the menu bar. Specify the center point of the pentagon. Move the pointer outward and click to specify the vertex point of the pentagon.

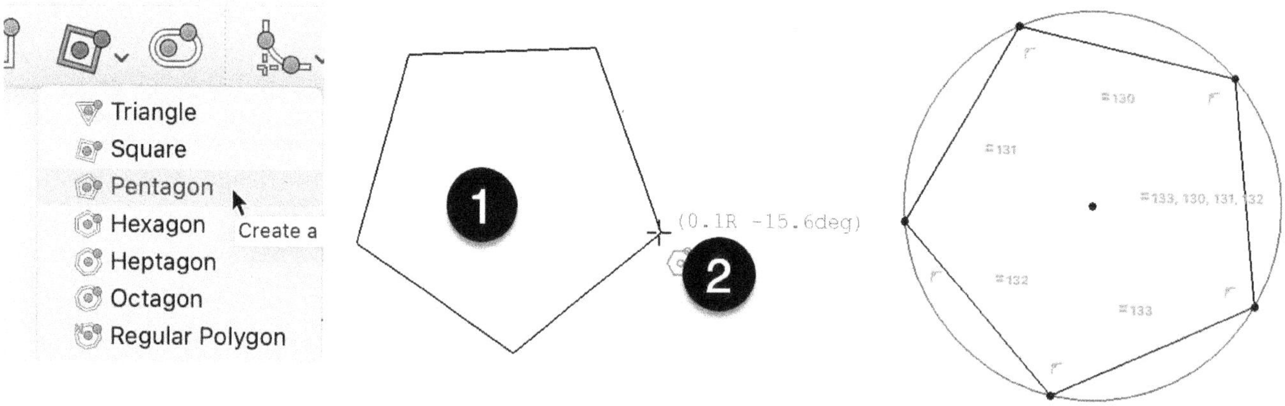

Hexagon

Click the **Polygon** drop-down > **Hexagon** on the **Sketcher geometries** toolbar (or) click **Sketch > Sketcher geometries > Create hexagon** on the menu bar. Specify the center point of the hexagon. Move the pointer outward and click to specify the vertex point of the hexagon.

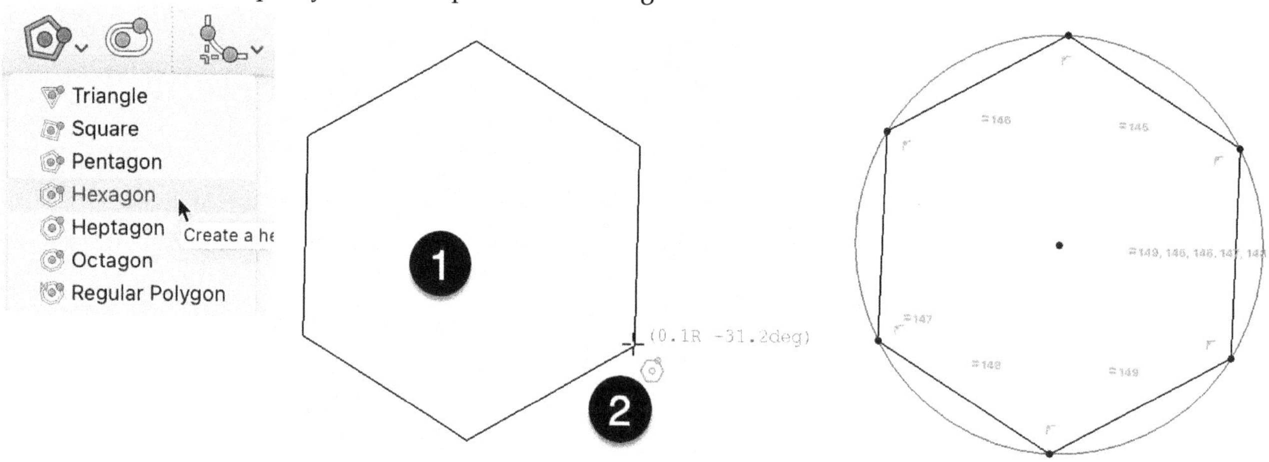

Heptagon

Click the **Polygon** drop-down > **Heptagon** on the **Sketcher geometries** toolbar (or) click **Sketch > Sketcher geometries > Create heptagon** on the menu bar. Specify the center point of the heptagon. Move the pointer outward and click to specify the vertex point of the heptagon.

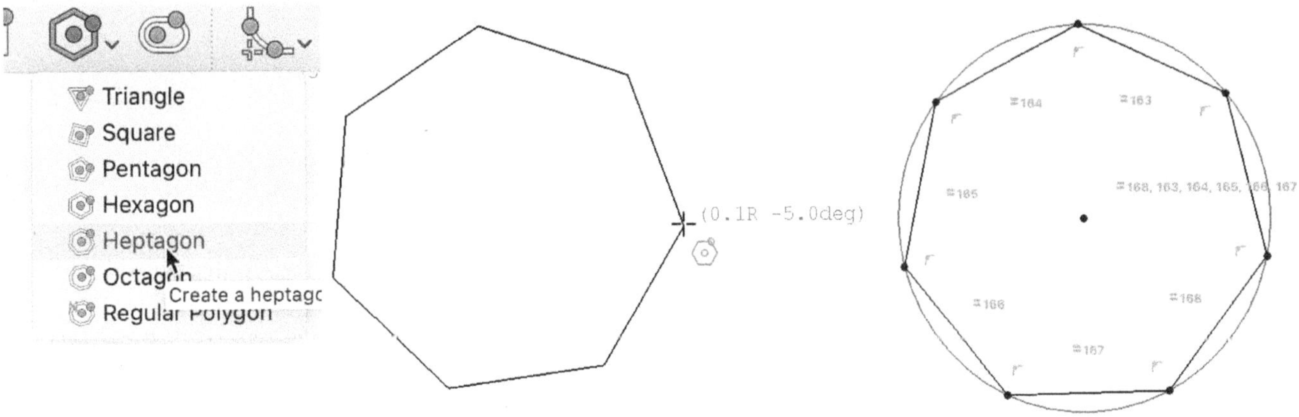

Octagon

Click the **Polygon** drop-down > **Octagon** on the **Sketcher geometries** toolbar (or) click **Sketch > Sketcher geometries > Create octagon** on the menu bar. Specify the center point of the octagon. Move the pointer outward and click to specify the vertex point of the octagon.

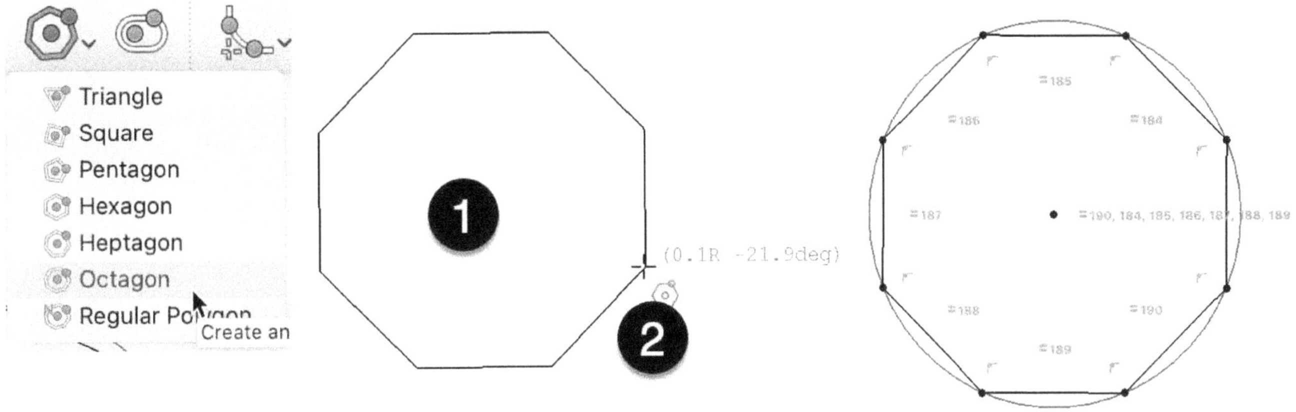

Regular Polygon

This command creates a polygon with the number of sides that you specify. Click the **Polygon** drop-down >

Regular Polygon 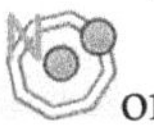 on the **Sketcher geometries** toolbar (or) click **Sketch > Sketcher geometries > Create regular polygon** on the menu bar. Next, specify the number of sides of the polygon in the **Create array** dialog. Click **OK** and click on the graphics window to define the center of the polygon. Move the pointer outward and click to the polygon. Notice that a circle is displayed touching the vertices of the polygon.

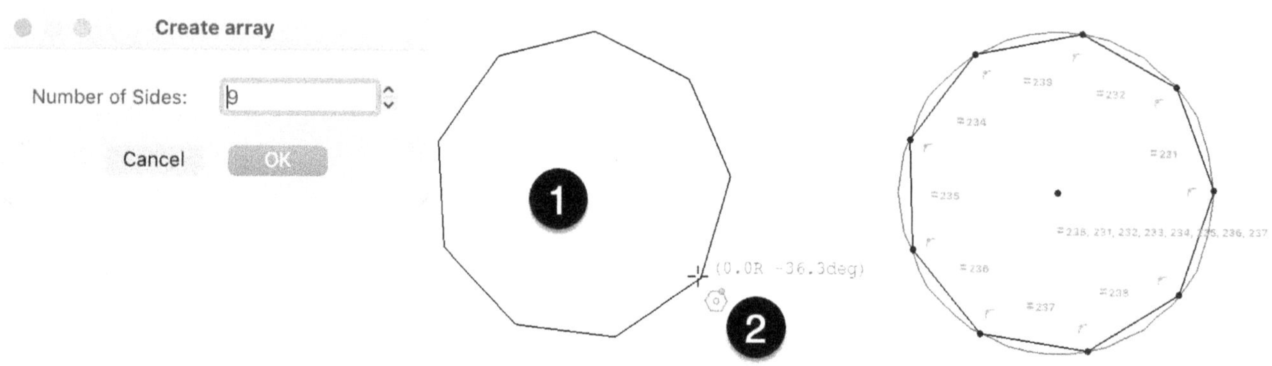

The Ellipse by center, major radius, point command

This command creates an ellipse using a center point, and major and minor axes. Click the **Conics** drop-down > **Ellipse by center, major radius, point** on the **Sketcher geometries** toolbar (or) click **Sketch > Sketcher geometries > Create ellipse by center** on the menu bar. In the graphics window, click to define the center point of the ellipse. Move the pointer away from the center point and click to define the distance and orientation of the first axis. Next, move the pointer in the direction perpendicular to the first axis and click; the ellipse is created.

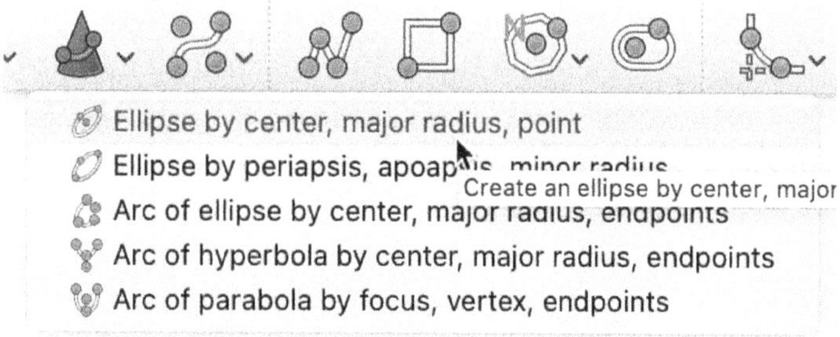

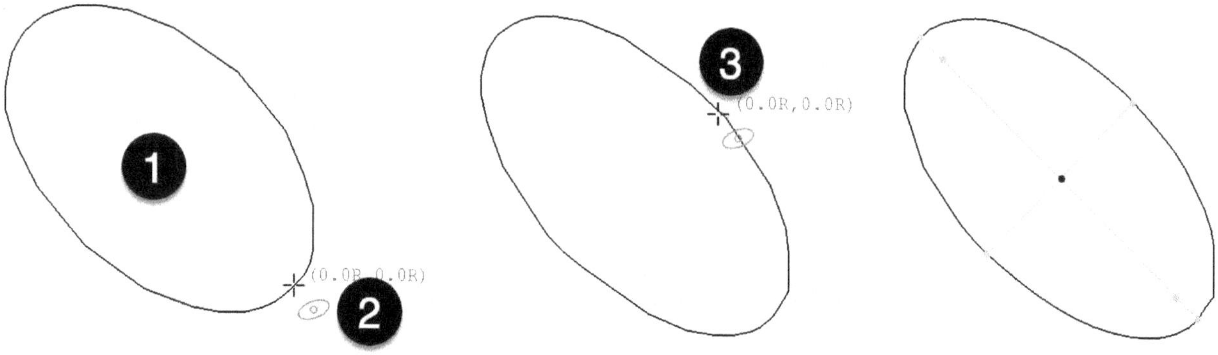

The Ellipse by periapsis, apoapus, minor radius command

This command creates an ellipse by specifying three points. The first two points define the location and length of the first axis. The third point defines the second axis of the ellipse. Click the **Conics** drop-down > **Ellipse by periapsis, apoapus, minor radius** on the **Sketcher geometries** toolbar (or) click **Sketch > Sketcher geometries > Create ellipse by 3 points** on the menu bar. In the graphics window, click to define the start point of the first axis.

Move the pointer away and click to define the distance and orientation of the first axis. Next, move the pointer in the direction perpendicular to the first axis and click; the ellipse is created.

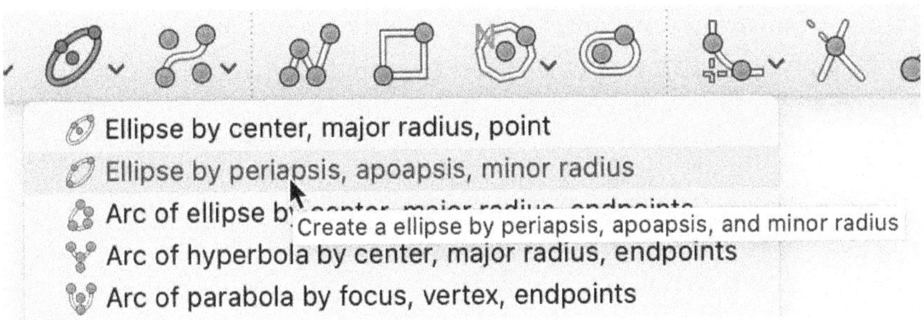

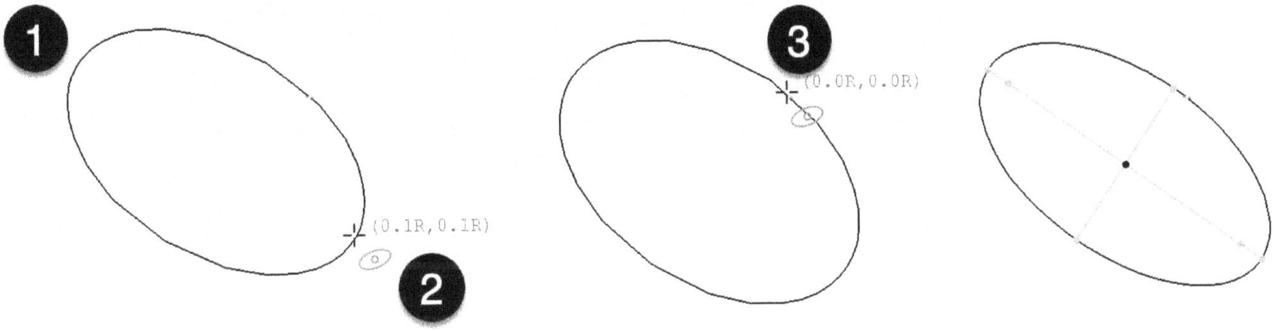

Toggle Construction geometry

This command converts a sketch element into a construction element. The construction elements support you in creating a sketch of the desired shape and size. To convert a sketch element to a construction element, select it and click the **Toggle Construction geometry** icon on the **Sketcher geometries** toolbar. You can also convert it back to a sketch element by selecting it and deactivating the **Construction** button on the **Sketch Palette**.

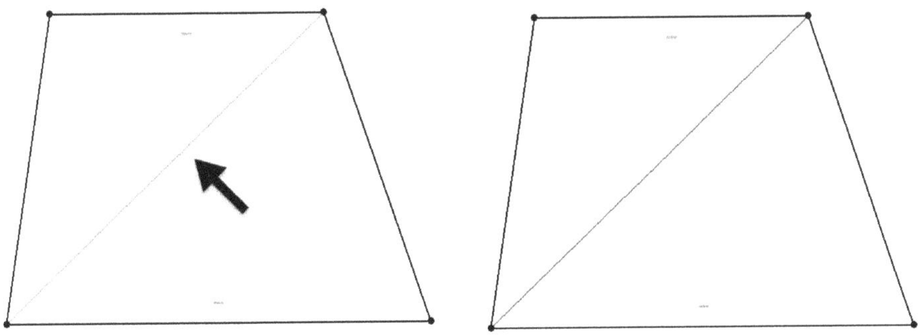

Adding Dimensions

It is generally considered a good practice to ensure that every sketch you create is fully constrained before creating solid features. The term 'fully-constrained' means that the sketch has a definite shape and size. You can fully-constrain a sketch by using dimensional and geometric constraints. You can add dimensional constraints to a sketch by using the commands available on the **Sketcher constraints** toolbar.

Horizontal Distance

To add a horizontal distance constraint, click the **Horizontal distance** icon on the **Sketcher constraints** toolbar (or) click **Sketch > Sketcher constraints > Constrain horizontal distance** on the menu bar. Select two points (or) a line from the graphics window. Next, type-in a value in the **Insert Length** dialog and click **OK**; the horizontal dimension is created.

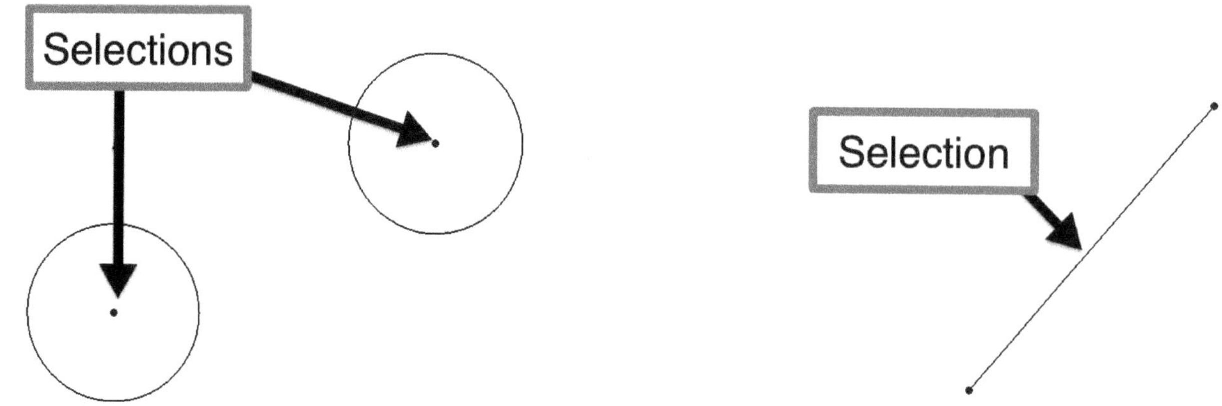

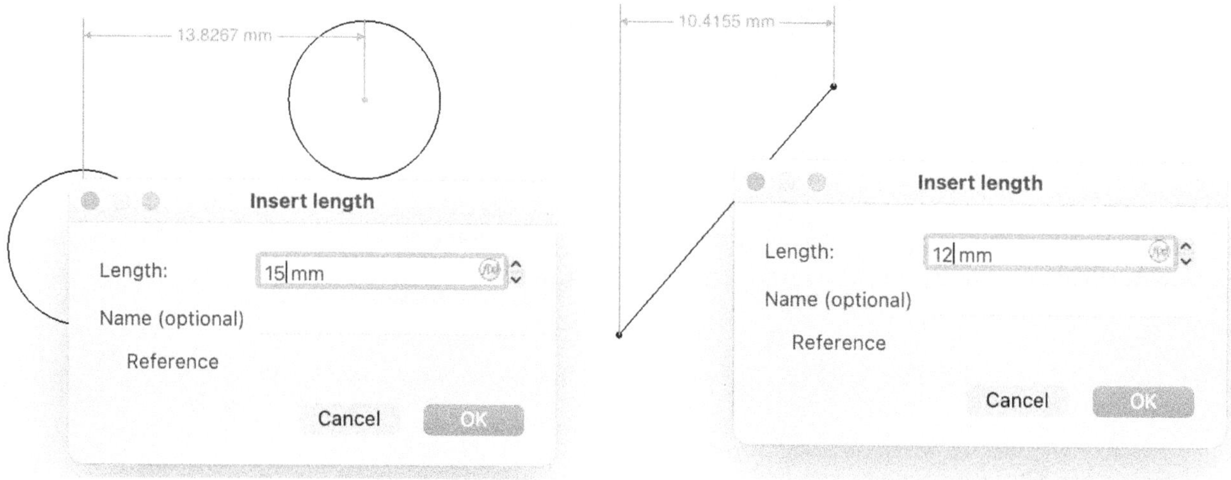

Vertical Distance Constraint

To create a vertical distance constraint, click the **Vertical distance** icon on the **Sketcher constraints** toolbar (or) click **Sketch > Sketcher constraints > Constrain vertical distance** on the menu bar. Select two points (or) a line from the graphics window. Next, type-in a value in the **Insert Length** dialog and click **OK**; the vertical distance constraint is created.

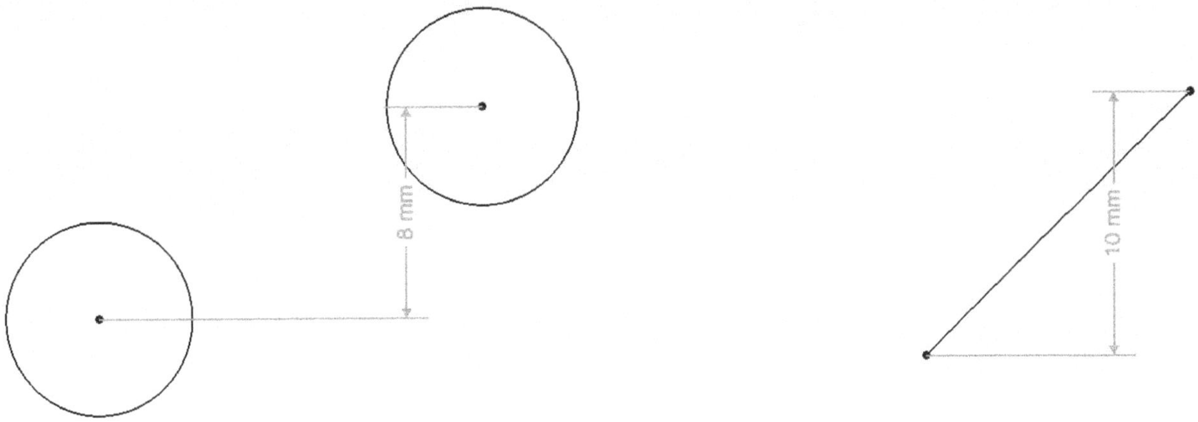

Distance Constraint

To create a dimension aligned between two points or a selected line, click the **Distance** icon on the **Sketcher constraints** toolbar (or) click **Sketch > Sketcher constraints > Constrain distance** on the menu bar. Select two points or a line from the graphics window. Next, position the dimension and edit its value.

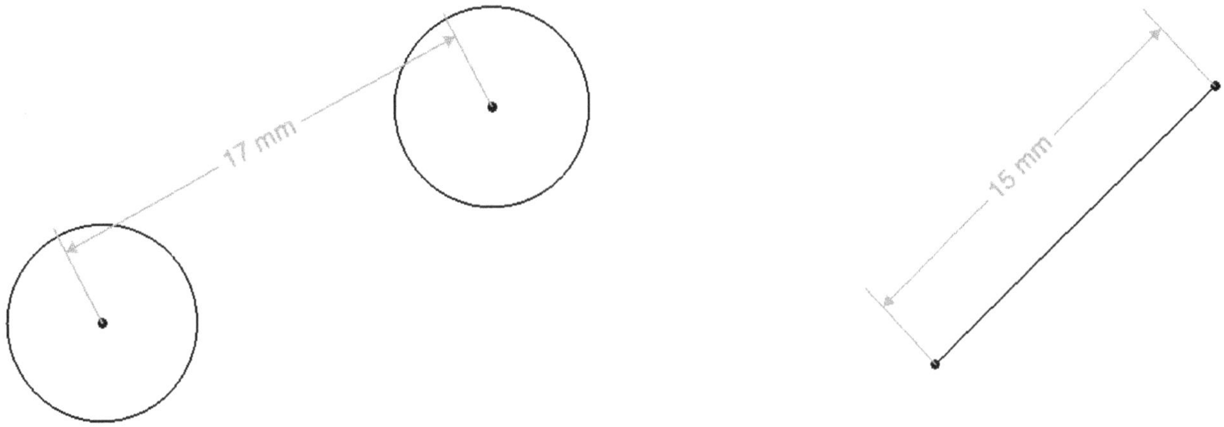

Angle Constraint

Click the **Angle** icon on the **Sketcher constraints** toolbar (or) click **Sketch > Sketcher constraints > Constrain angle** on the menu bar. Next, select two lines that are positioned at an angle to each other. Move the pointer between the selected lines and click to position the dimension. Next, type in a value and click the **OK** button.

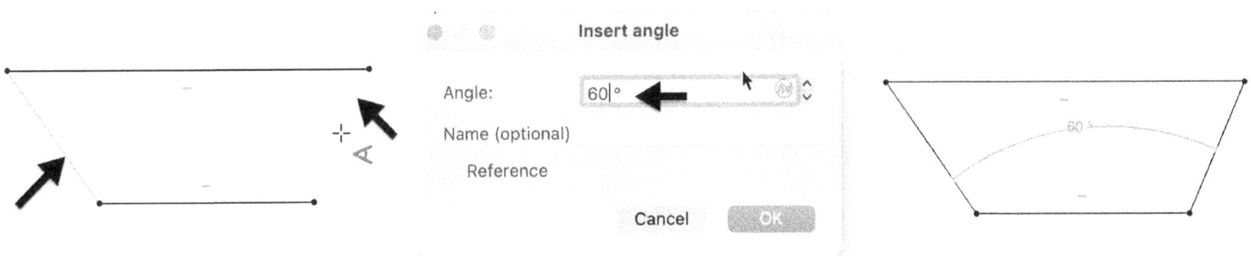

Diameter Constraint

This command adds a diameter constraint to a circle or arc. To do this, click the **Constrain Diameter** icon on the **Sketcher constraints** toolbar (or) click **Sketch > Sketcher constraints > Constrain diameter** on the menu bar.

Next, select a circle or arc; the **Change diameter** dialog pops up. Type-in a value in this dialog, and then click **OK** to update the dimension.

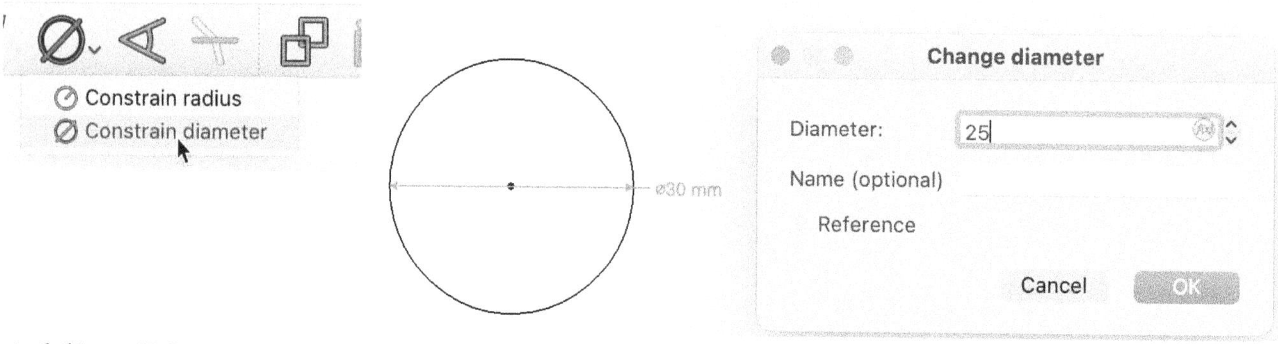

Adding Dimensions to an Arc

FreeCAD 0.19 allows you to add five types of dimensions to an arc: Radius, Diameter, and Linear dimension.

Adding a Radius

To add a radius to an arc, click the **Constrain radius** icon on the **Sketcher constraints** toolbar (or) click **Sketch > Sketcher constraints > Constrain radius** on the menu bar. Next, select the arc and type the radius value in the **Change radius** dialog. Click **OK** to create the radius dimension.

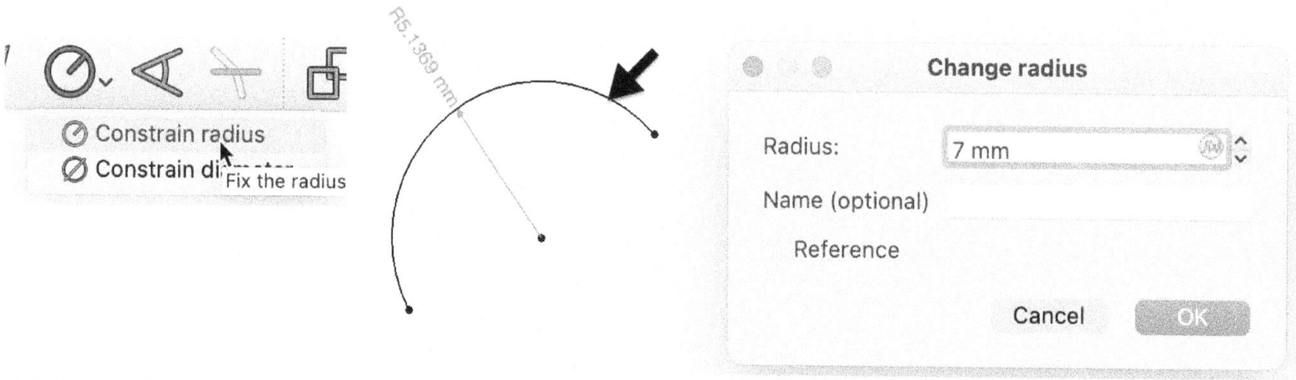

Adding a Linear dimension to the Arc

To add a linear dimension to an arc, activate the **Distance** command and select its endpoints. Type the distance value and click **OK**.

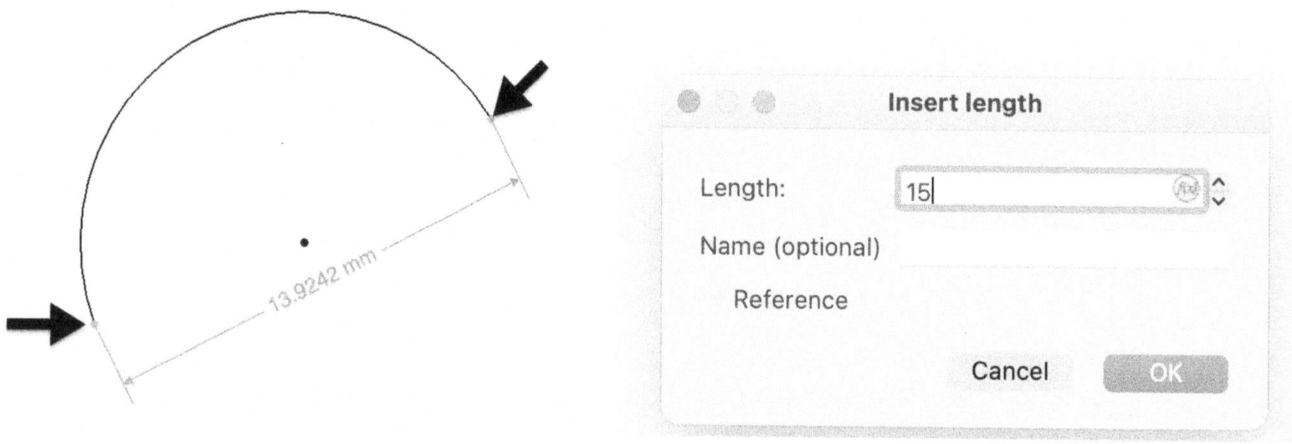

Lock

This command is used to lock a selected point at its location. Click the **Lock**  icon on the **Sketcher constraints** toolbar (or) click **Sketch > Sketcher constraints > Constrain lock** on the menu bar. Select the point to be locked; dimensions are created between the selected point and the sketch origin.

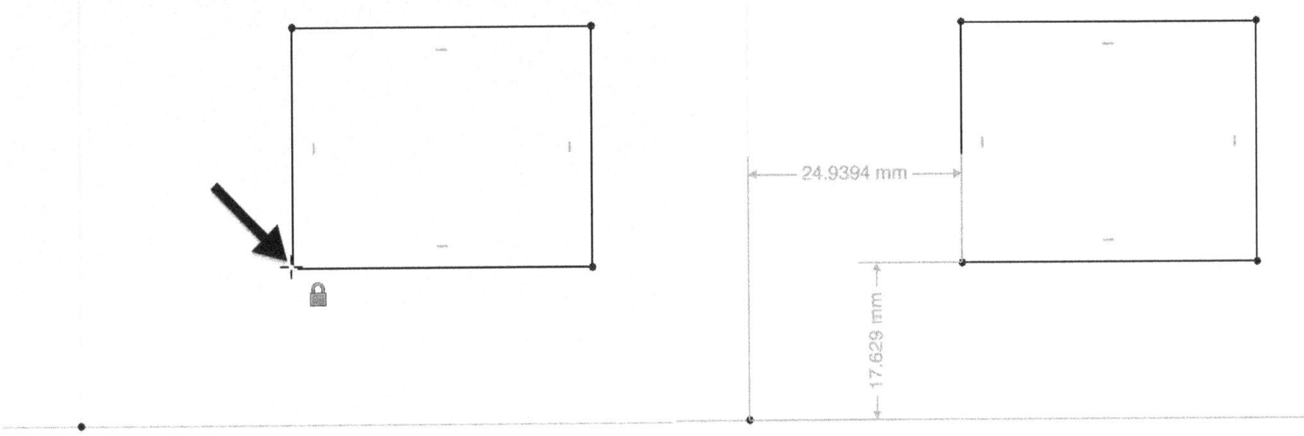

Over-constrained Sketch

When creating sketches for solid or surface features, FreeCAD 0.19 will not allow you to over-constrain the geometry. The term 'over-constrain' means adding more dimensions than required. The following figure shows a fully constrained sketch. If you add another dimension to this sketch (e.g., diagonal dimension), a message appears in the **Tasks** tab of the **Combo View** panel. It shows that the sketch contains a redundant constrain. Now, check the **Reference** option on the Insert Length dialog and click **OK**; the new dimension will be displayed in blue color.

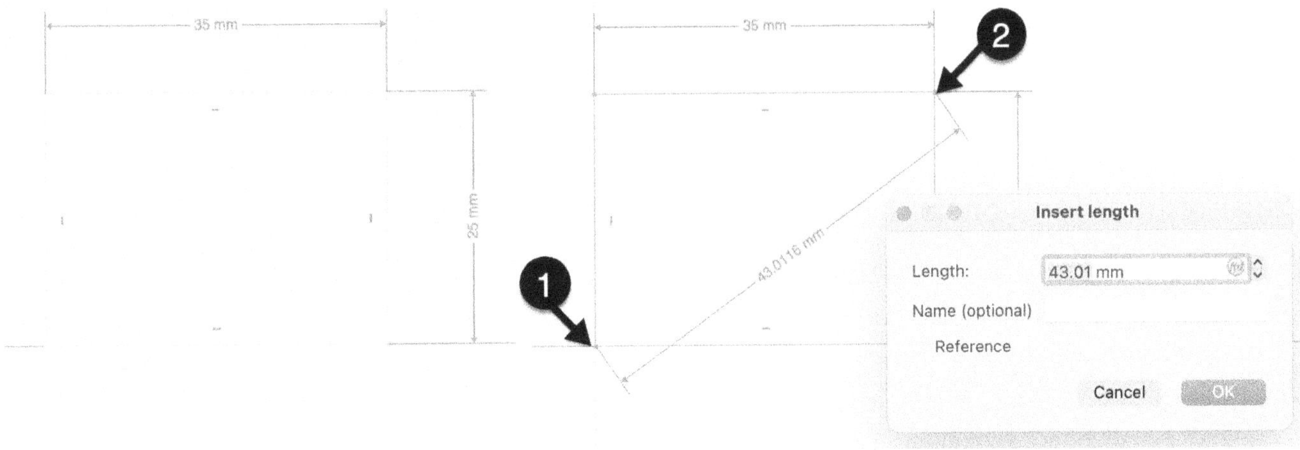

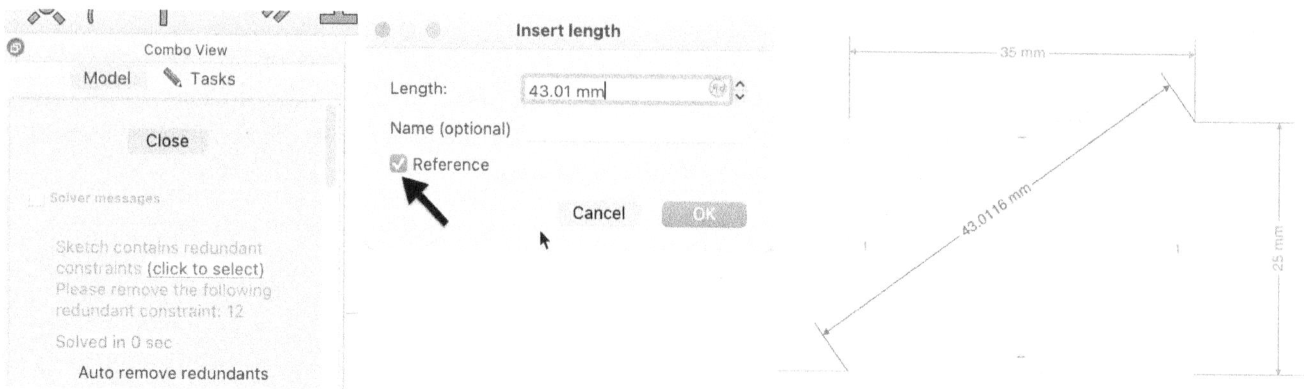

Geometric Constraints

The constraints are used to control the shape of a drawing by establishing relationships between the sketch elements. You can apply constraints to a sketch using the commands displayed on the **Sketcher constraints** toolbar.

Coincident Constraint

This constraint connects a point with another point. On the **Sketcher constraints** toolbar, click **Coincident constraint** and select the points to be made coincident to each other. The selected points will be connected.

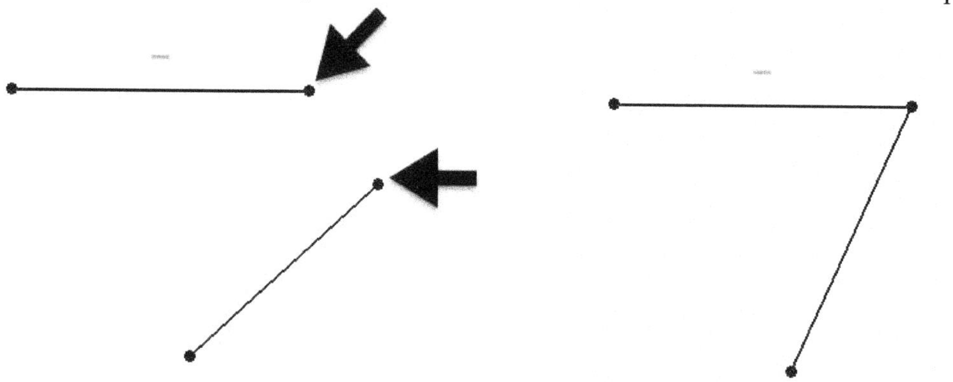

Point onto object constraint

This constraint makes a vertex or a point to be on a line, curve, arc, or circle. Click the **Point onto object** icon on the **Sketcher constraints** toolbar (or) click **Sketch > Sketcher constraints > Constrain point onto object**. Next, select a line, circle, arc, or curve. Select the point to be made coincident. The point will lie on the selected entity or its extension.

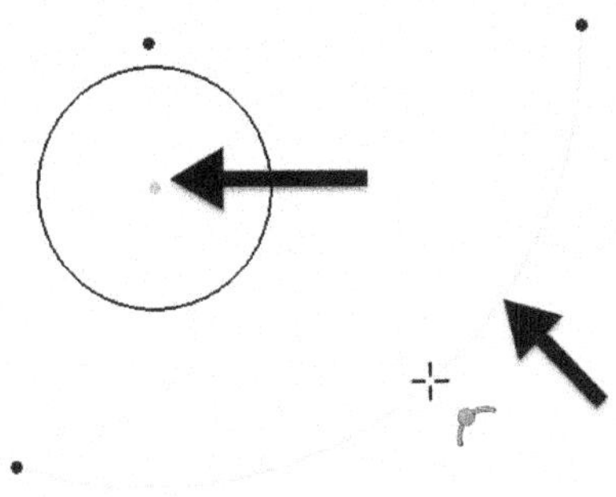

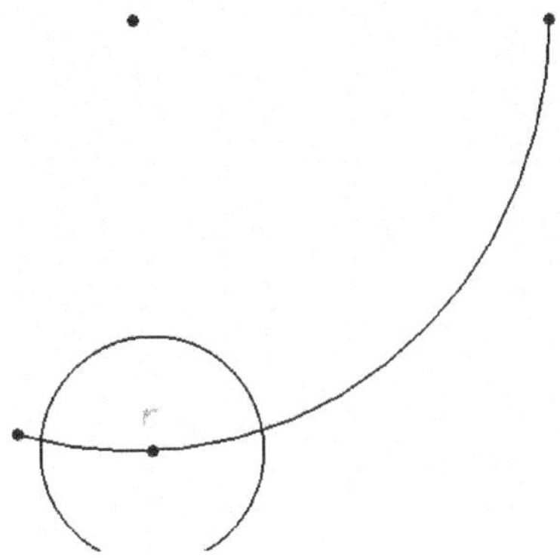

Horizontal Constraint

This constraint makes a line horizontal. Click the **Horizontal Constraint** icon on the **Sketcher constraints** toolbar (or) click **Sketch > Sketcher constraints > Constrain horizontally**. Next, select a line positioned; the line is made horizontal.

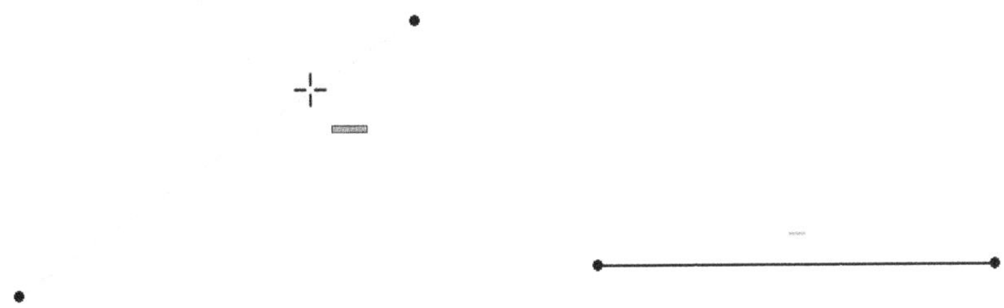

The **Horizontal Constraint** also aligns the two selected points horizontally. Click the Horizontal Constraint icon and select the points to be aligned horizontally.

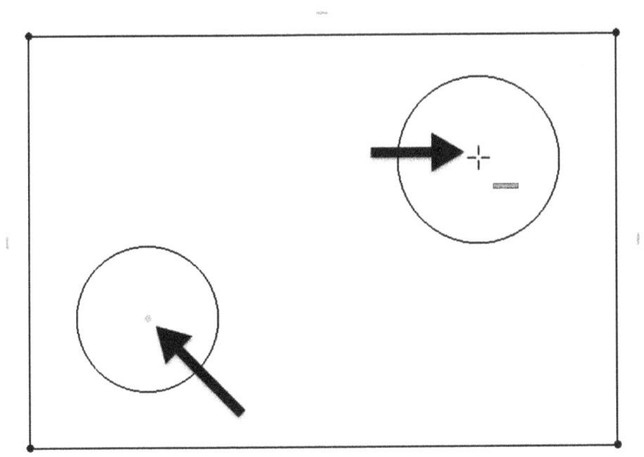

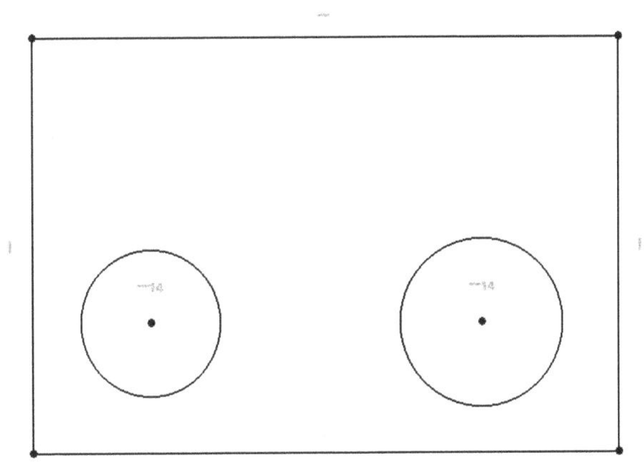

Vertical Constraint

This constraint makes a line vertical. Click the **Vertical Constraint** icon on the **Sketcher constraints** toolbar (or) click **Sketch > Sketcher constraints > Constrain vertically**. Select a line; the line is made vertical.

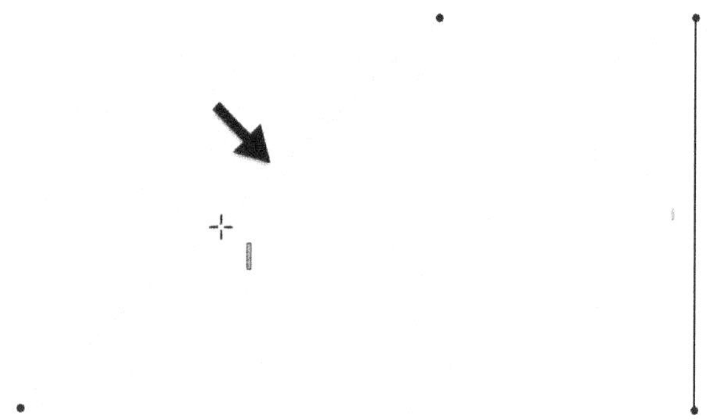

The **Vertical Constraint** also aligns the two selected points vertically. Click the **Vertical Constraint** button on the **Sketcher constraints** toolbar and then select the points to align them vertically.

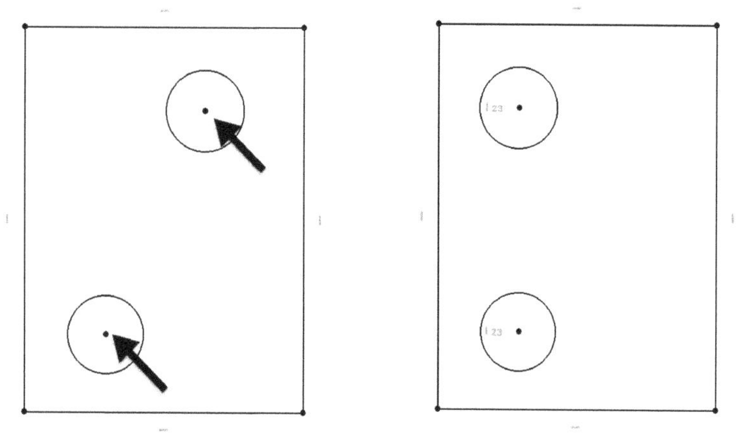

Equal

The **Equal**  constraint makes two lines equal in length.

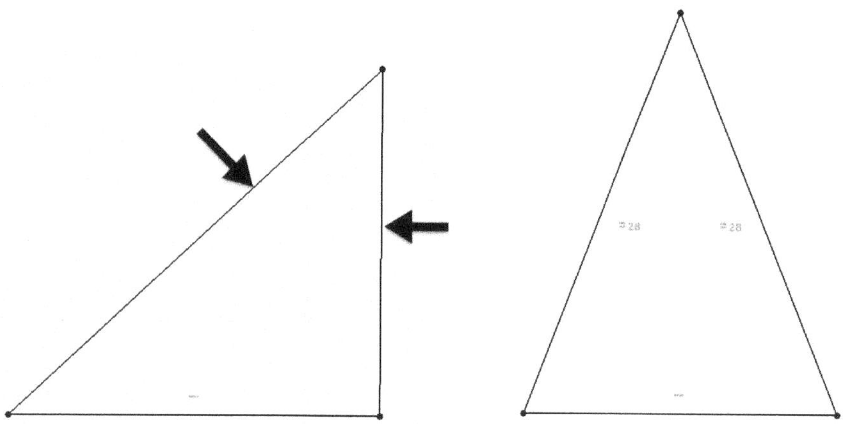

In addition to that, this constraint makes two circles or arcs equal in size.

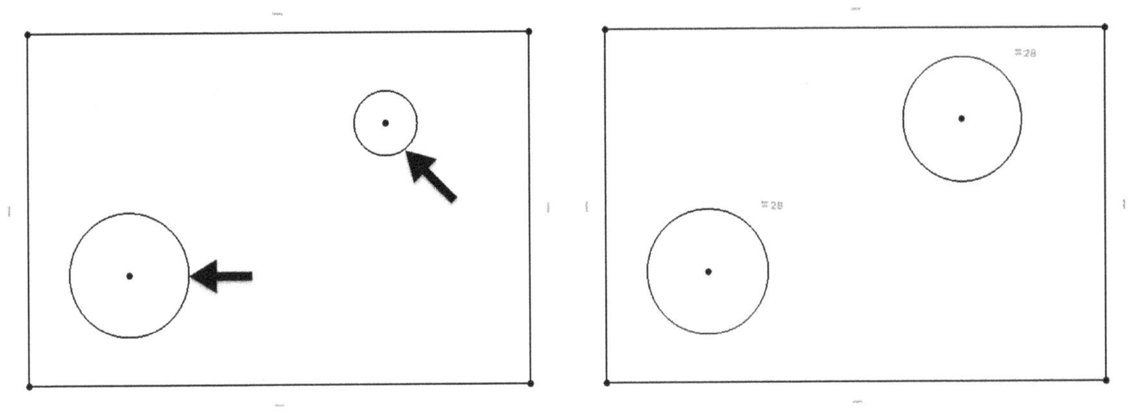

Tangent

This constraint makes an arc, circle, or line tangent to another arc or circle. Click the **Tangent constraint** icon on the **Sketcher constraints** toolbar (or) click **Sketch > Sketcher constraints > Constrain tangent**. Next, select a circle, arc, or line. Select another circle, arc, or line; both the elements become tangent to each other.

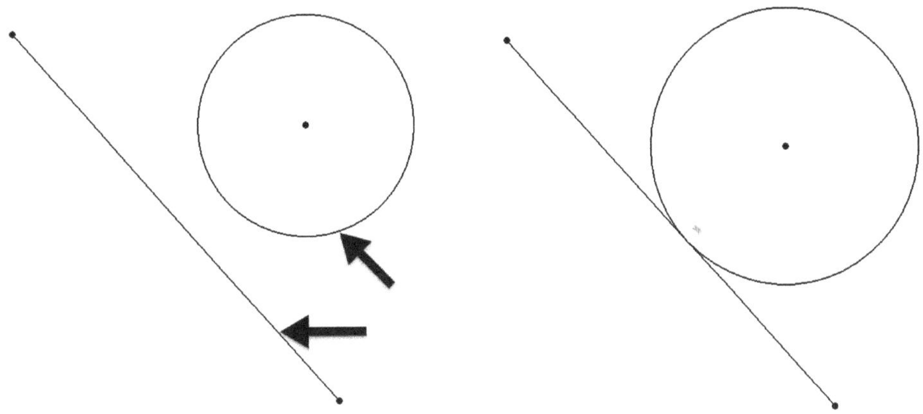

Parallel Constraint

This constraint makes two lines parallel to each other. Click the **Parallel constraint** icon on the **Sketcher constraints** toolbar (or) click **Sketch > Sketcher constraints > Constrain parallel.** Next, select two lines from the sketch. The under-constrained line is made parallel to the constrained line. For example, if you select a line with the **Horizontal** constraint and a free to move line, the free-to-move line becomes parallel to the horizontal line.

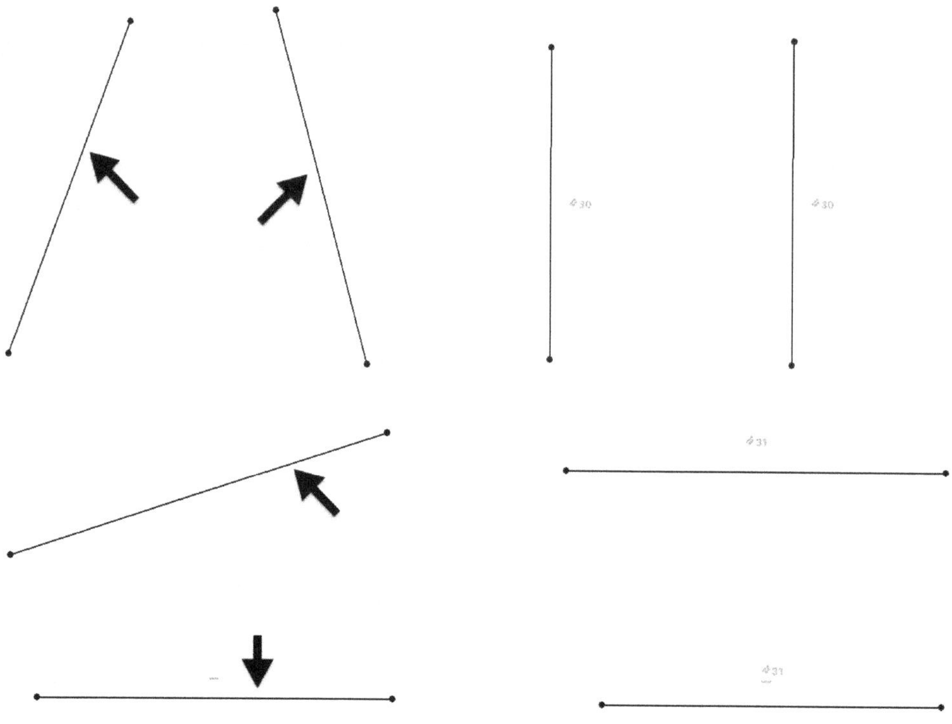

Perpendicular Constraint

This constraint makes two lines perpendicular to each other. Click the **Perpendicular constraint** icon on the **Sketcher constraints** toolbar (or) click **Sketch > Sketcher constraints > Constrain perpendicular**. Next, select two lines from the sketch. The two lines will be made perpendicular to each other.

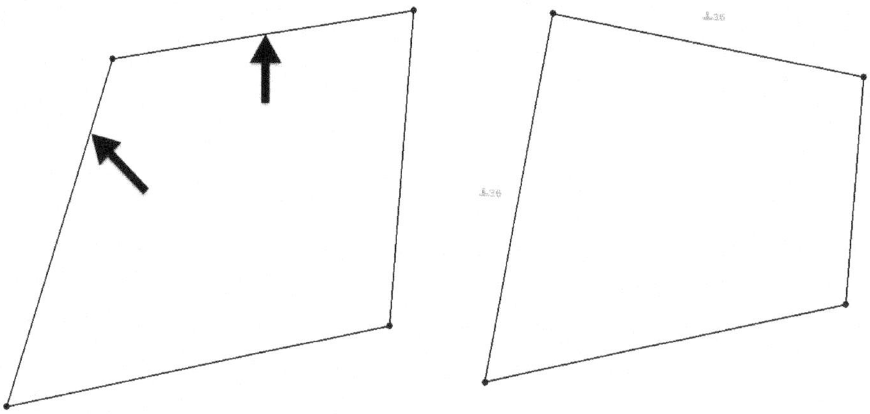

Symmetrical Constraint

This constraint makes two points symmetric about a line or third point. The objects will have the same position about the symmetry line. Click the **Symmetrical constraint** icon on the **Sketcher constraints** toolbar (or) click **Sketch > Sketcher constraints > Constrain symmetrical**, and click on the first point. Next, click on the second point, and then select the symmetry line. The two points will be made symmetric about the symmetry line.

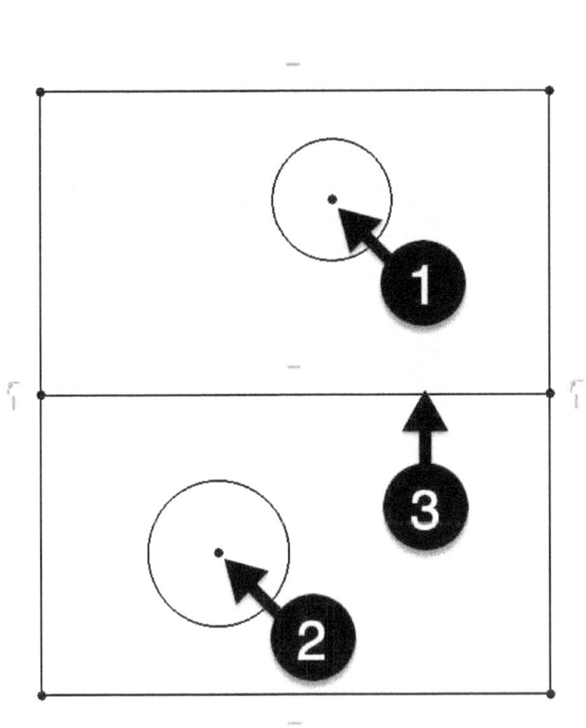

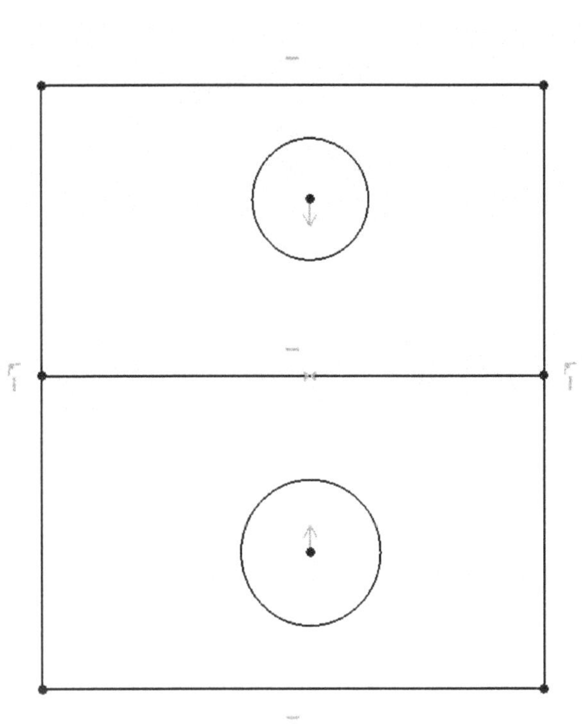

Block Constraint

This constraint fixes the selected objects at their current location. Click the **Block constraint** icon on the **Sketcher constraints** toolbar (or) click **Sketch > Sketcher constraints > Constrain block** on the menu bar. Next, select single or multiple objects from the graphics window. Now, click on the fixed objects and try to drag them; the objects are unmovable.

The Sketch Fillet command

This command rounds a sharp corner created by the intersection of two lines, arcs, circles, and rectangle. Click the **Fillet > Sketch Fillet** icon on the **Sketcher geometries** toolbar (or) click **Sketch > Sketch geometries > Create fillet** on the menu bar. Next, select the elements to be filleted. The elements to be filleted are not required to touch each other. Keep on selecting the elements of the sketch; the fillets are added at the corners at which two selected elements intersect. Note that the existing constraints associated with the intersection edge will be deleted.

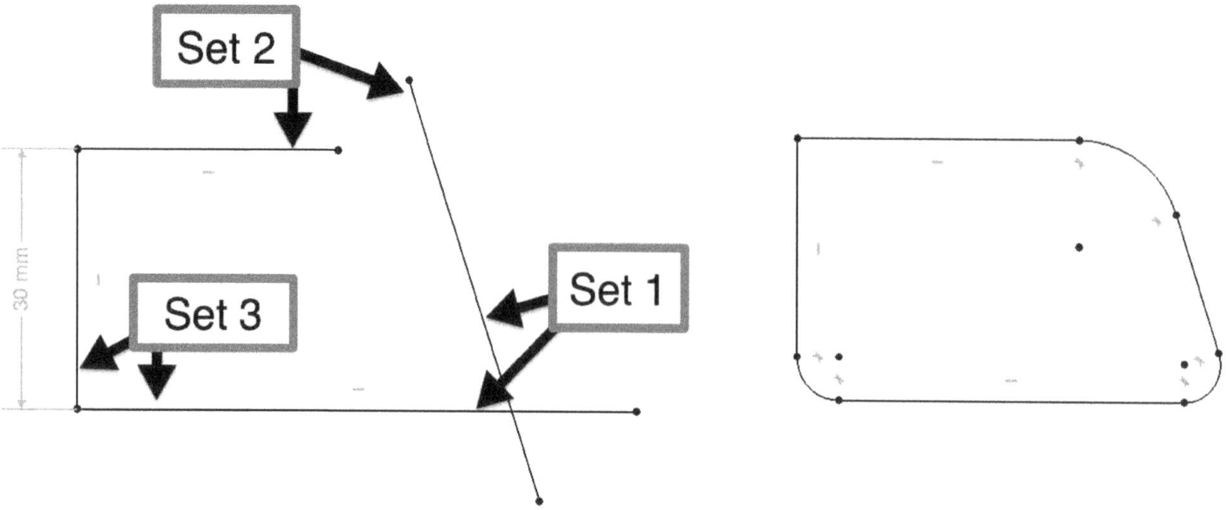

The Corner-preserving fillet command

This command rounds a sharp corner created by the intersection of two elements without deleting the existing constraint. Click the **Fillet > Corner-preserving Fillet** icon on the **Sketcher geometries** toolbar (or) click **Sketch > Sketch geometries > Create corner-preserving fillet** on the menu bar. Next, select the intersecting elements with an existing constraint.

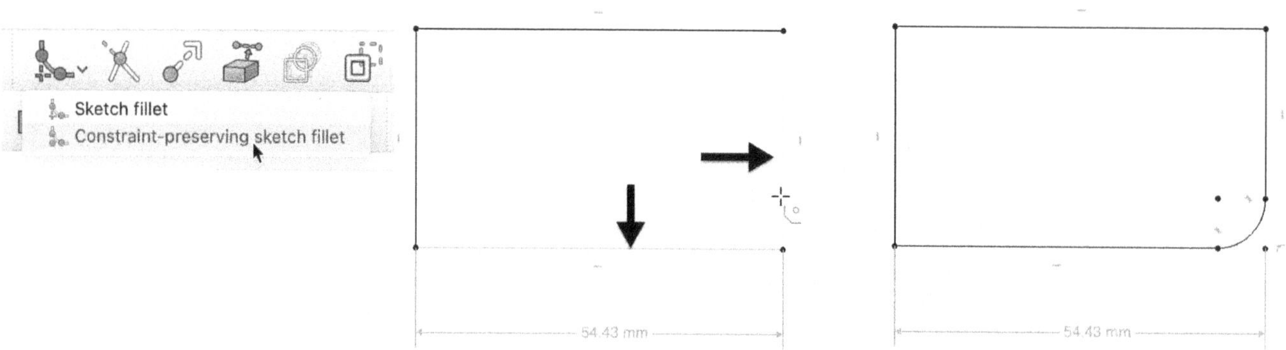

The Extend edge command

This command extends or trims elements such as lines, arcs, and curves until they touch another element called the boundary edge. Click the **Extend edge** icon on the **Sketcher geometries** toolbar (or) click **Sketch > Sketch geometries > Extend edge** on the menu bar. Next, click on the element to extend. Select the boundary edge.

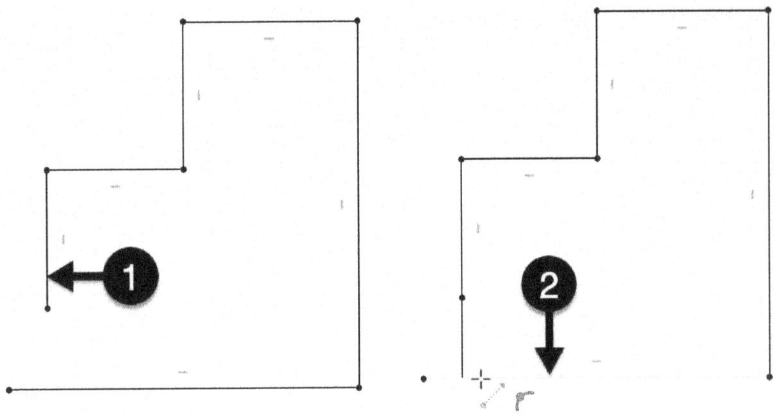

To trim an element up to a boundary element, select the element to be trimmed. Next, select the boundary edge.

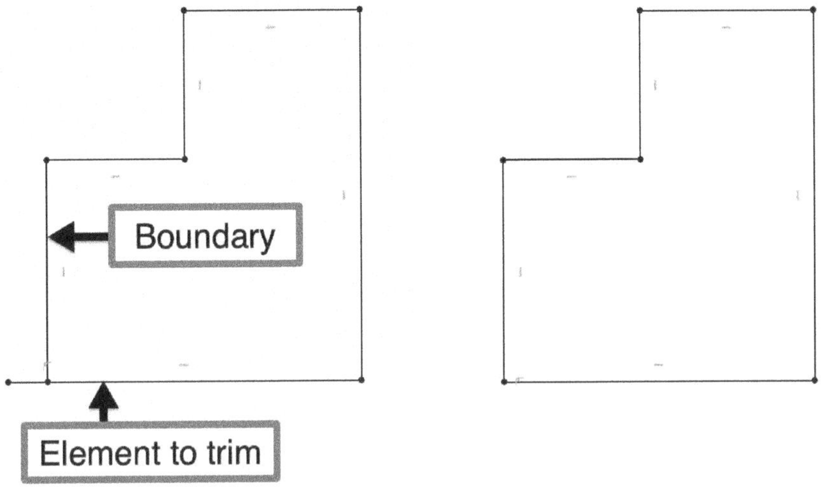

The Trim Edge command

This command trims the end of an element back to the intersection of another element. Click the **Trim edge** icon on the **Sketcher geometries** toolbar (or) click **Sketch > Sketch geometries > Trim edge** on the menu bar. Click on the element or elements to trim.

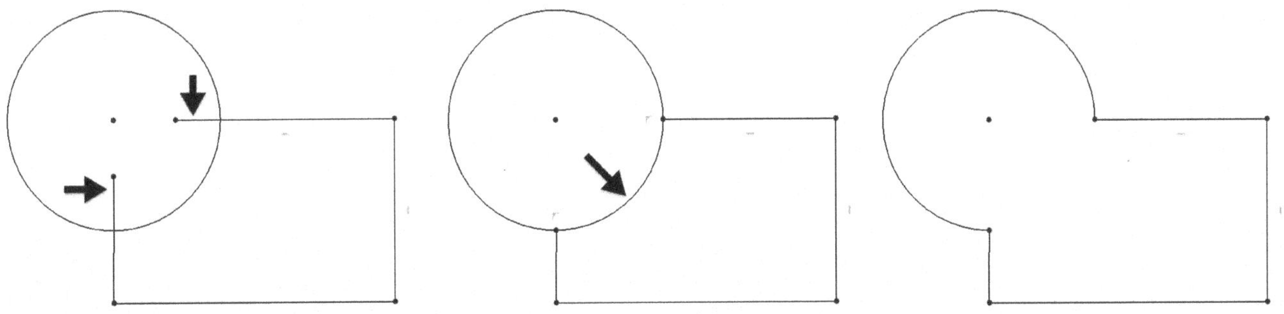

The Select unconstrained DOF command

This command selects the under-constrained element of a sketch. Click the **Select unconstrained DOF** icon on the **Sketcher tools** toolbar (or) click **Sketch > Sketch tools > Select unconstrained DOF** on the menu bar. The under-constrained sketch element is highlighted in green.

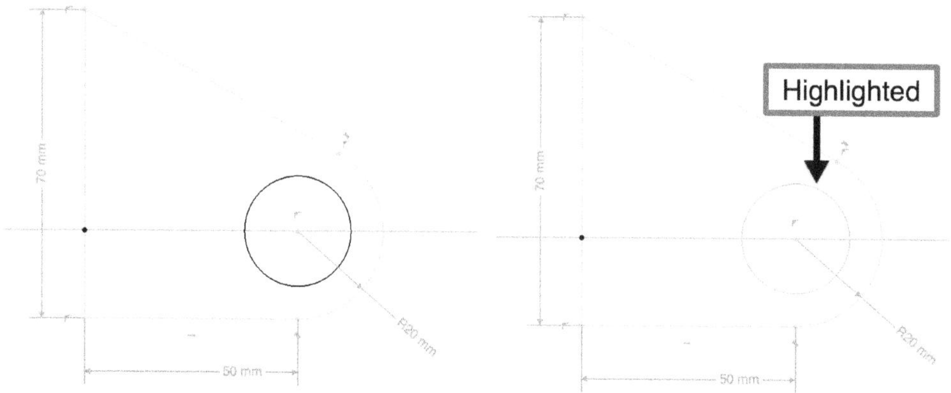

The Close shape command

This command creates a closed shape by connecting the end points of the two sketch elements. Select the open sketch elements to be connected. Next, click the **Close Shape** icon on the **Sketcher tools** toolbar (or) click **Sketch > Sketch tools > Close shape** on the menu bar.

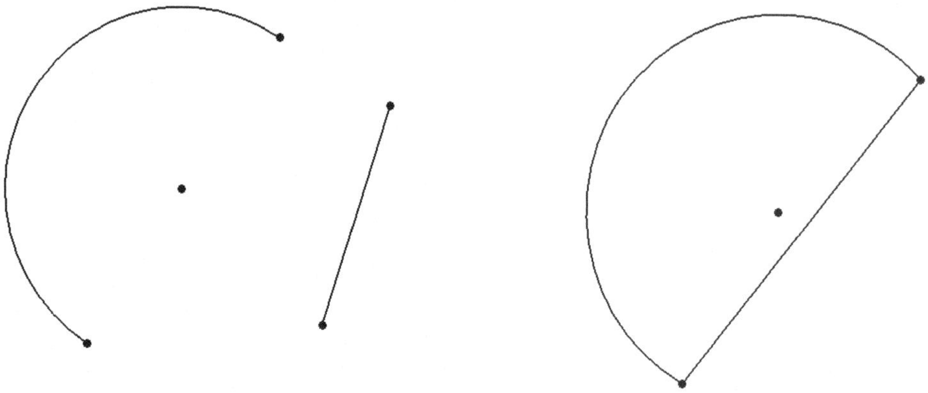

The Connect edges command

This command connects the edges of the two selected sketch elements. Select the open sketch elements to be connected. Next, click the **Connect edges** icon on the **Sketcher tools** toolbar (or) click **Sketch > Sketch tools > Connect edges** on the menu bar.

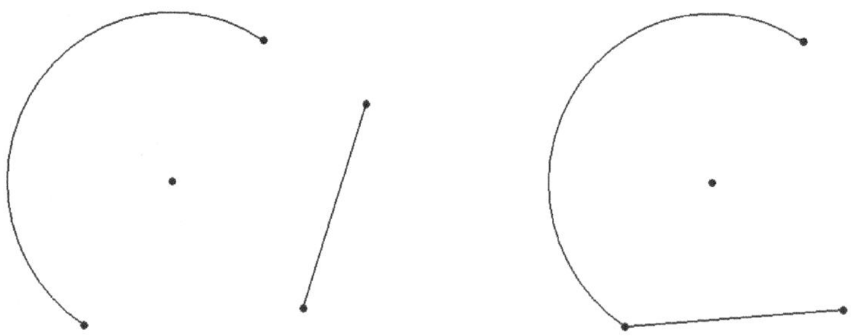

Rectangular array

This command creates a rectangular array of the selected sketch elements. Select the objects to array and click the **Rectangular array** icon on the **Sketcher tools** toolbar (or) click **Sketch > Sketch tools > Rectangular array** on the menu bar. On the **Create array** dialog, type-in values in the **Columns** and **Rows** boxes. Check the **Constrain inter-element separation** option if you want to create constraints between the elements of the array. The **Equal vertical/horizontal spacing** option creates an array with equal spacing between the rows and columns. The **Clone** option creates the clones of the base object. All the objects of the array will be modified if you modify the base object. Click the **OK** button on the dialog. Move the pointer horizontally or vertically and click to specify the spacing between the array objects.

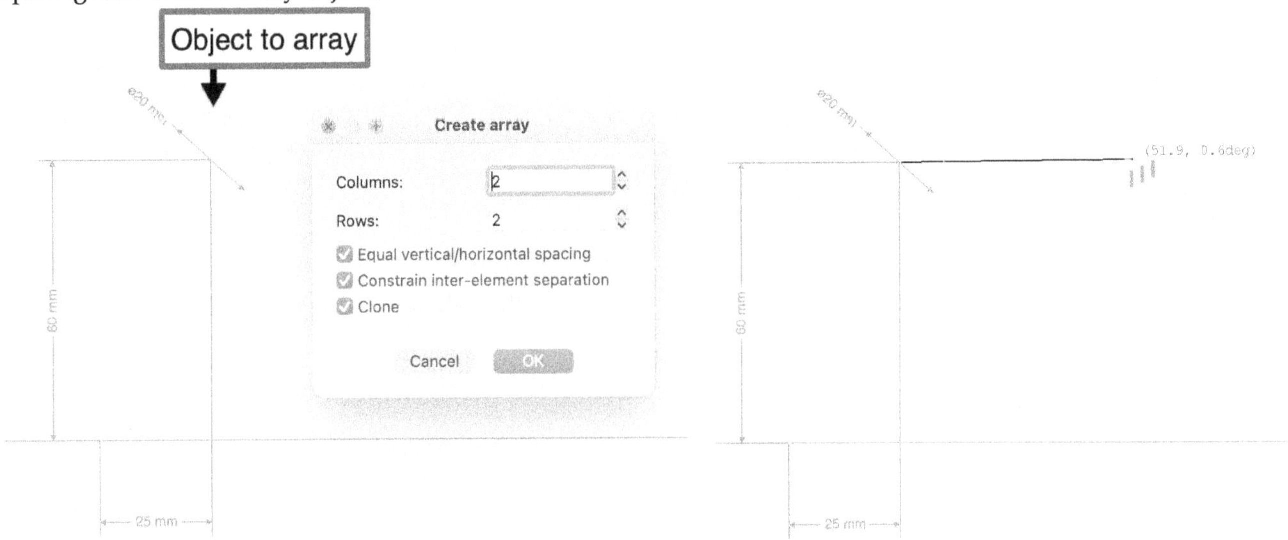

Next, modify the distance and angular constraints of the array.

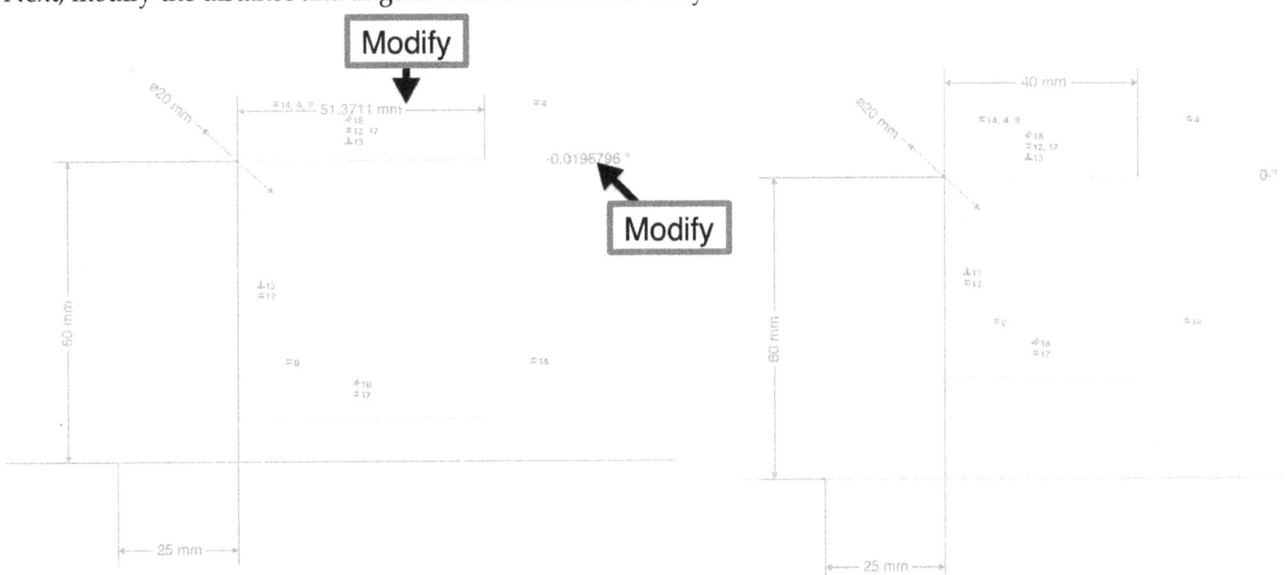

Delete all Constraints

This command is used to delete all constraints of a sketch. To do this, click the **Delete all constraints** icon on the **Sketcher tools** toolbar (or) click **Sketch > Sketcher tools > Delete all constraints** on the menu bar. Next, click **Yes** on the message box.

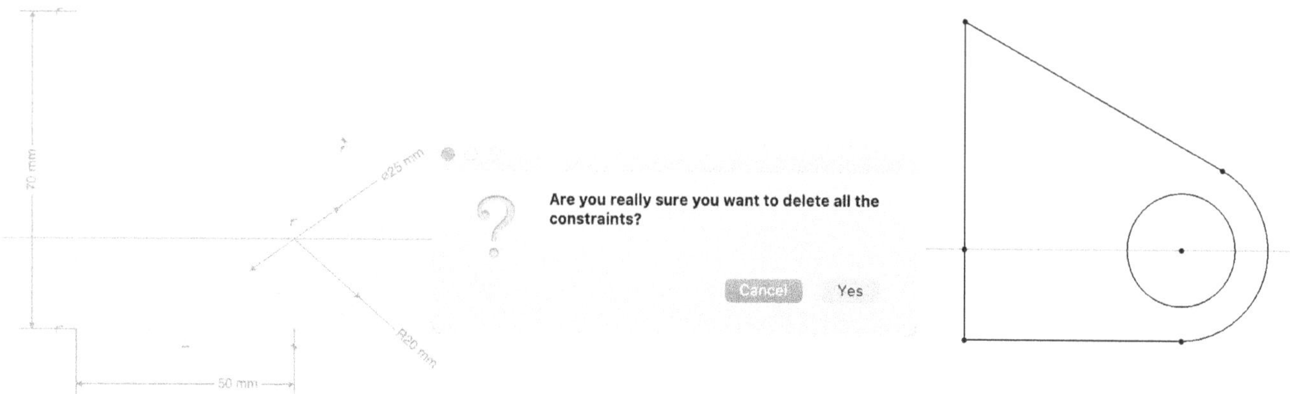

Delete all geometry

This command is used to delete all elements of a sketch. To do this, click **Sketch > Sketcher tools > Delete all geometry** on the menu bar. Next, click **Yes** on the message box.

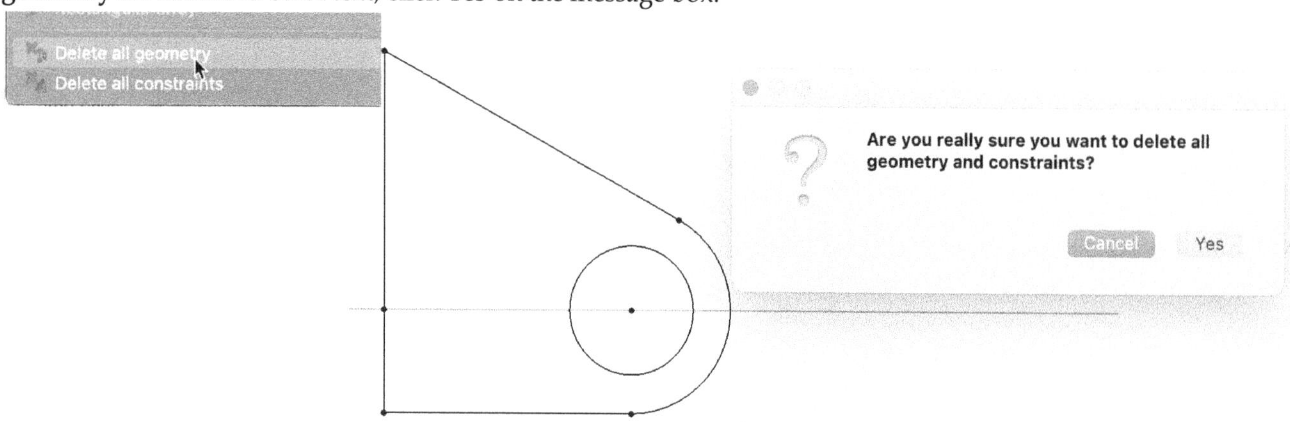

The Mirror Sketch command

This command creates a mirror copy of the selected elements using the X, or Y axes, or origin point. Select the objects to mirror and click the **Mirror sketch** icon on the **Sketcher** toolbar (or) click **Sketch > Mirror sketch** on the menu bar. Next, select the **X-Axis** option and click **OK** on the **Select Mirror Axis/Point** dialog to mirror the objects about X axes. Click the **Model** tab on the **Combo View** panel and notice that a new sketch for the mirror object is created.

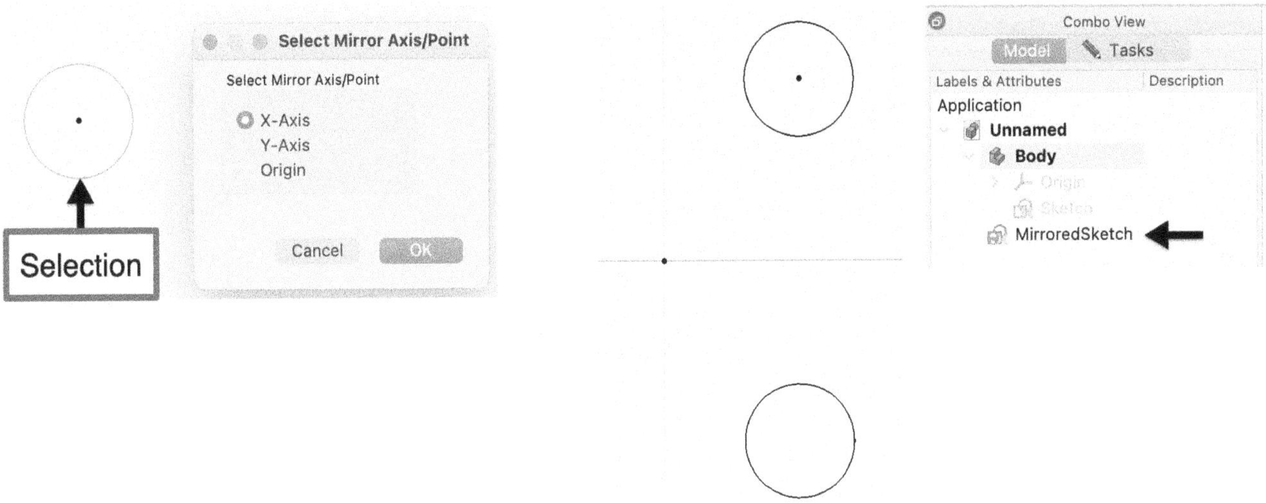

Creating Splines

Splines are non-uniform curves, which are used to create smooth shapes. In FreeCAD 0.19, you can create a smooth spline curve using two commands: **B-spline by control points** and **Periodic B-spline by control points**.

B-spline by control points

The **B-spline by control points** command helps you to create a spline by defining various points called as control points. Click the **Spline** drop-down > **B-spline by control points** 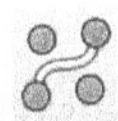 icon on the **Sketcher geometries** toolbar (or) click **Sketch > Sketcher geometries > Create B-spline** on the menu bar. In the graphics window, click to specify the first control point. Move the pointer and specify the second point. Likewise, specify the other control points. As you define the control points, the construction lines are created connecting them. Also, the spline will be created. Right-click to complete the spline. You can control the shape of the spline using the control points. Indeed, you can add dimensions and constraints to the control point.

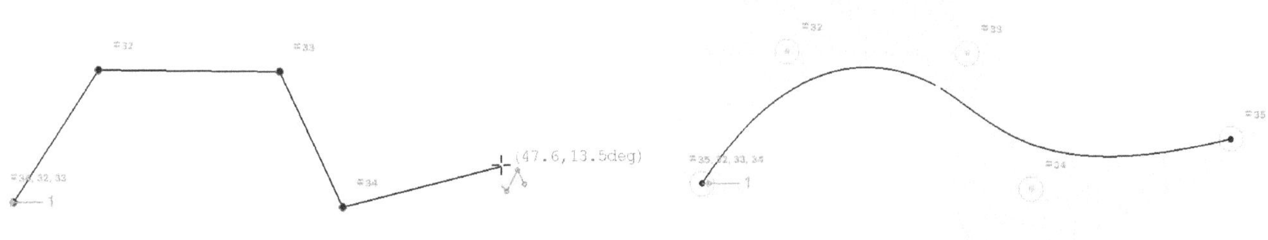

Periodic B-spline by control points

The **Periodic B-spline by control points** command creates a closed B-spline. Click the **Spline** drop-down > **Periodic B-spline by control points**  icon on the **Sketcher geometries** toolbar (or) click **Sketch > Sketcher geometries > Create periodic B-spline** on the menu bar. Next, specify the control points and right-click.

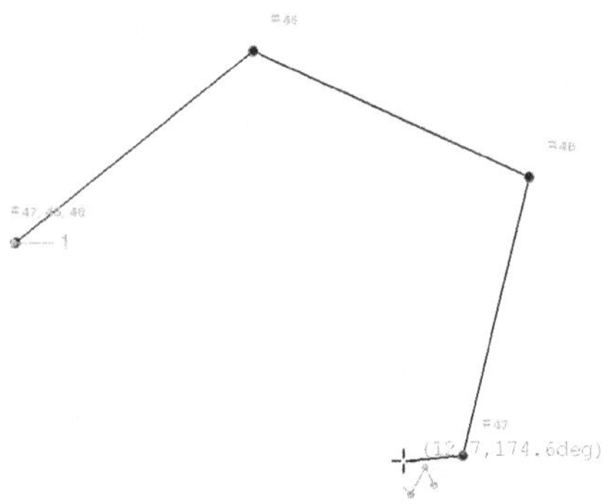

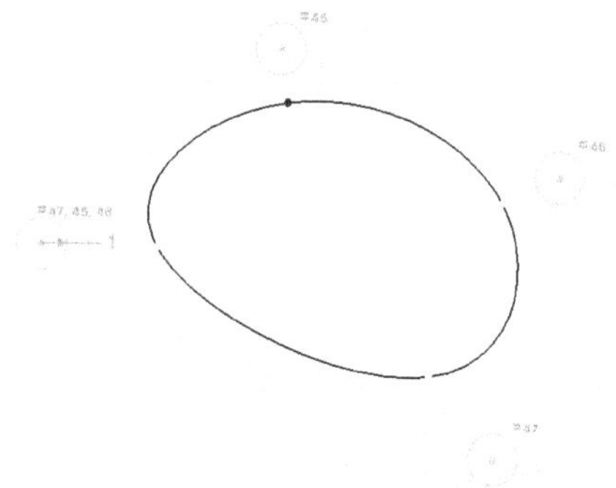

Tutorial 1 (Millimetres)

In this example, you will draw the sketch shown below.

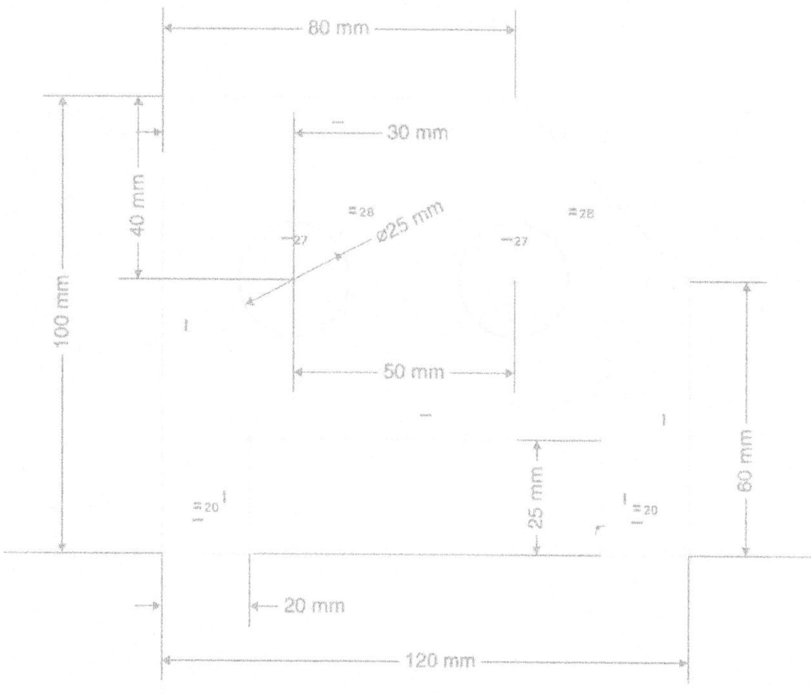

Starting a new part file

1. Click **FreeCAD** on the desktop.
2. Click the **New** icon on the **File** toolbar; it creates a new document.
3. On the **Workbench** toolbar, select **Workbench** drop-down **> Part Design** (or) select **View > Workbench > Part Design** on the Menu bar.
4. Click **Edit > Preferences** on the Menu bar; the **Preferences** dialog appears on the screen.
5. Click **Units** tab and select **Unit System > Standard (mm/kg/s/degree)**.
6. Select **Number of decimals > 2** and click **OK** on the **Preferences** dialog.

Starting a sketch

1. To start a new sketch, click **Part Design Helper** toolbar > **Create a new sketch** .
2. On the **Combo View** panel, select the **XZ_Plane** from the **Tasks** tab and click **OK**.
3. On the **Combo View** panel, click the **Tasks** tab and expand the **Edit Controls** section. Next, make sure that the **Show grid** checkbox is selected.
4. On the **Sketcher geometries** toolbar, click the **Polyline** icon (or) click **Sketch > Sketcher geometries > Create polyline**.
5. Click on the origin point to define the first point of the line.
6. Move the pointer horizontally toward the right and notice that the horizontal constraint symbol appears below the cursor.
7. Click to define the endpoint of the line.
8. Move the pointer vertically upward and notice the vertical constraint symbol below the cursor. Click to draw the vertical line.

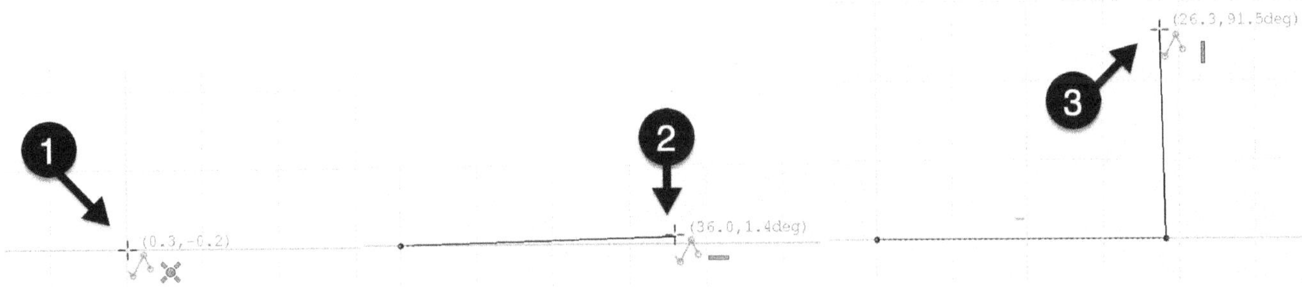

9. Move the pointer horizontally toward the right. Click when the horizontal constraint glyph appears.
10. Move the pointer vertically downward. Click when the vertical constraint and coincident constraint glyphs appear.
11. Move the pointer horizontally toward right up to a short distance. Click when the horizontal constraint glyph appears.

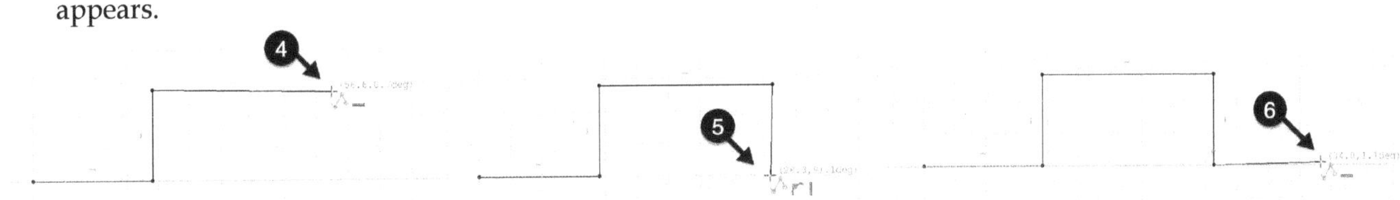

12. Move the pointer vertically upward and click.
13. Move the pointer in the top-left direction and click to create an inclined line.
14. Move the pointer horizontally towards left and click.

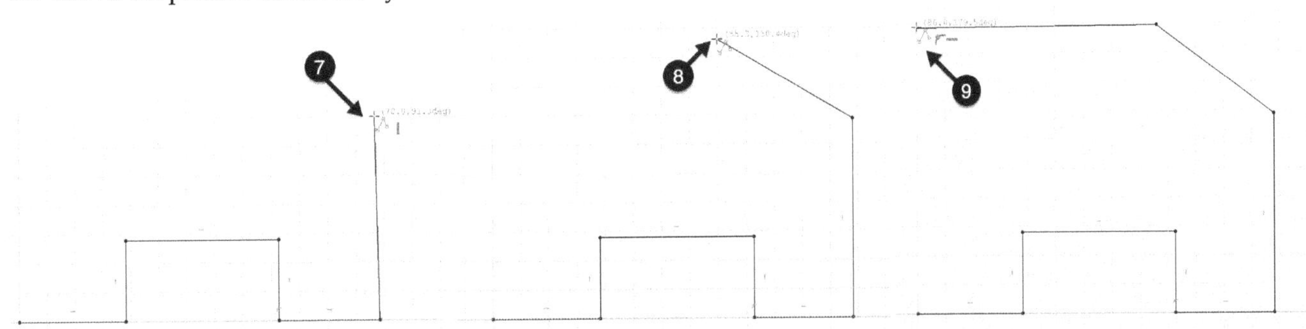

15. Move the pointer vertically downward and click on the origin point.
16. Right-click twice to deactivate the **Polyline** tool.

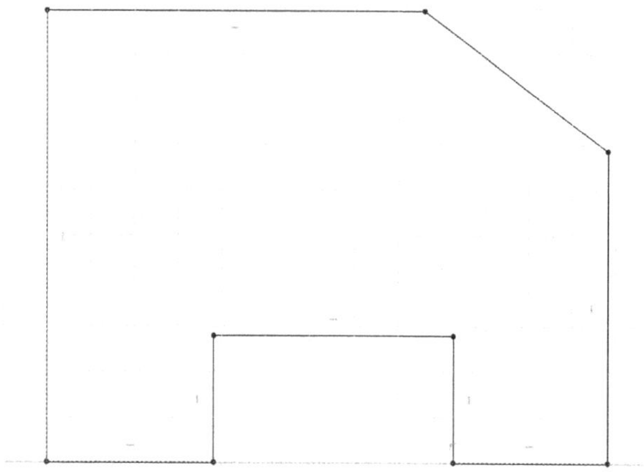

Adding Constraints

1. On the **Sketcher constraints** toolbar, click **Equal** . Next, select the two horizontal lines at the bottom; the selected lines become equal in length.

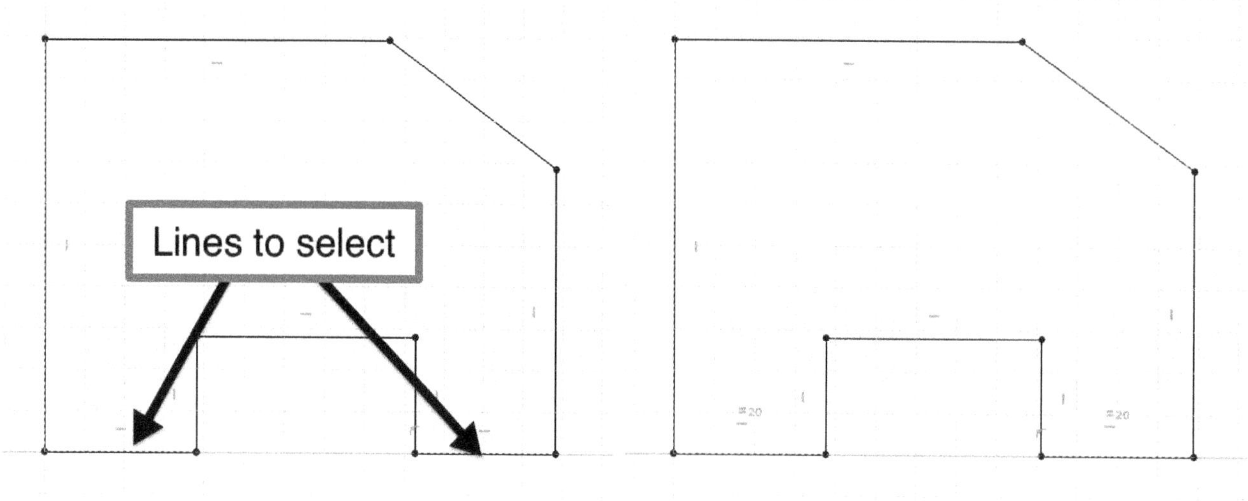

2. Select the small vertical lines; the selected lines are made equal in length.

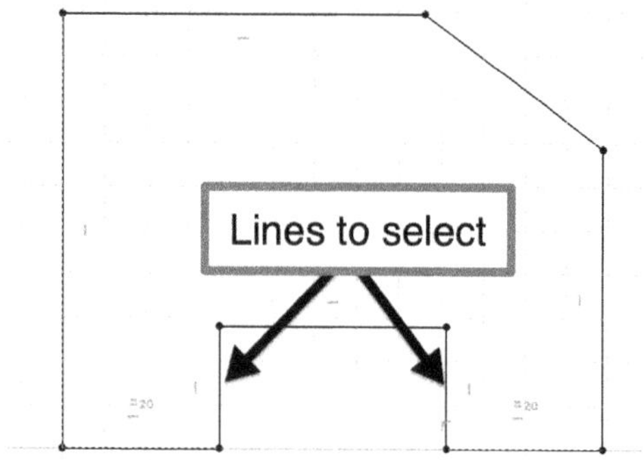

Adding Dimensions

1. On the **Sketcher constraints** toolbar, click the **Constrain horizontal distance** icon (or) click **Sketch > Sketcher constraints > Constrain horizontal distance** on the menu bar.
2. Select the bottom horizontal line of the rectangle.
3. Type-in **20** in the **Length** box of the **Insert Length** dialog and click the **OK** button.

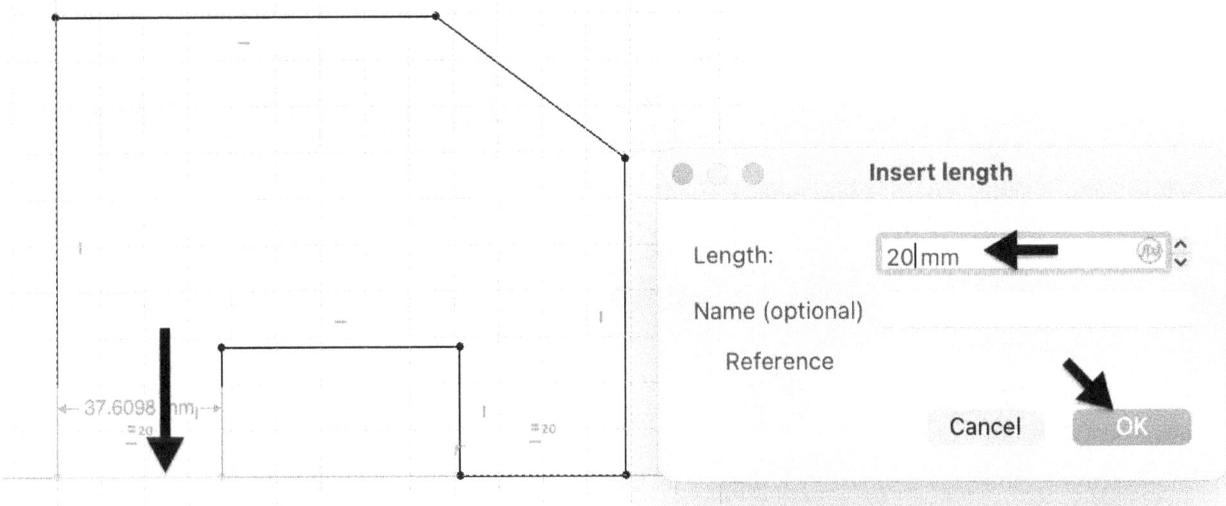

4. On the **Sketcher constraints** toolbar, click the **Constrain vertical distance** icon (or) click **Sketch > Sketcher constraints > Constrain Vertical distance** on the menu bar.
5. Click on the small vertical line located at the right side. Move the mouse pointer towards the left and click to position the dimension.
6. Type-in **25** in the **Length** box of the **Insert Length** dialog and click the **OK** button.

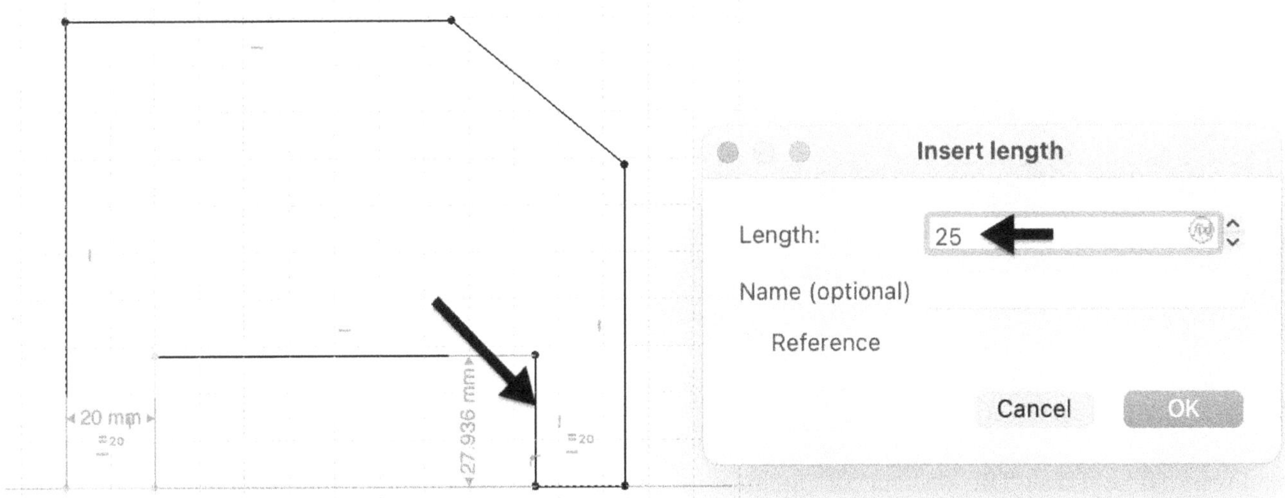

7. Create other distance constraints in the sequence, as shown below. Use the **Constrain horizontal distance** and **Constrain vertical distance** commands.

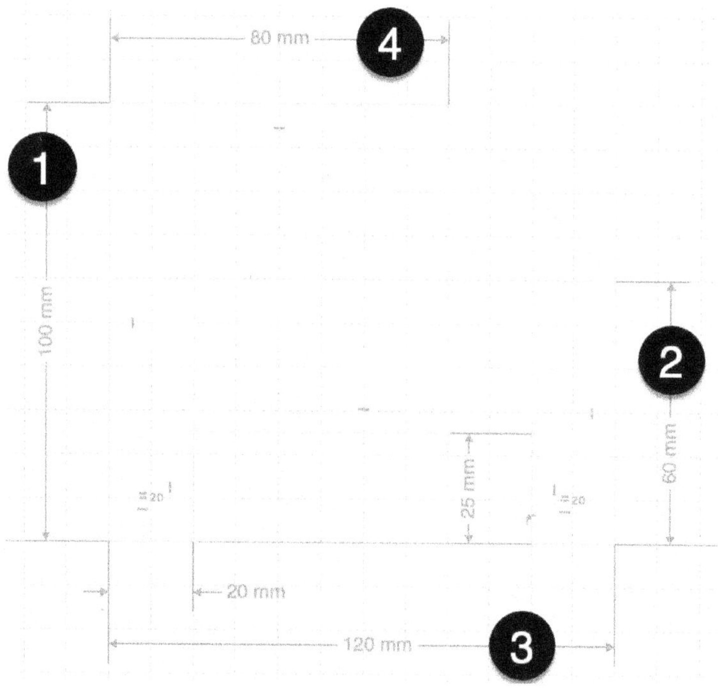

8. On the **Sketcher geometries** toolbar, click the **Create Circle** icon (or) click **Sketch > Sketcher geometries > Create Circle**.
9. Click inside the sketch region to define the center point of the circle. Move the mouse pointer and click to define the diameter. Likewise, create another circle.

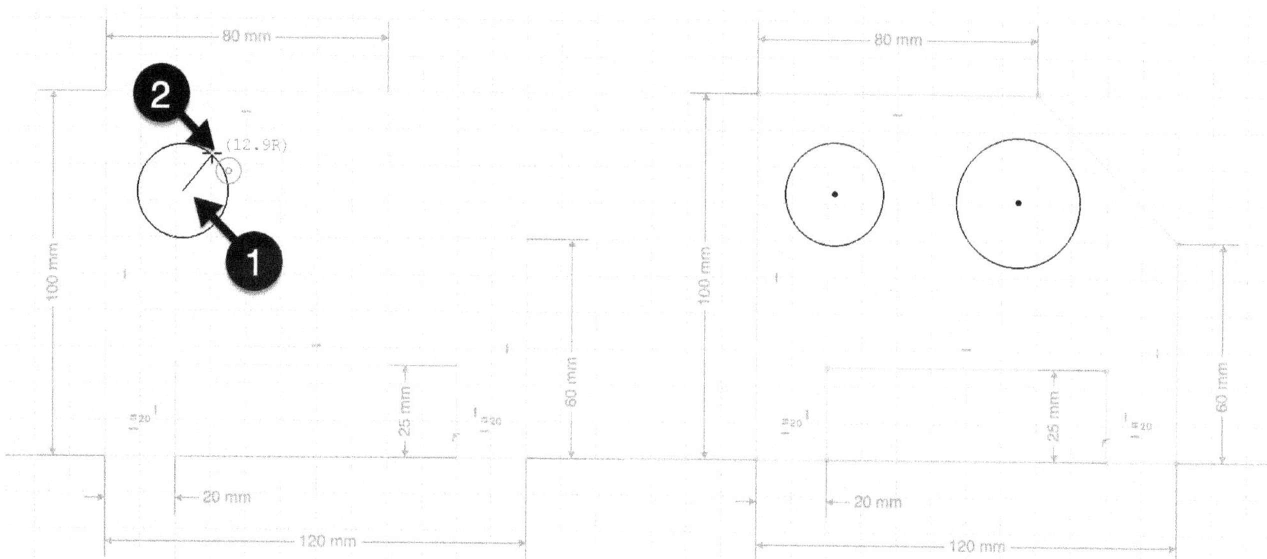

10. Press and hold the CTRL key and click on the center points of the two circles. On the **Sketcher constraints** toolbar, click **Constrain horizontally** ; the centers of the circles are aligned horizontally.

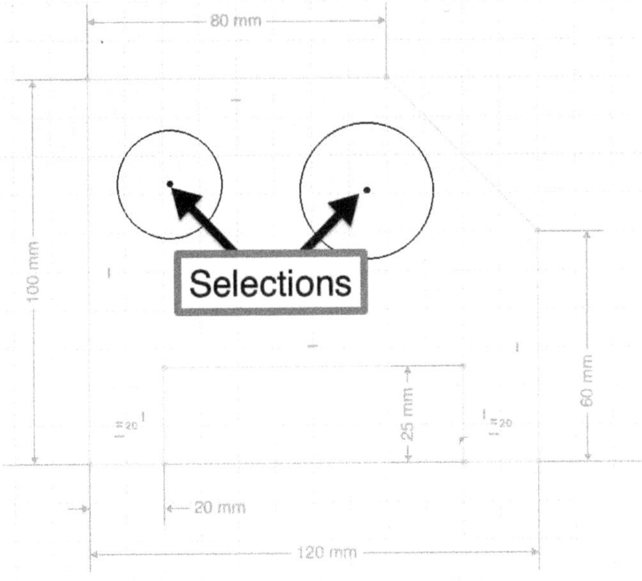

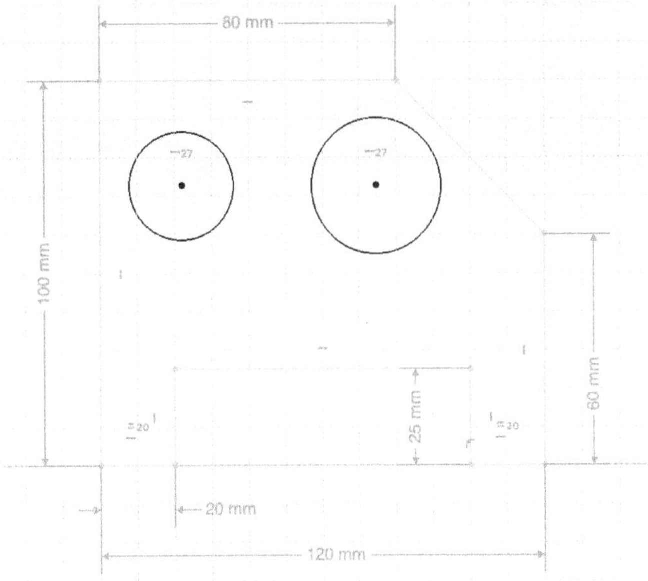

11. On the **Sketcher constraint** toolbar, click the **Constrain equal** icon. Next, click on the two circles; the diameters of the circles become equal.

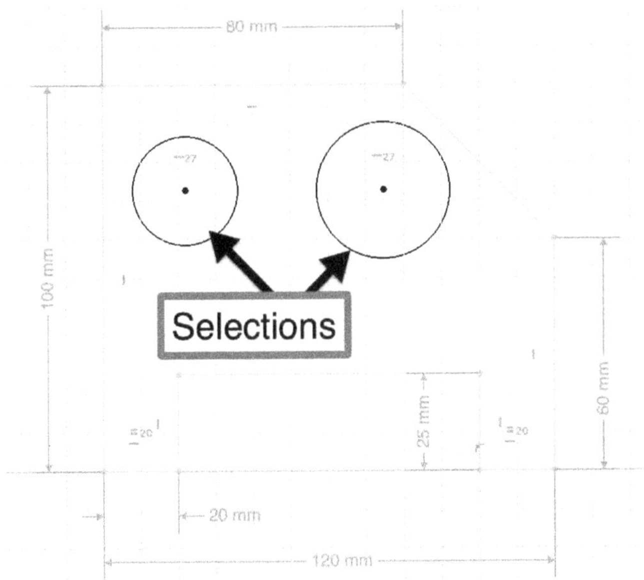

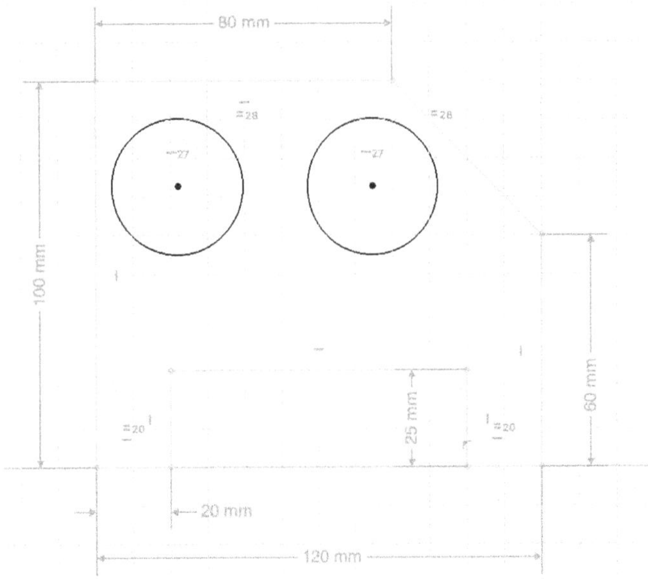

12. On the **Sketcher constraints** toolbar, Click **Constraint on arc or circle** drop-down > **Constrain diameter**.
13. Click on any one of the circles. Move the mouse pointer and click to position the dimension. Type **25** in the **Diameter** box and press Enter.

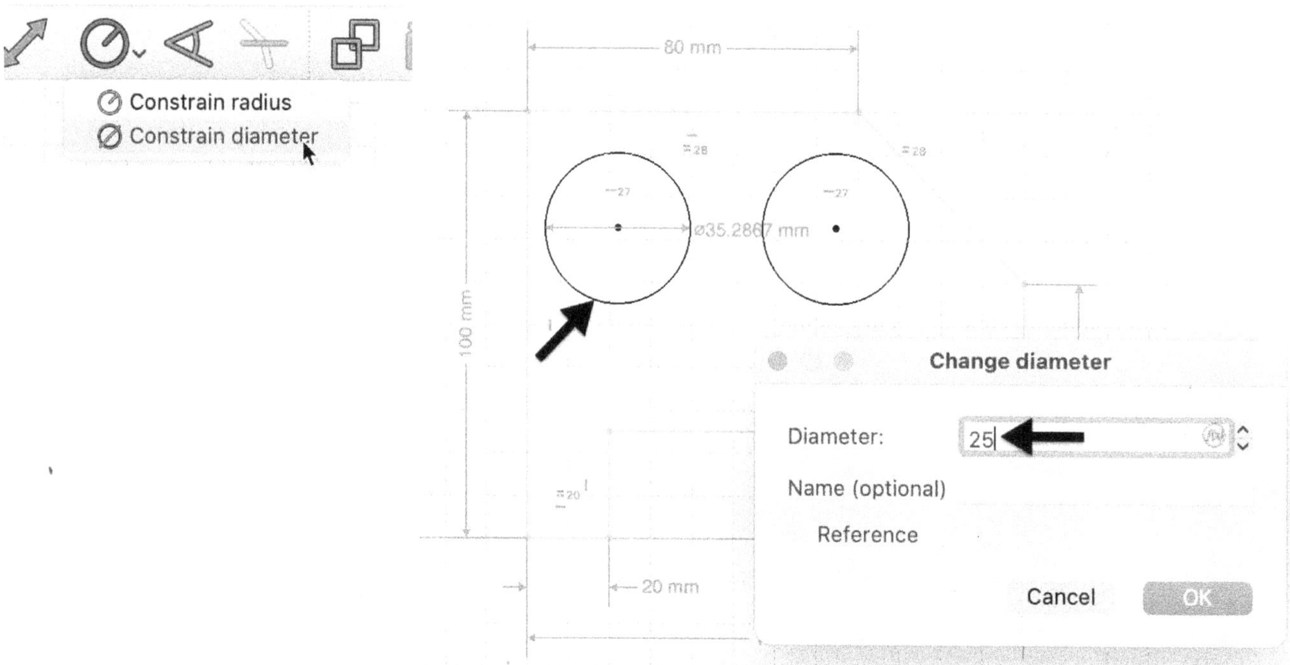

14. Create the remaining horizontal and vertical distance constraints between the circles and the adjacent lines, as shown below.

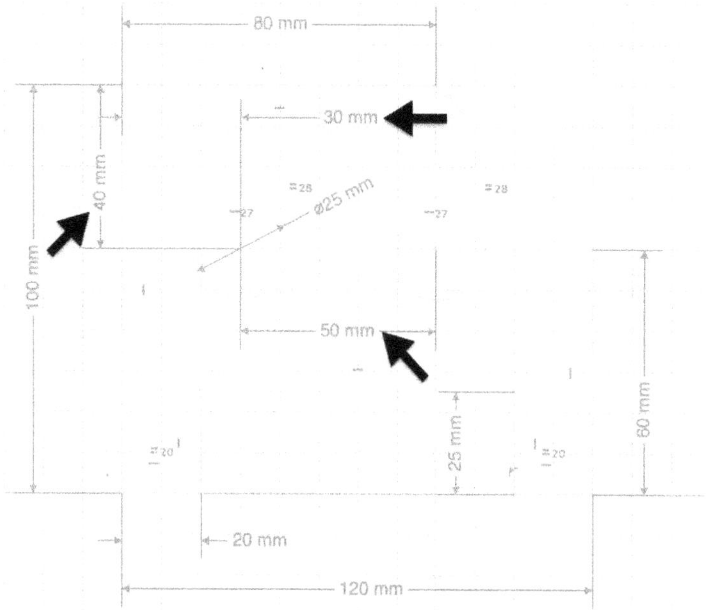

15. Click the **Close** button on the **Tasks** tab of the **Combo View** panel.
16. Click **File > Save** on the Menu bar. Next, browse to the required location on your computer and type **C2_example1** in the **File name** box. Click **Save**.
17. Close the file tab on the bottom of the window.

Tutorial 2 (Inches)

In this example, you draw the sketch shown below.

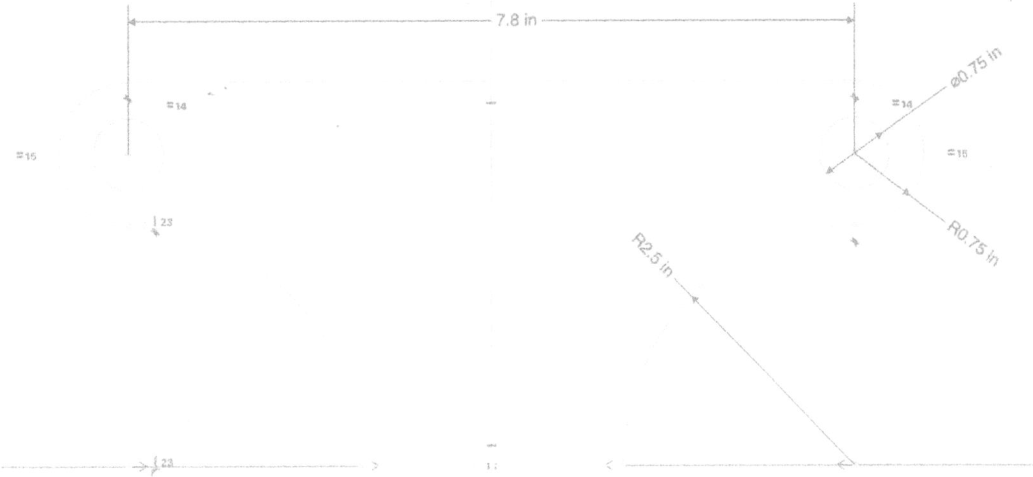

Creating a New document

1. Start FreeCAD 0.19.
2. Click the **New** icon on the **File** toolbar; it creates a new document.
3. Select the **Part Design** option from the **Workbenches** drop-down.
4. On the Menu bar, click **Edit > Preferences**. Next, click the **Units** tab on the **Preferences** dialog.
5. Select **Unit System > Imperial decimal**. Next, type **3** in the **Number of decimals** box, and then click **OK**.

Creating a Sketch

1. Click the **Create sketch** icon on the **Part Design Helper** toolbar.
2. Select the XZ Plane and click **OK** on the **Combo View** to start the sketch.
3. On the **Sketcher geometries** toolbar, click the **Create line** icon.
4. Click on the left side of the graphics window.
5. Move the pointer horizontally towards the right and click to draw a line.

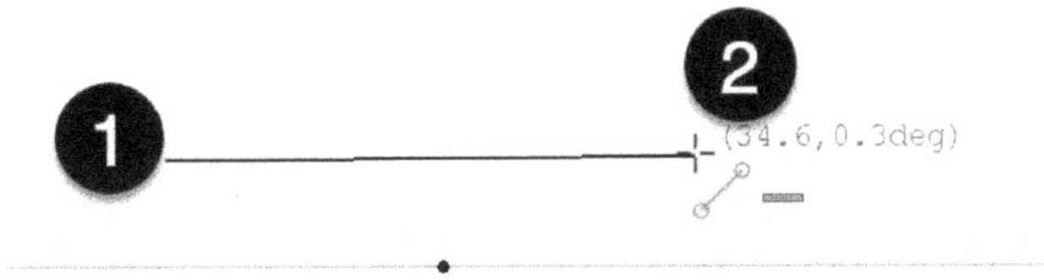

6. Right-click to deactivate the **Create line** tool.
7. Click **Arcs** drop-down **> End points and rim point arc** on the **Sketcher geometries** toolbar and click on the endpoint of the line.
8. Move the pointer in the top-right direction, and then click to define the endpoint of the arc.
9. Move the pointer and click to define the radius of the arc.

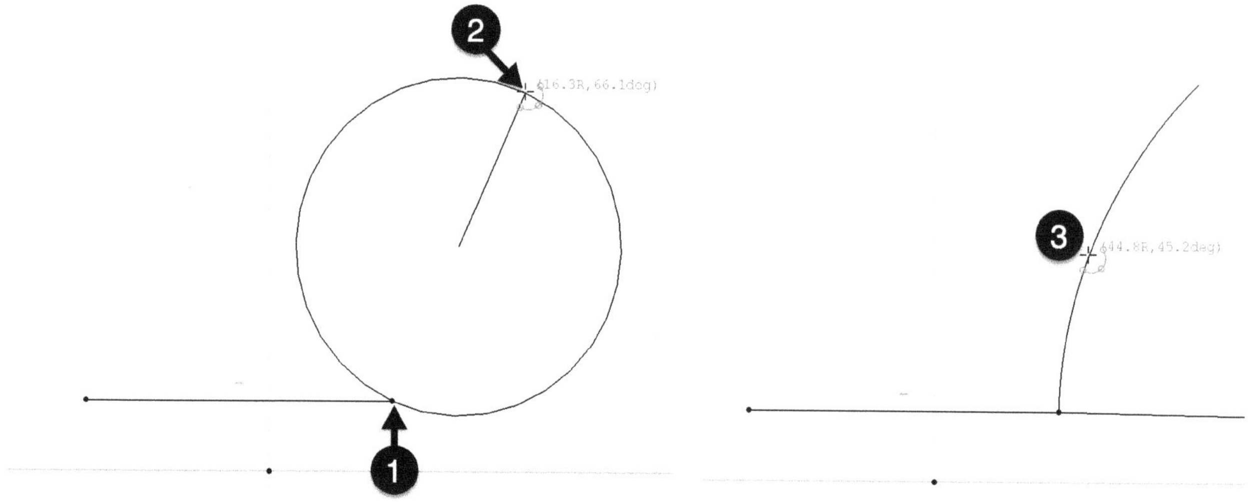

10. Click on the endpoint of the arc. Next, move the pointer upward and click to define the endpoint of the arc.
11. Move the pointer and click to define the radius of the arc.

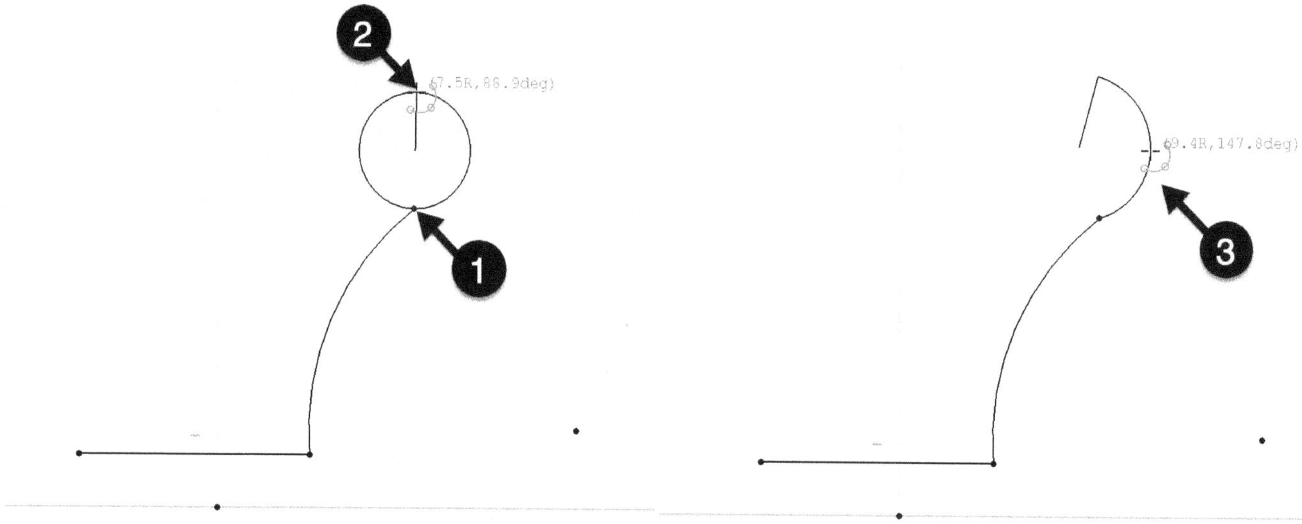

12. Activate the **Create line** tool (click the **Create line** icon on the **Sketcher geometries** Toolbar).
13. Select the endpoint of the arc.
14. Move the mouse pointer towards the left and click to create a horizontal line. Note that the length of the new line should be greater than that of the lower horizontal line.

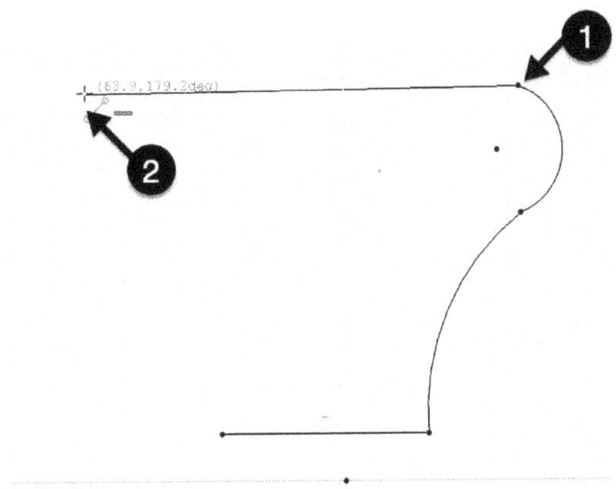

15. Press **Esc** to deactivate the **Create Line** tool.
16. Activate the **End points and rim point arc** tool and click on the endpoint of the line.
17. Move the pointer downwards and click. Next, move the pointer towards the left and click to define the radius of the arc.

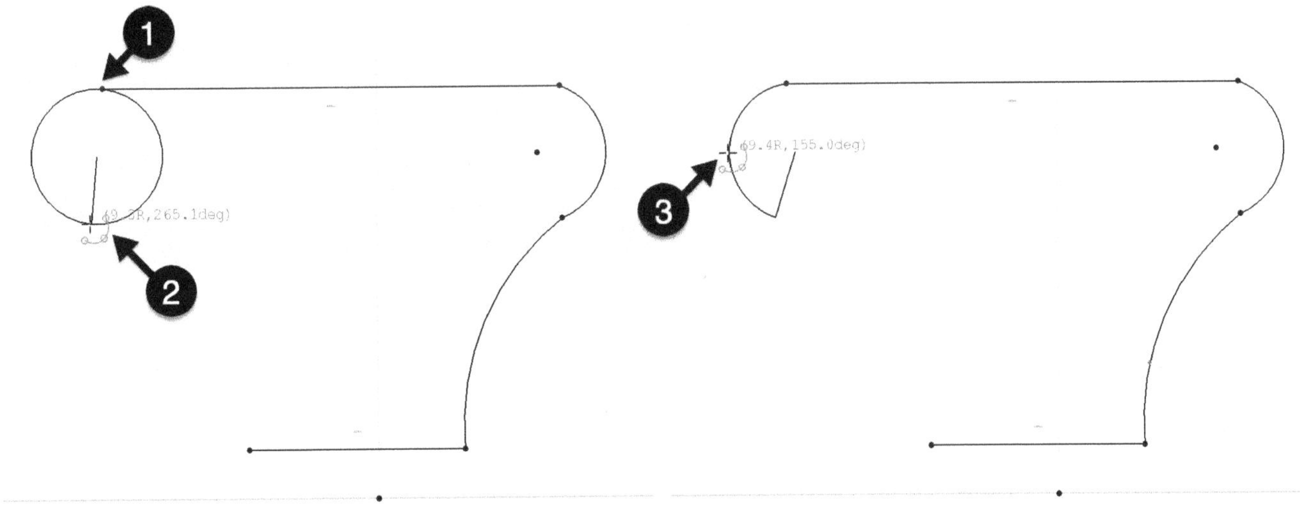

18. Click on the endpoint of the previous arc. Next, select the start point of the sketch.
19. Move the pointer and click to define the radius of the arc.

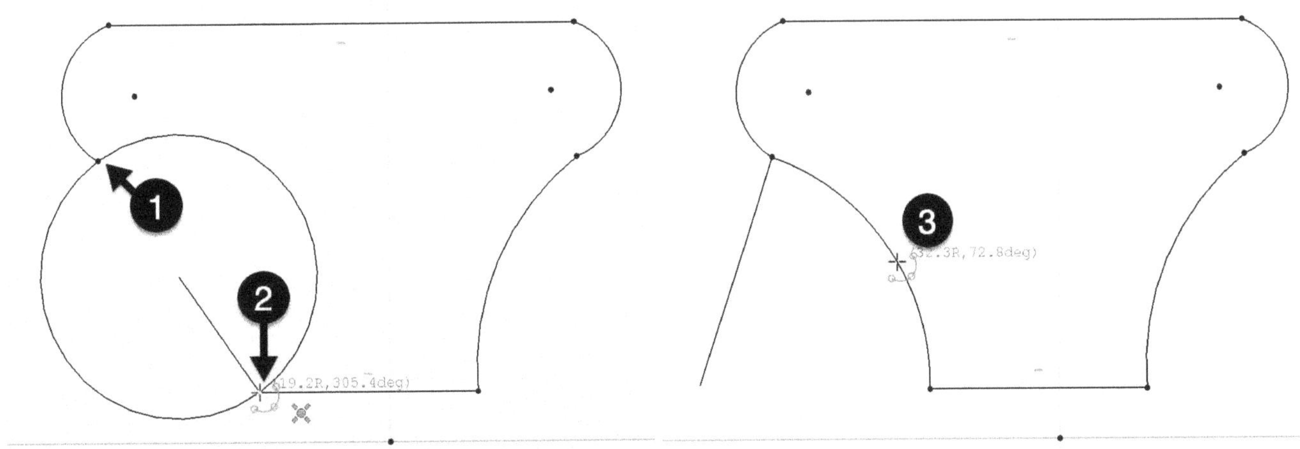

20. Click the **Constrain symmetrical** icon on the **Sketcher constraints** toolbar.

21. Select the left endpoint of the lower horizontal line, the right endpoint of the lower horizontal line, and the origin point. The endpoints of the horizontal line are made symmetric about the origin point.

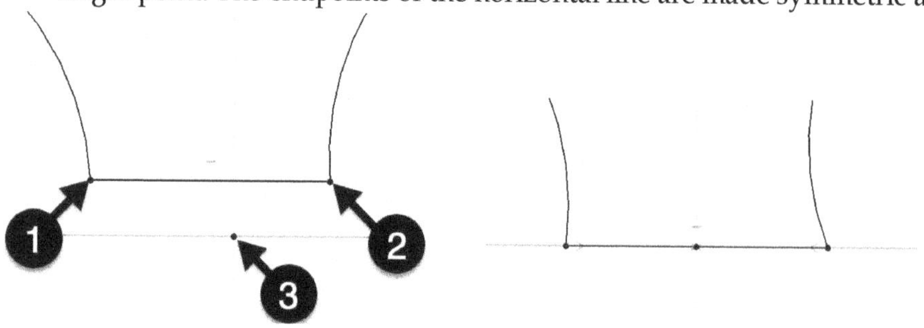

22. Click the **Constrain point onto object** icon on the **Sketcher constraints** toolbar.
23. Select the center point of any one of the large arcs. Next, select the horizontal axis.

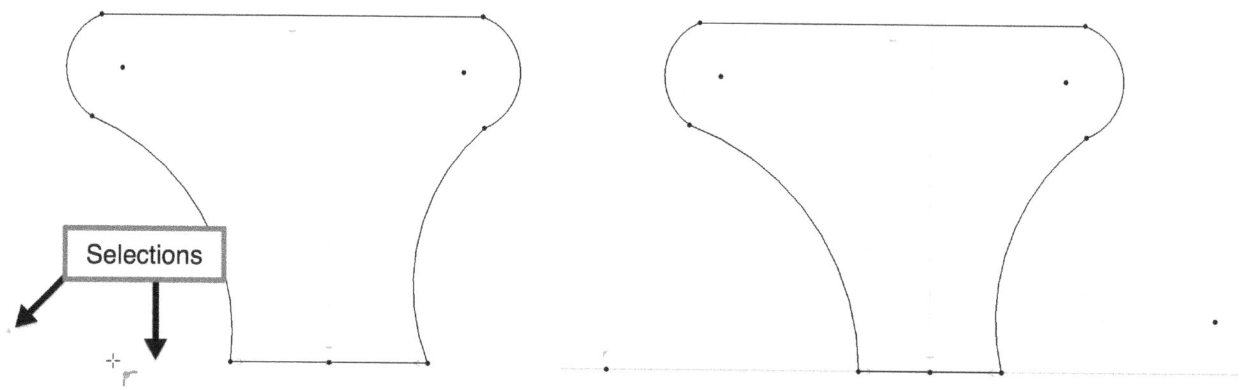

24. Click the **Constrain symmetrical** icon on the **Sketcher constraints** toolbar.
25. Select the center point of the left large arc, the center point of the large right arc, and the origin point. The centerpoints of the arcs are made symmetric about the origin point.

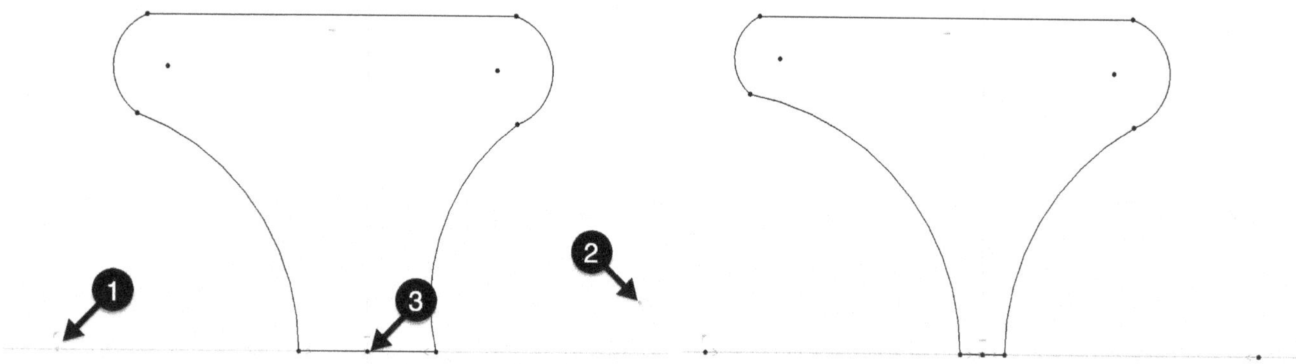

26. On the menu bar, click **Sketch > Sketcher geometries > Create Circle**.
27. Select the center point of any one of the small arcs. Next, move the pointer outward and click to create the circle.
28. Select the center point of another small arc, move the pointer outward, and then click.

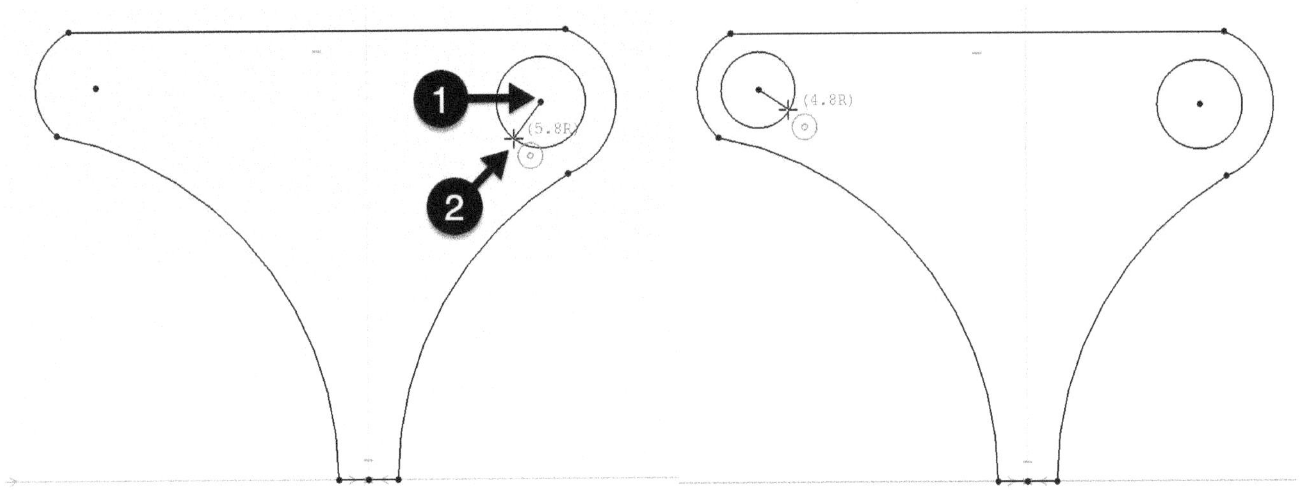

29. Click the **Constrain equal** icon on the **Sketcher constraints** toolbar. Next, select the two circles to make them equal in size.
30. Select the two small arcs to make them equal in radius.

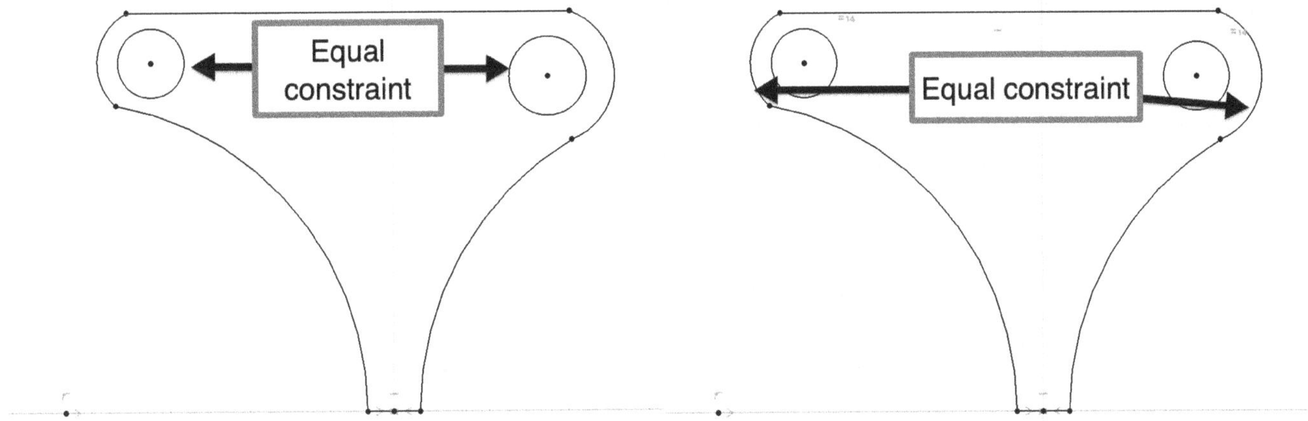

31. Click the **Constrain tangent** icon on the **Sketcher constraints** toolbar. Next, select the large and small arcs on the left side to make them tangent to each other.
32. Likewise, create the tangent constraints by selecting the other elements, as shown. Press ESC.

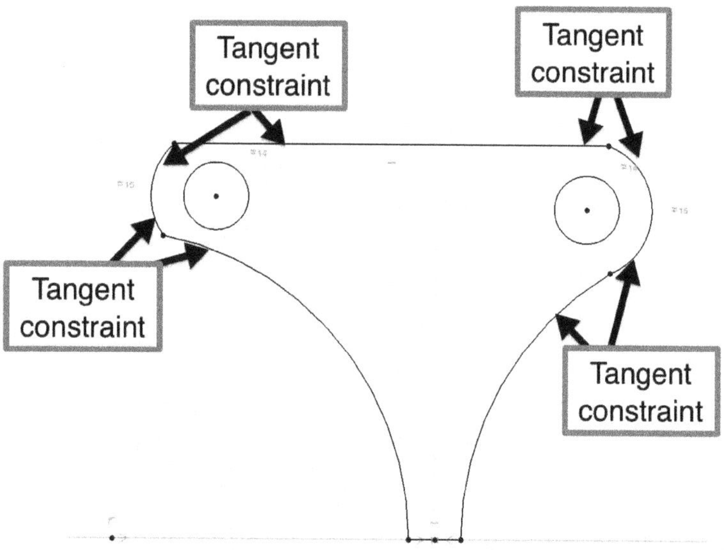

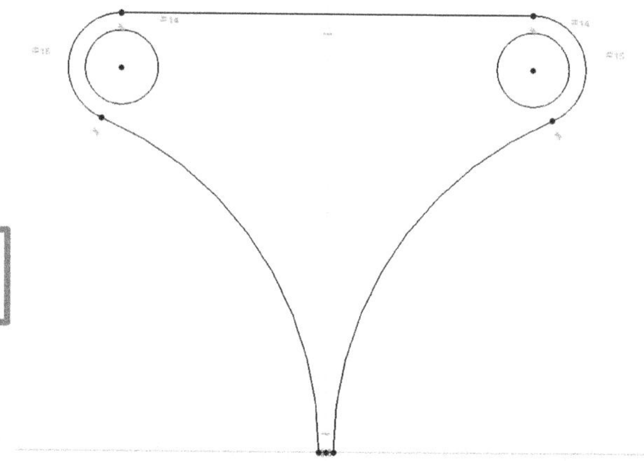

33. Click the **Constrain radius** icon on the **Sketcher constraints** toolbar.
34. Select any one of the small arcs. Type **0.75** in the **Radius** box and click **OK**.

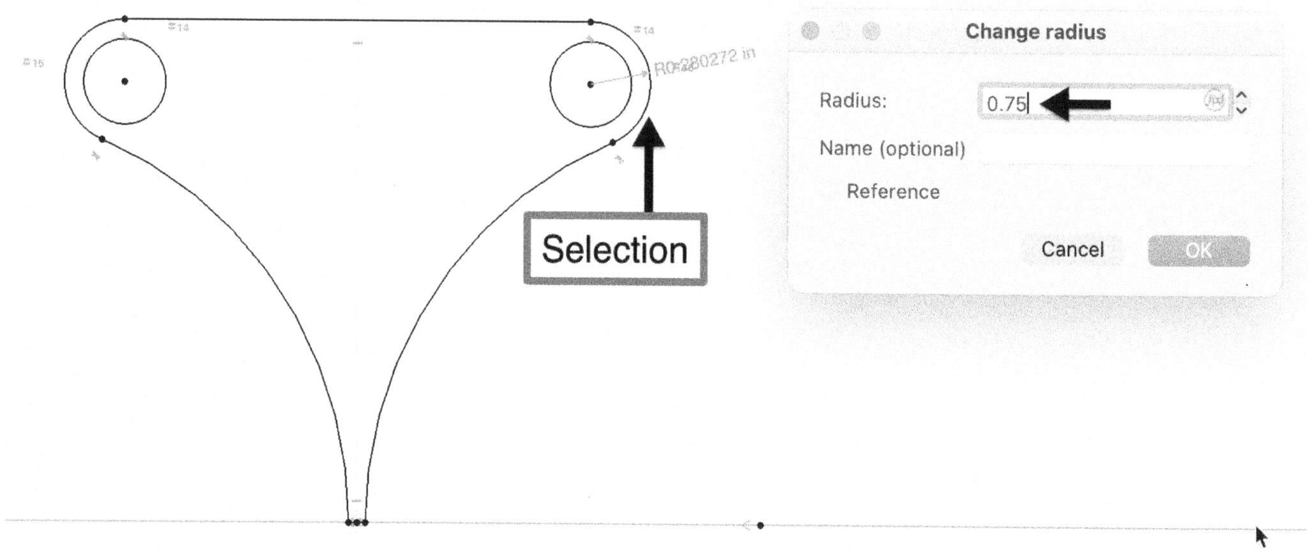

35. On the **Sketcher constraints** toolbar, click **Constrain radius > Constrain diameter**.
36. Select any one of the circles. Type 0.75 in the **Diameter** box and click **OK**.

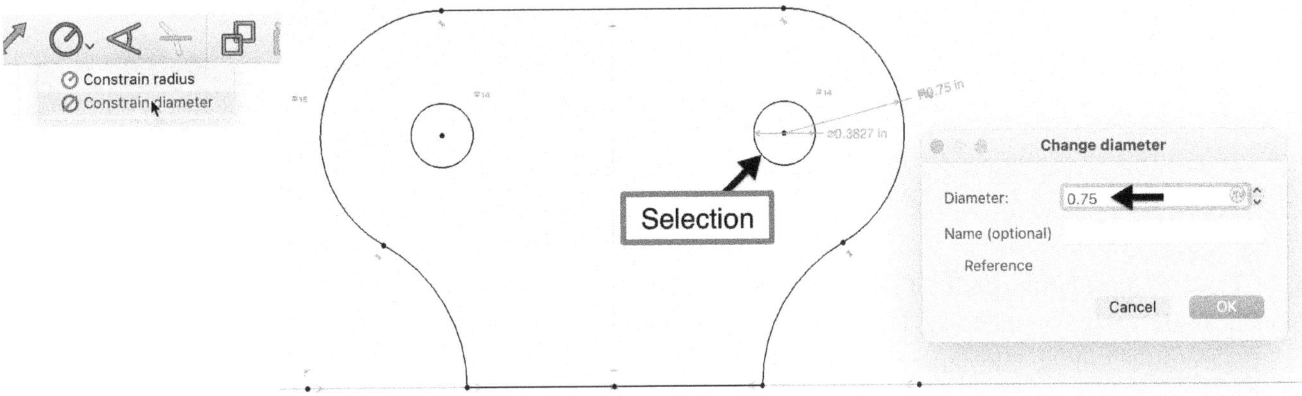

37. Click the **Constrain horizontal distance** icon on the **Sketcher constraints** toolbar.
38. Select the centerpoints of the two circles. Type 7.8 in the **Length** box and click **OK**.

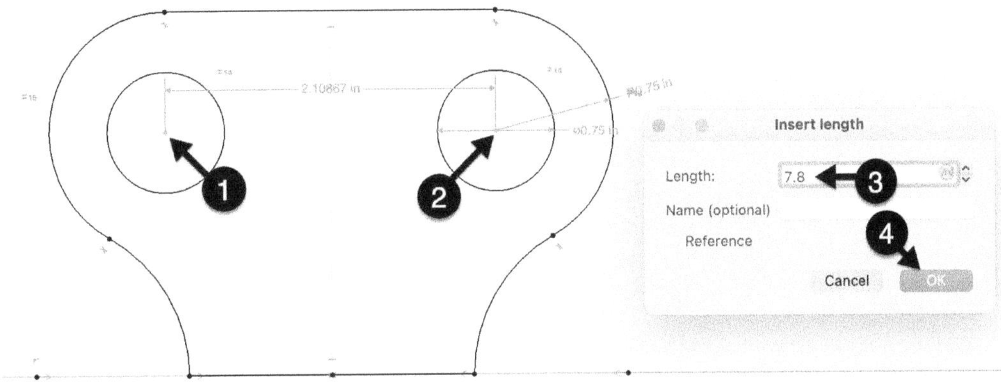

39. Press and hold the Ctrl key and select the centerpoint and endpoint of the left large arc.
40. Click the **Constrain vertically** icon on the **Sketcher constraints** toolbar; the selected points are aligned vertically.

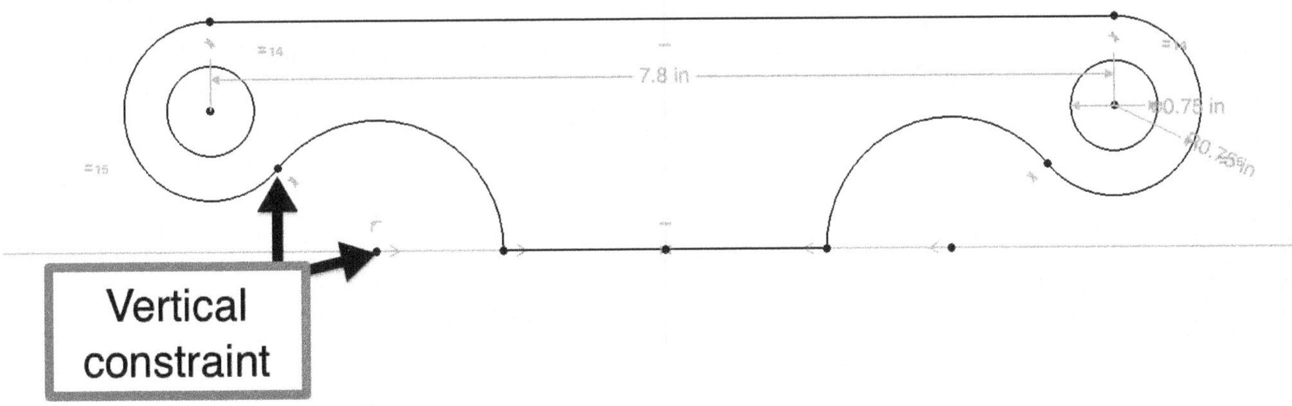

41. Click the **Constrain radius** icon on the **Sketcher constraints** toolbar.
42. Select any one of the large arcs. Type **2.5** in the **Radius** box and click **OK**.

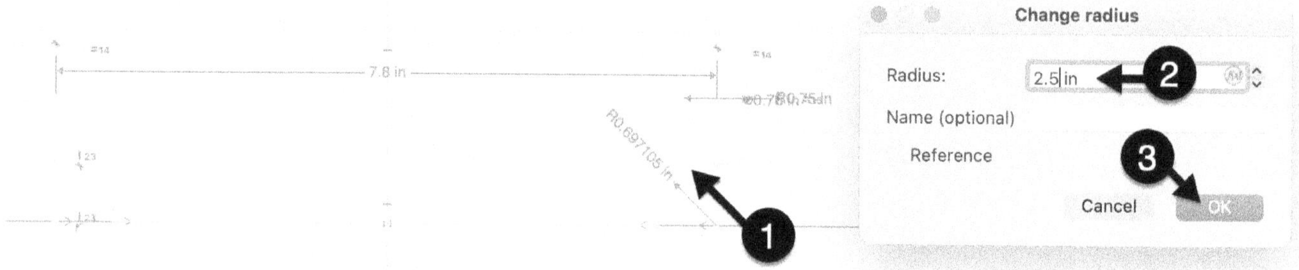

43. Click the **Close** button on the **Tasks** tab of the **Combo View** panel.
44. Click **File > Save** on the Menu bar. Next, browse to the required location on your computer and type **C2_example2** in the **File name** box. Click **Save**.
45. Close the file tab on the bottom left corner of the window.

Tutorial 3 (Millimetres)

In this example, you draw the sketch shown below.

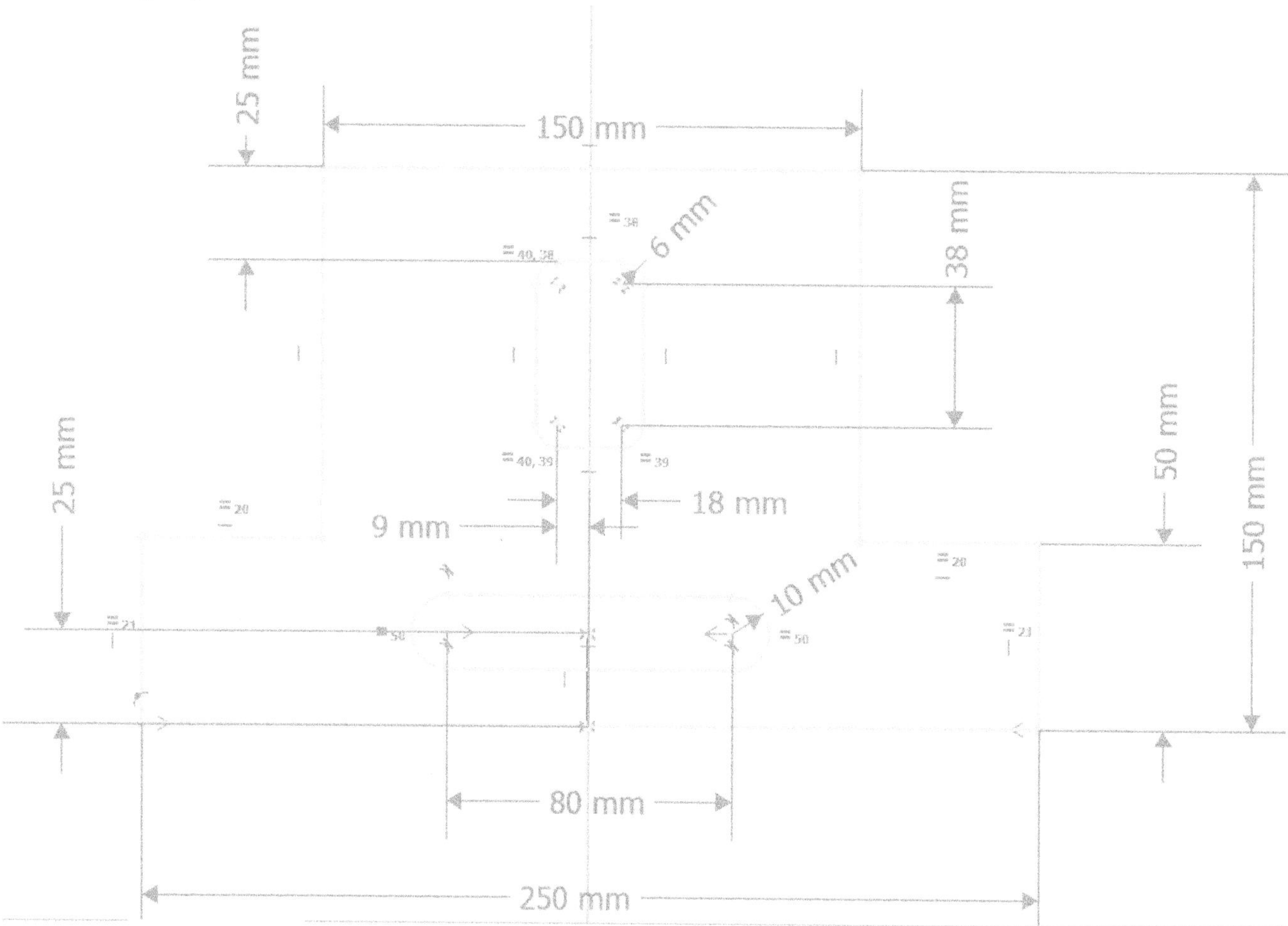

Creating a New Document

1. Start **FreeCAD 0.19**.
2. Click the **New** icon on the **File** toolbar; it creates a new document.
3. Click **Edit > Preferences** on the menu bar; the **Preferences** dialog appears on the screen.
4. Click **Units** tab on the dialog and select **Unit System > Standard(mm/kg/s/degree)**.

5. Select **Number of decimals > 2** and click **Apply**.
6. Click **OK** to close the **Preferences** dialog.
7. On the **Workbench** toolbar, select **Workbench** drop-down > **Part Design** (or) select **View > Workbench > Part Design** on the Menu bar.

Creating a Sketch

1. To start a new sketch, click **Create Sketch** tool on the **Part Design Helper** toolbar.
2. Select **XZ_Plane** from the **Select feature** section on the **Combo View** panel.
3. Click **OK** on the **Combo View** panel.
4. Activate the **Create line** tool (click the **Create line** icon on the **Sketcher geometries** toolbar).
5. Click on the origin point and move the pointer vertically up. Next, click to create a vertical line.

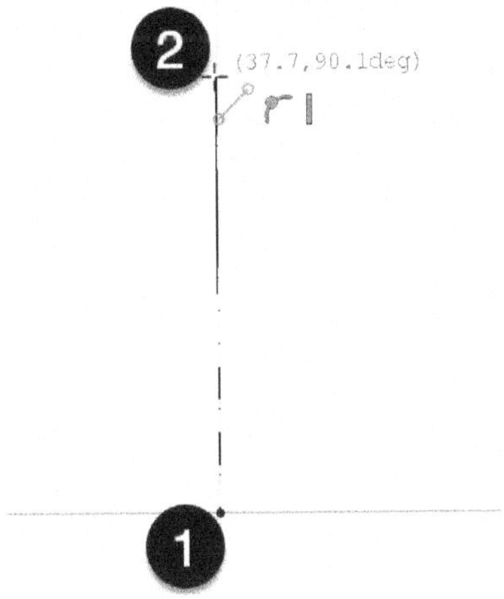

6. Press **Esc** to deactivate the **Create line** tool.
7. Select the vertical line and click the **Construction Mode** tool on the **Sketcher geometries** toolbar.
8. Activate the **Create polyline** tool and click the horizontal axis of the sketch on the left side of the construction line.
9. Move the mouse pointer vertically up and click to define the second point.
10. Move the pointer horizontally toward the right and click to create a horizontal line.
11. Move the pointer vertically up and click to create a vertical line. Press ESC.

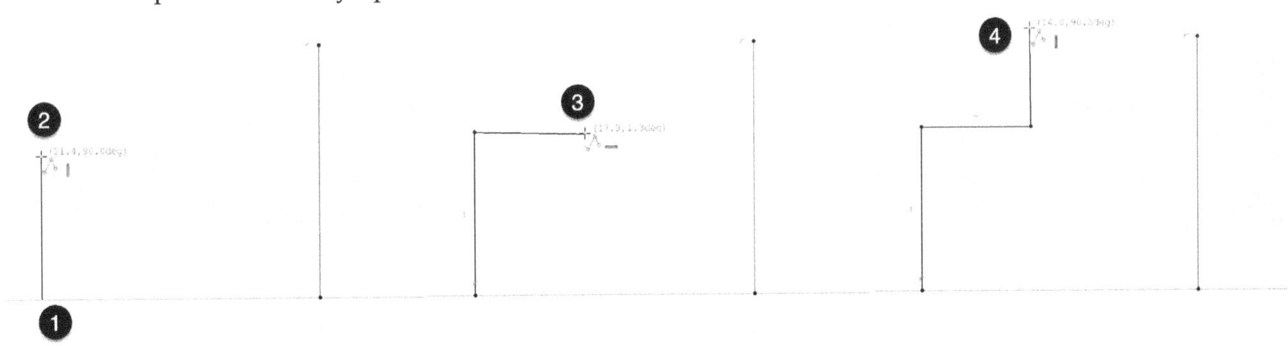

12. Press and hold the Ctrl key and select the lines and the construction line in the sequence given in the figure.

13. Click the **Symmetry** icon on the **Sketcher tools** toolbar (or) click **Sketch > Sketcher tools > Symmetry** on the menu bar; the lines are mirrored about the construction line.

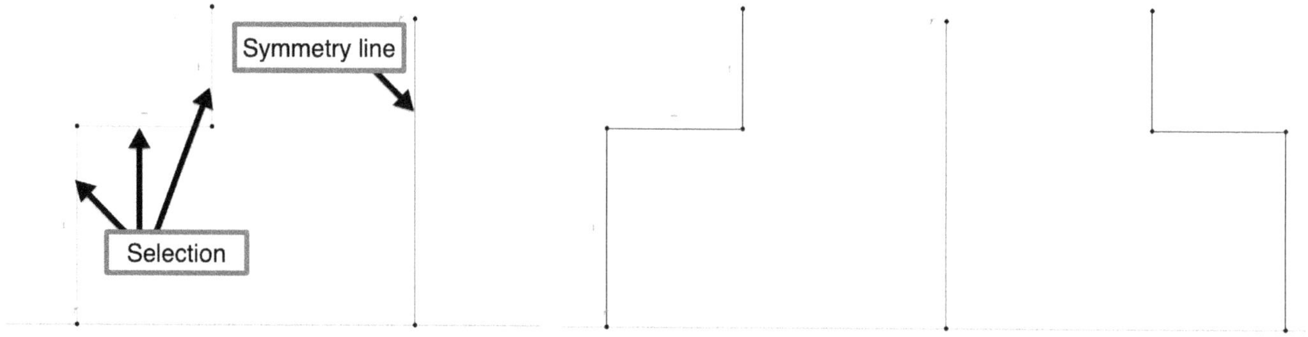

14. Activate the **Create line** tool (click the **Create line** icon on the **Sketcher geometries** toolbar).
15. Select the endpoints of the lower vertical lines displayed on both sides of the construction line.
16. Select the endpoints of the upper vertical lines. Next, press ESC.

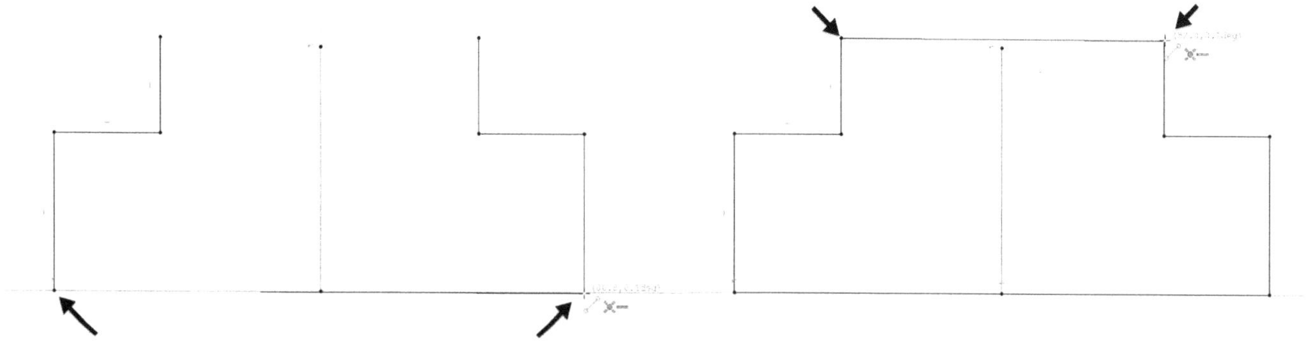

17. Click the **Constrain symmetrical** icon on the **Sketcher constraints** toolbar.
18. Select the horizontal line and origin point, as shown in the figure. The endpoints of the horizontal line are made symmetric about the origin point. Also, notice that the **Solver messages** section on the **Combo View** panel displays an error.

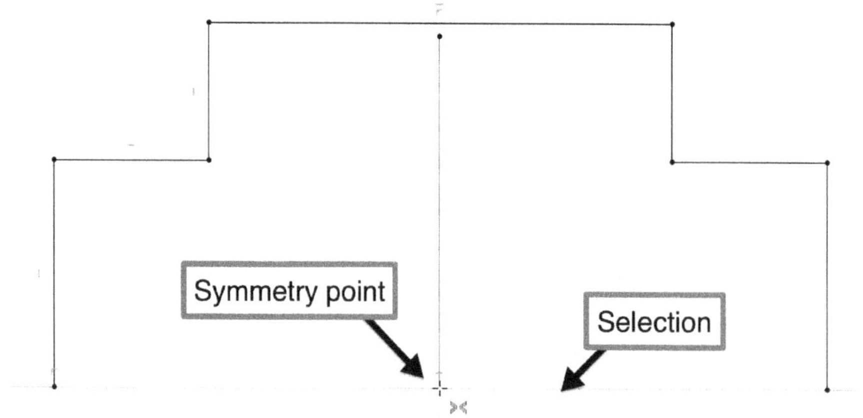

19. Click the **(click to select)** link in the Solver messages section. Next, scroll down in the **Constraints** section and notice that the **Horizontal Constraint** is highlighted.

20. Right click on the **Horizontal Constraint** and select **Delete**. (Skip Step 19 and 20, if there is no error)

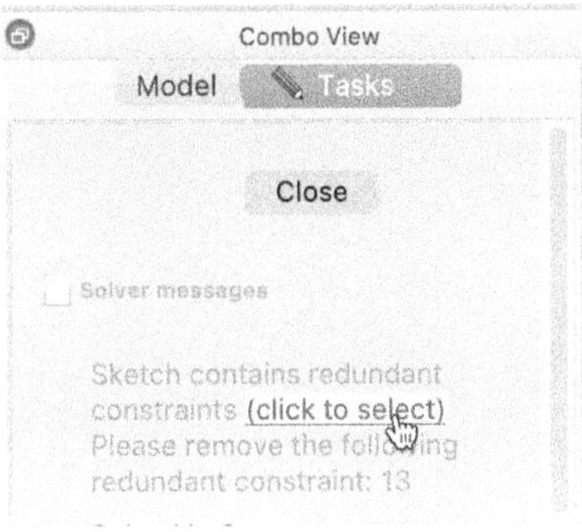

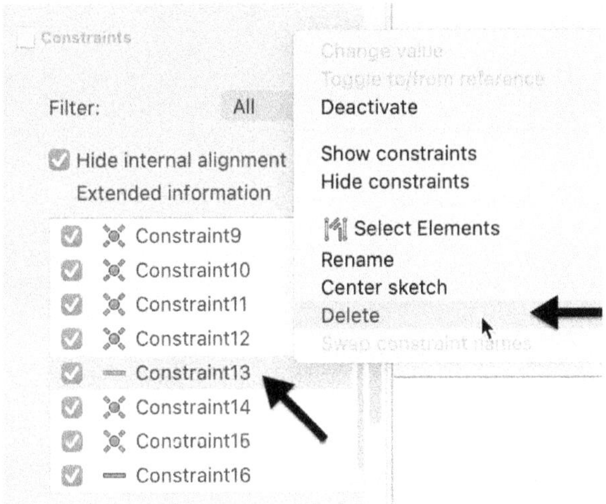

21. Click the **Constrain equal** icon on the **Sketcher constraints** toolbar
22. Select the two horizontal lines, as shown.
23. Select the two vertical lines to make them equal in length.

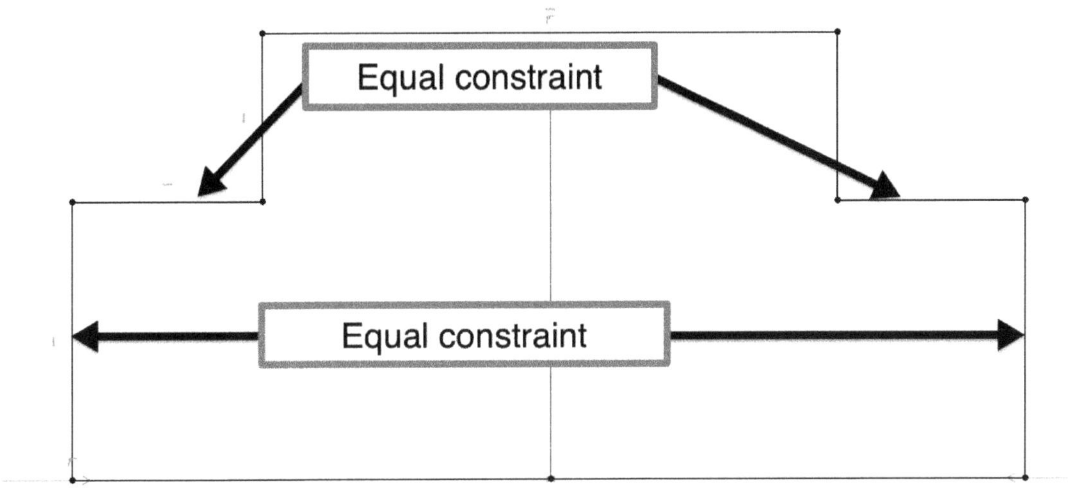

24. Apply the Vertical and Horizontal constraints to the lines, as shown.
25. Create horizontal distance and vertical distance constraints, as shown.

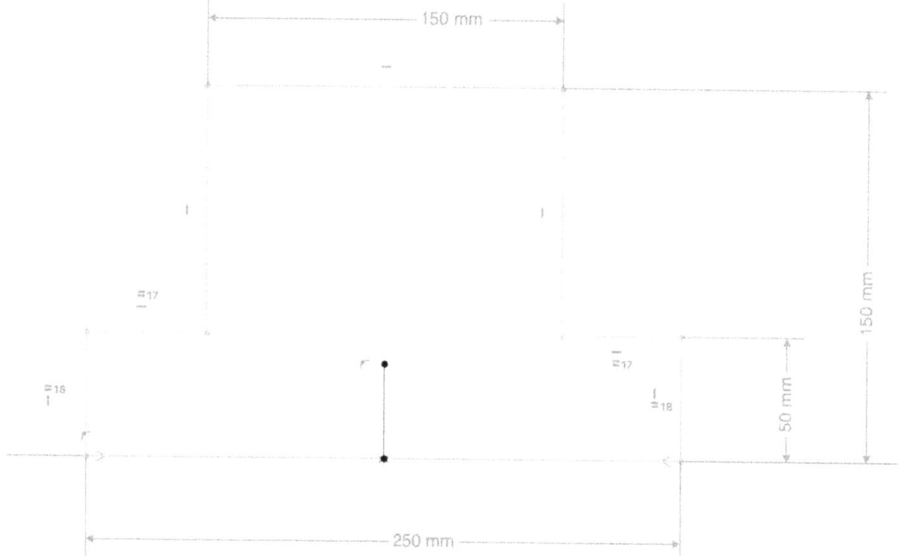

26. Click the **Create rectangle** icon on the **Sketcher geometries** toolbar.
27. Click on the left side to define the first corner of the rectangle. Next, move the cursor diagonally and click to define the second corner of the rectangle.
28. Click the **Create fillet** icon on the **Sketcher geometries** toolbar. Next, select the corner points of the rectangle.

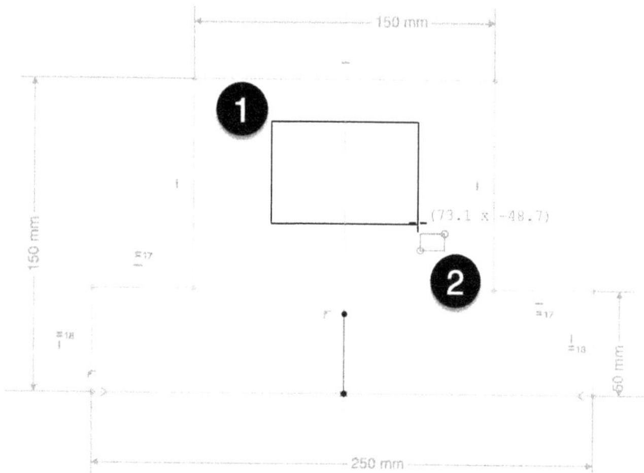

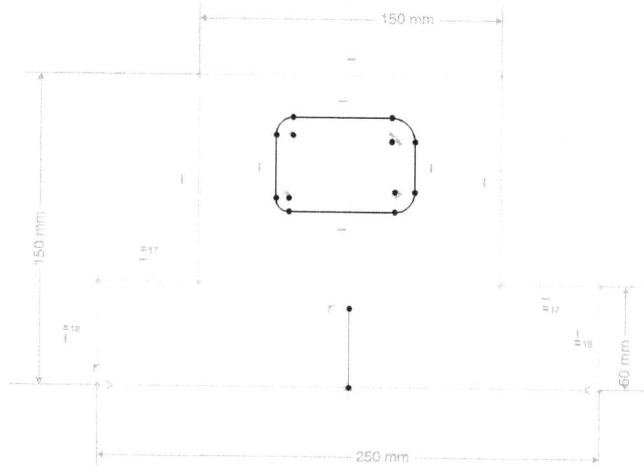

29. Click the **Constrain equal** icon on the **Sketcher constraints** toolbar.
30. Select all the fillets of the rectangle in the clockwise direction. Next, select the first and last fillet to make them equal.
31. Click the **Constrain radius** icon on the **Sketcher constraints** toolbar.
32. Select any one of the fillets. Type **6** in the **Radius** box and click **OK**.

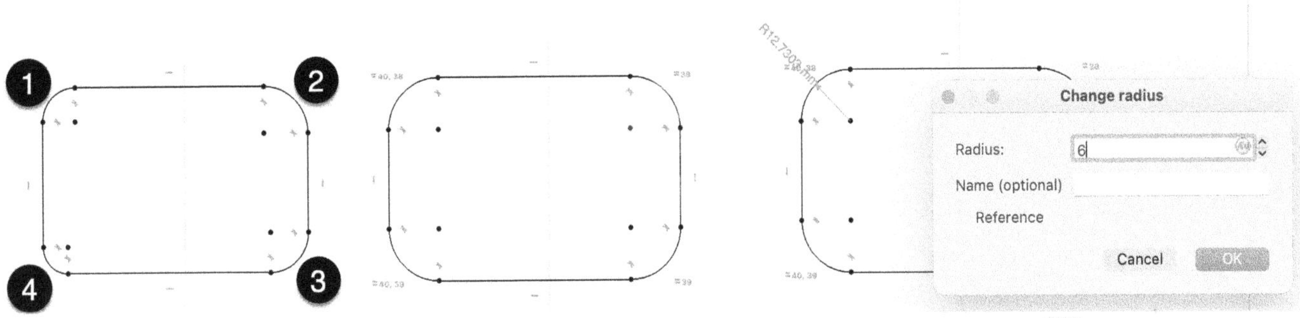

33. On the menu bar, click **Sketch > Sketcher constraints > Constrain horizontal distance** .
34. Select the centerpoints of the two lower fillets. Next, type 18 in the **Length** box and click **OK**.

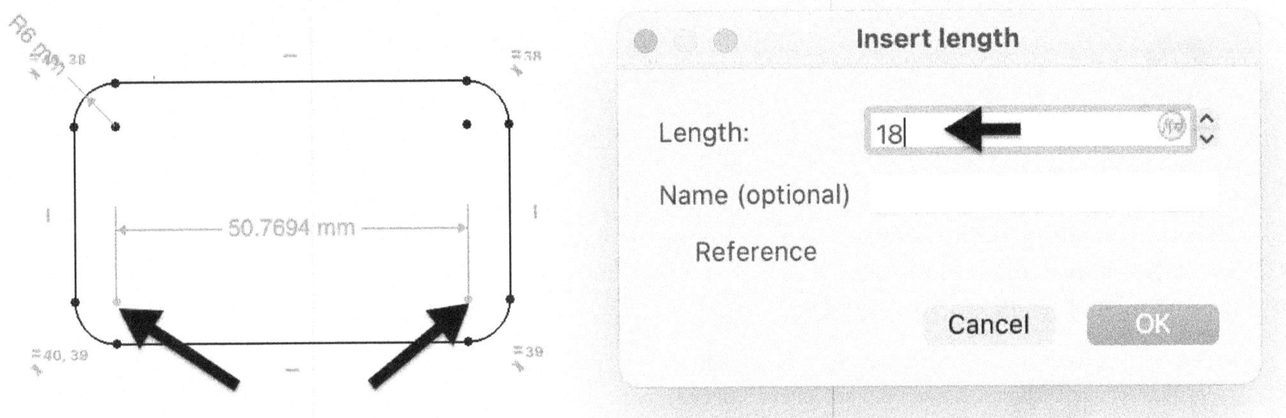

35. On the menu bar, click **Sketch > Sketcher constraints > Constrain vertical distance** .
36. Select the centerpoints of the two right fillets. Next, type **38** in the **Length** box and click **OK**.

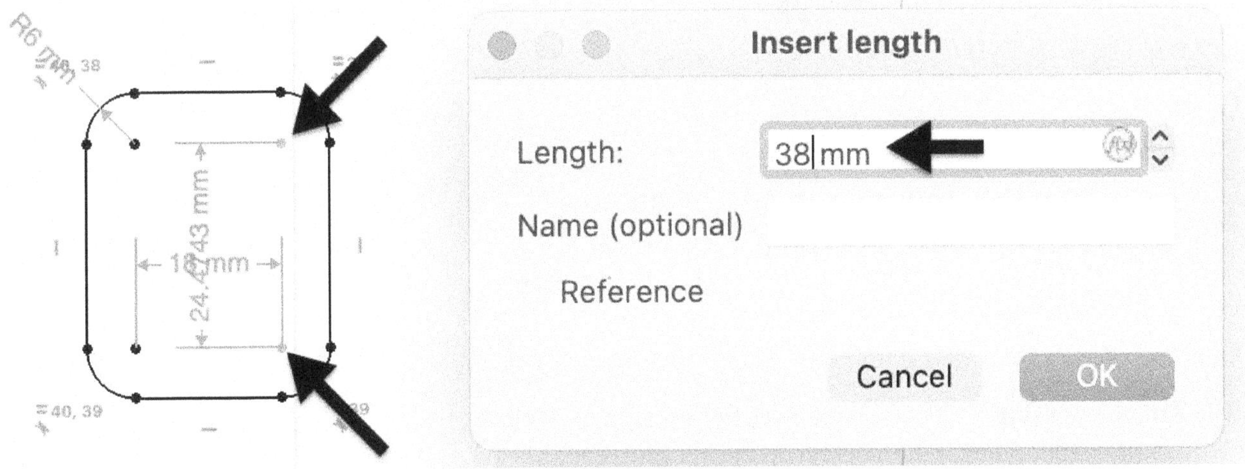

37. Click the **Constrain horizontal distance** icon on the **Sketcher constraints** toolbar.
38. Select the origin point of the sketch and the center point of the bottom-left fillet of the rectangle.
39. Type **9** in the **Length** box and click **OK**.

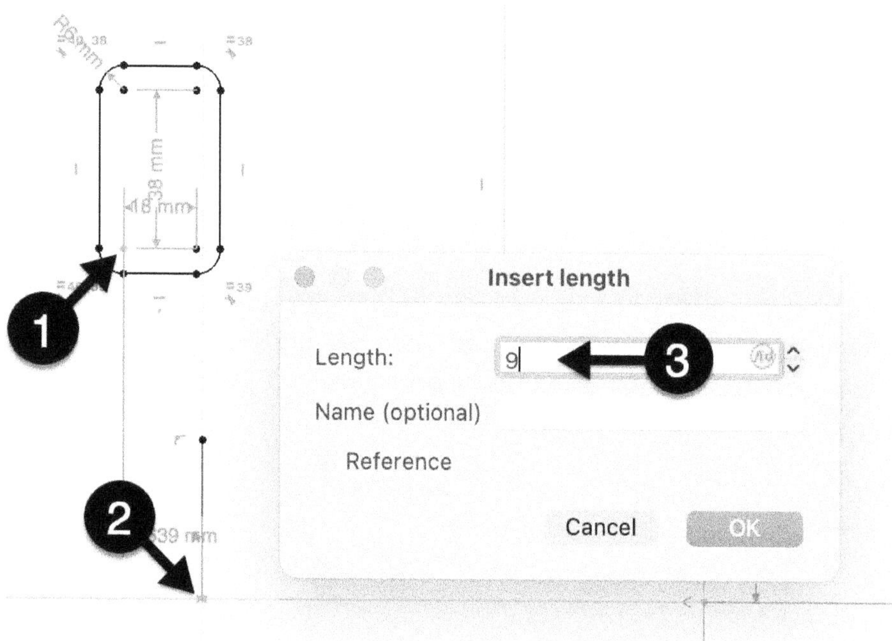

40. Click the **Constrain vertical distance** icon on the **Sketcher constraints** toolbar.
41. Select the top-left corner of the outer loop. Next, select the top left corner of the rectangle.
42. Type 25 in the **Length** box and click **OK.**

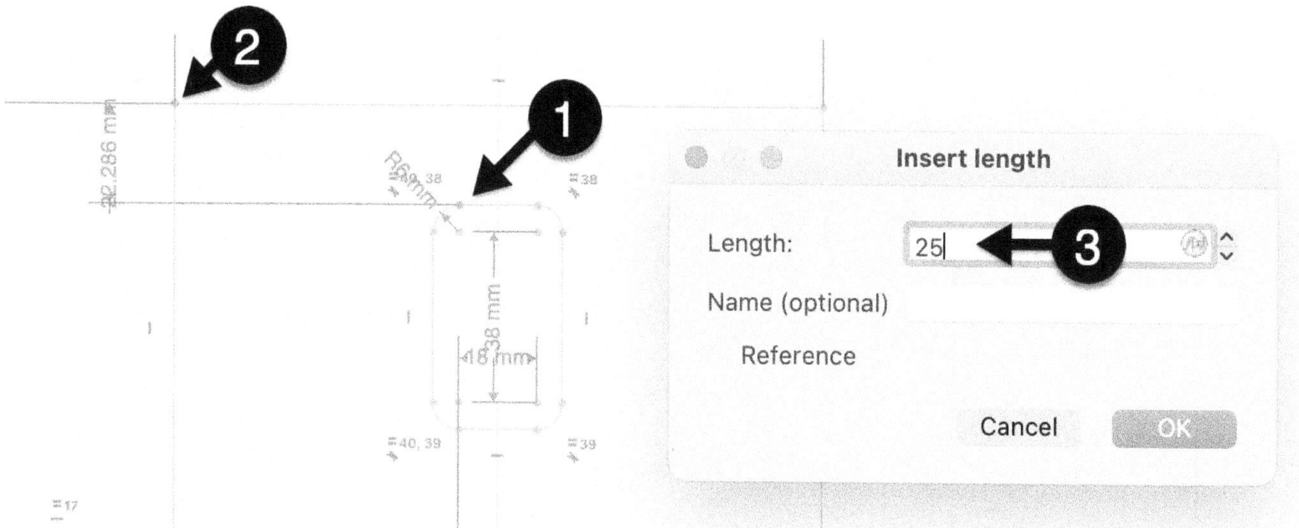

43. Click the **Create slot** icon on the **Sketcher geometries** toolbar.
44. Specify the center point of the slot end on the left side of the sketch.
45. Move the pointer toward the top right corner and click to specify the other end of the slot.

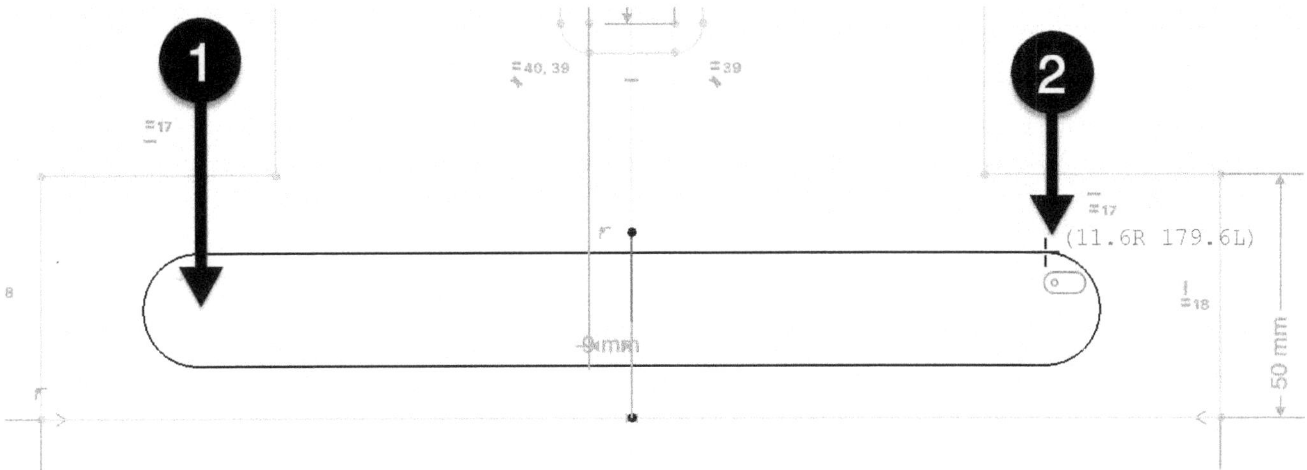

46. Click the **Constrain symmetrical** icon on the **Sketcher constraints** toolbar.
47. Select the left endpoint of the slot, right endpoint of the slot, and the top endpoint of the vertical construction line. The two endpoints are made symmetric about the top endpoint of the vertical construction line.

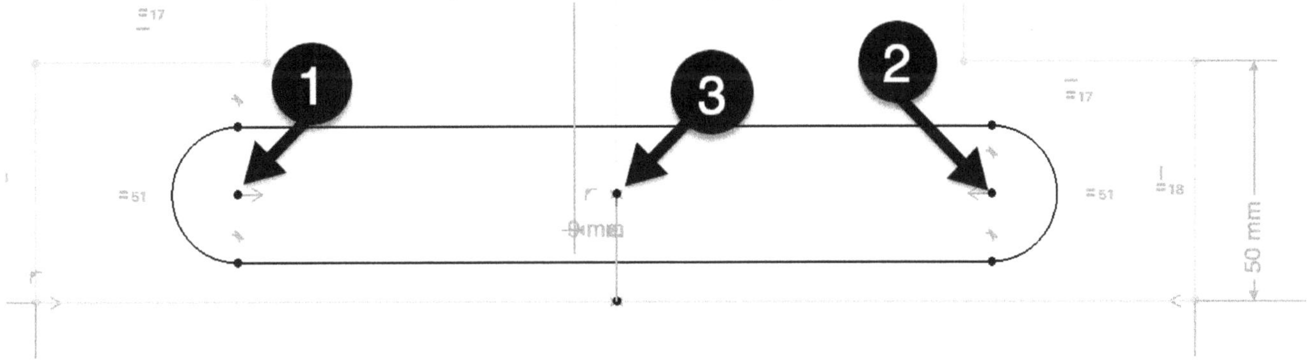

48. Add constraints to the slot and the vertical construction line, as shown.

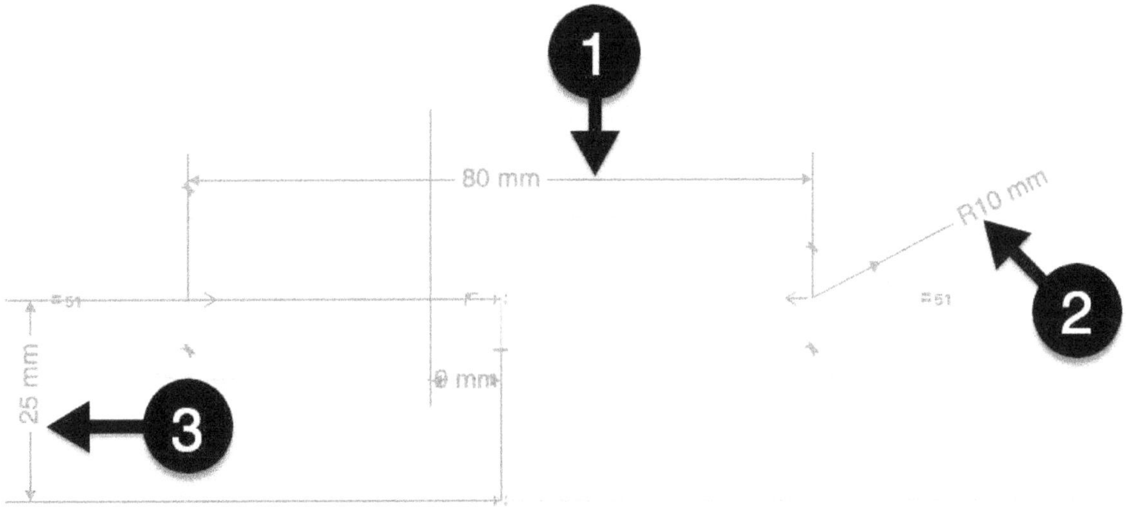

49. Click the **Close** button on the **Tasks** tab of the **Combo View** panel.
50. Click **File > Save** on the Menu bar. Next, browse to the required location on your computer and type **C2_example3** in the **File name** box. Click **Save**.
51. Close the file tab on the bottom left corner of the window.

Questions

1. What is the procedure to create sketches in FreeCAD?
2. List any two sketch *Constraints* in FreeCAD.
3. List any two distance constraints.
4. Describe two methods to create circles.
5. How to define the shape and size of a sketch?
6. How to create a three point arc?
7. Which command is used to create multiple lines?
8. List the commands to create arcs?
9. What is the command to create slots?

Exercises

Exercise 1

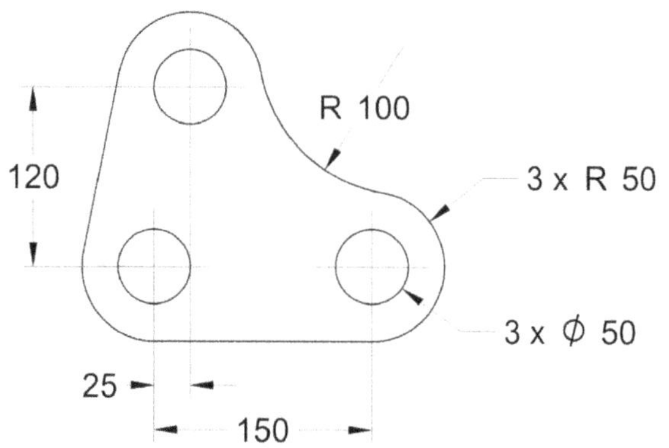

Exercise 2

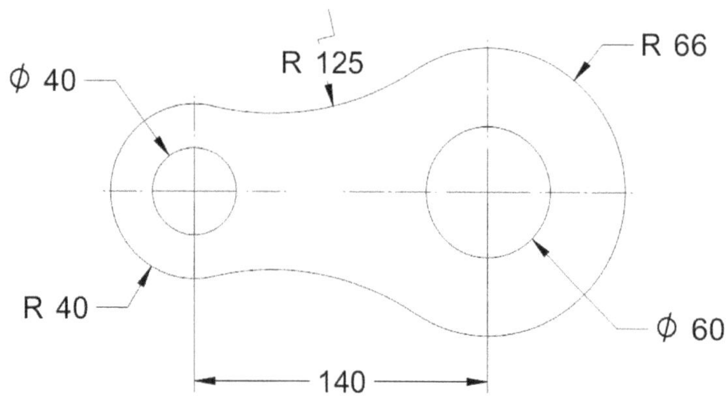

Exercise 3

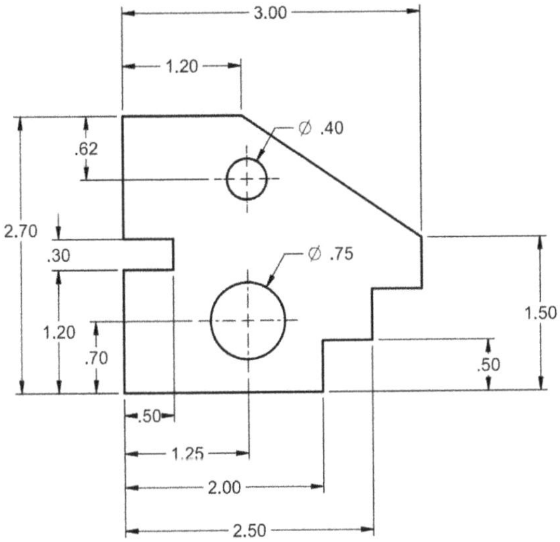

Chapter 3: Pad and Revolve Features

Pad and revolve features are used to create basic and simple parts. Most of the time, they form the base for complex parts, as well. These features are easy to create and require a single sketch. Now, you will learn the commands to create these features.

In this chapter, you will learn to:

- Create the *Pad* and *Revolve* features
- Create Datum Planes
- Work with additional options in the *Pad* and *Revolve* commands

Padded Features

Padding is the process of taking a two-dimensional profile and converting it into a 3D model by giving it some thickness. A simple example of this would be taking a circle and converting it into a cylinder. Once you have created a sketch profile or profiles you want to *Pad*, click the **Pad** icon on the **Part Design Modeling** toolbar; the sketch is selected automatically. Type-in a value in the **Length** box to specify the Pad length.

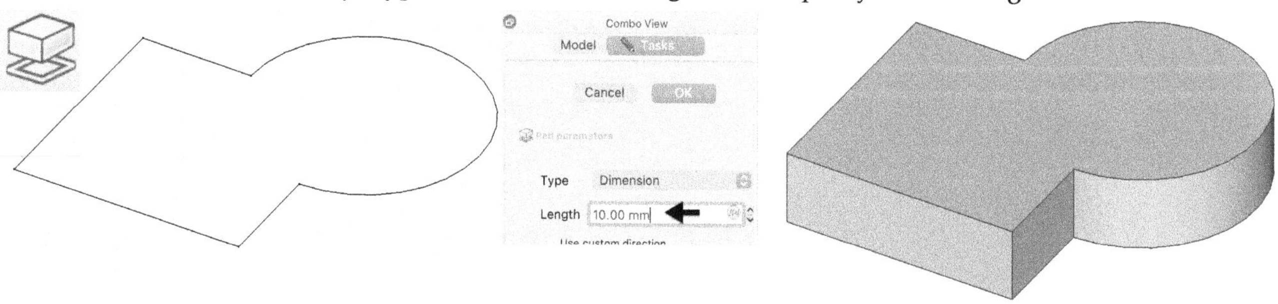

You can select **Symmetric to plane** option on the **Combo View** panel to Pad the sketch symmetrically about the sketch plane.

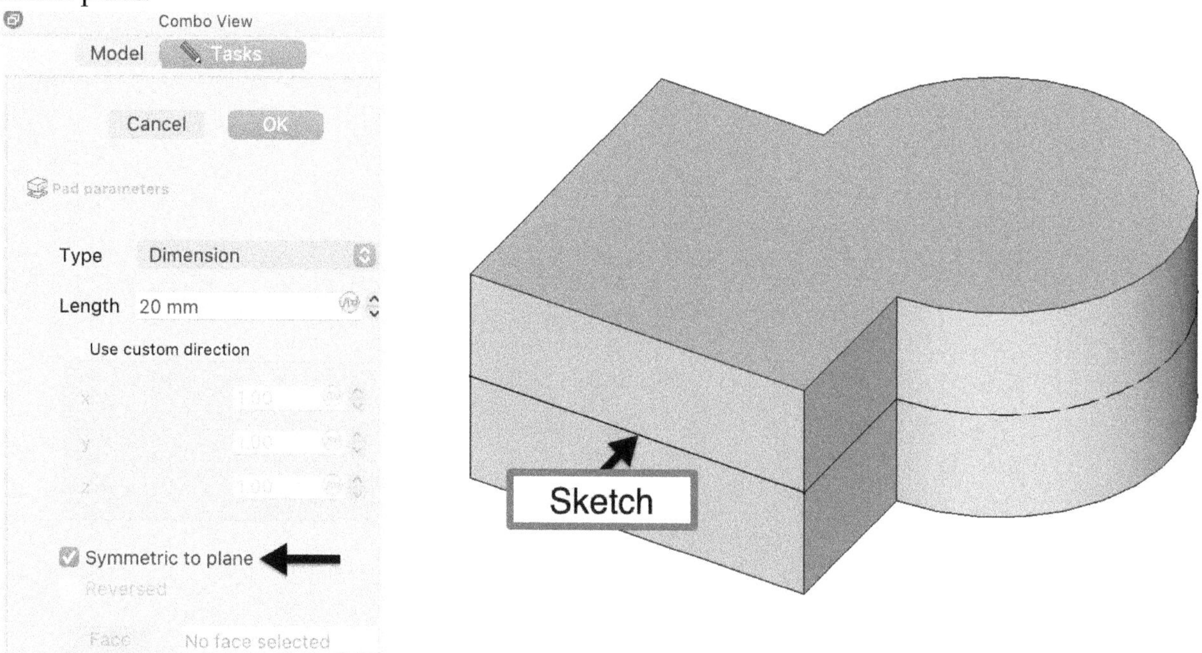

Click **OK** on the **Tasks** tab of the **Combo View** panel to complete the *Padded* feature.

Revolved Features

Revolving is the process of taking a two-dimensional profile and revolving it about a centerline to create a 3D geometry (shapes that are axially symmetric). While creating a sketch for the *Revolved* feature, it is vital to think about the cross-sectional shape that will define the 3D geometry once it is revolved around an axis. For instance, the following geometry has a hole in the center. This could be created with a separate *Pocket* or *Hole* feature. But in order to make that hole part of the *Revolved* feature, you need to create the profile at a distance from the axis of revolution. The distance value should be the radius of the hole to be created at the center. You can define the axis of revolution using the vertical or horizontal axis, or by creating a construction line in the sketch.

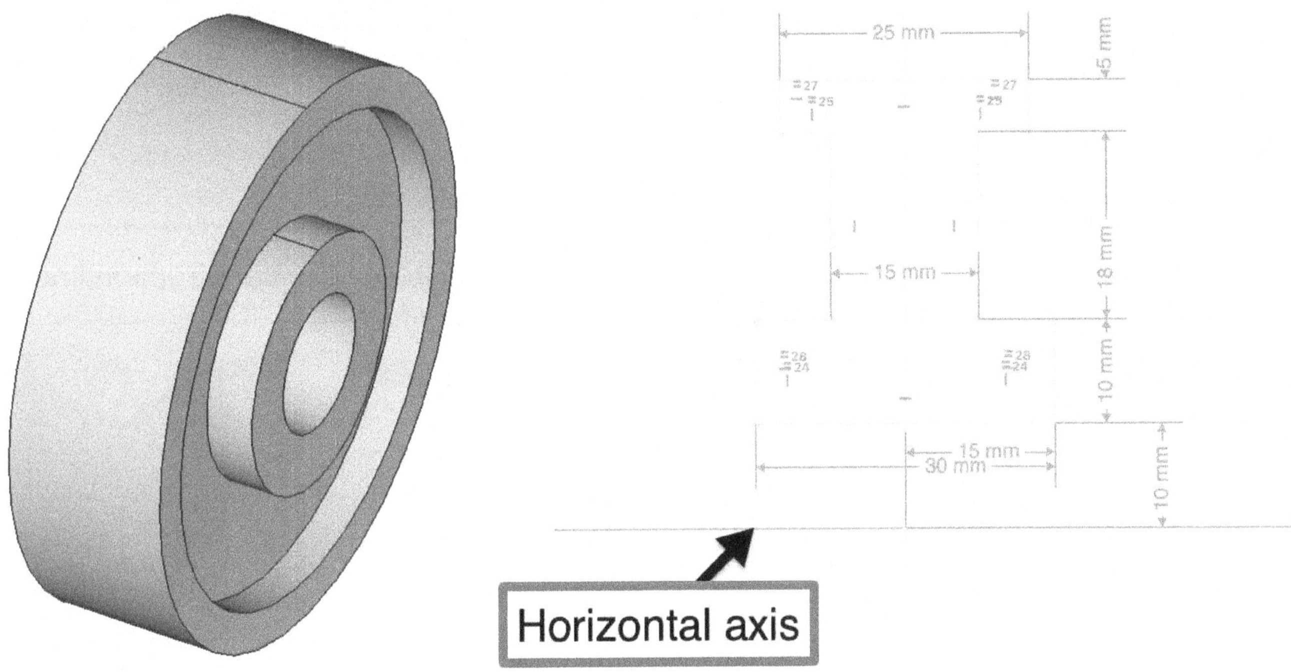

After completing the sketch, click the **Revolution** icon on the **Part Design** toolbar (or) click **Part Design > Create an additive feature > Revolution**; the sketch profile is selected automatically. Next, you have to select the axis of revolution from the **Axis** drop-down available on the **Tasks** tab of the **Combo View** panel. You can select the Vertical sketch axis, Horizontal sketch axis, Construction line, or Base X, Y, or Z axes. You can also select a part edge or other reference element using the **Select reference** option available in the **Axis** drop-down.

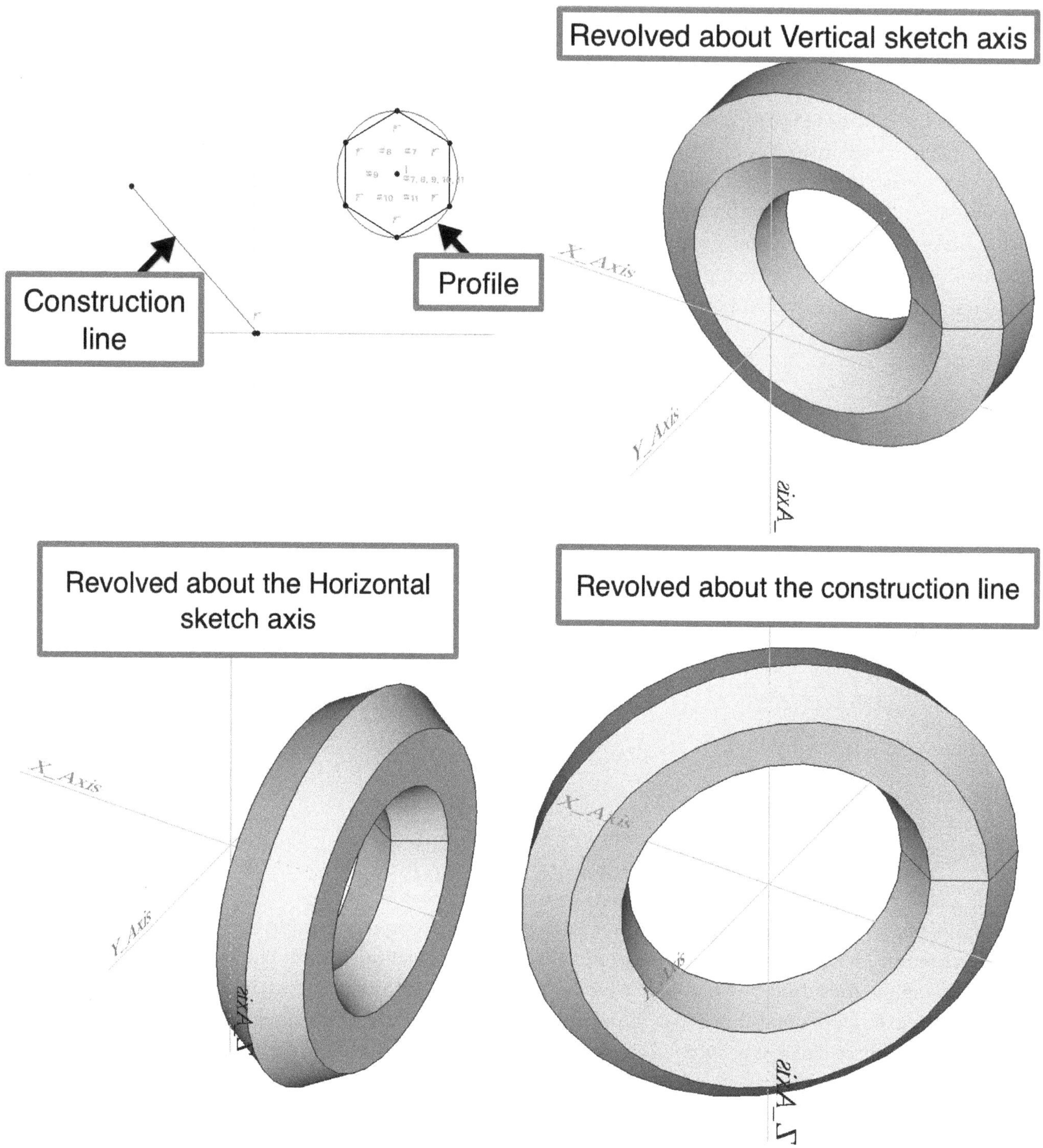
Revolved about Vertical sketch axis
Construction line
Profile
Revolved about the Horizontal sketch axis
Revolved about the construction line

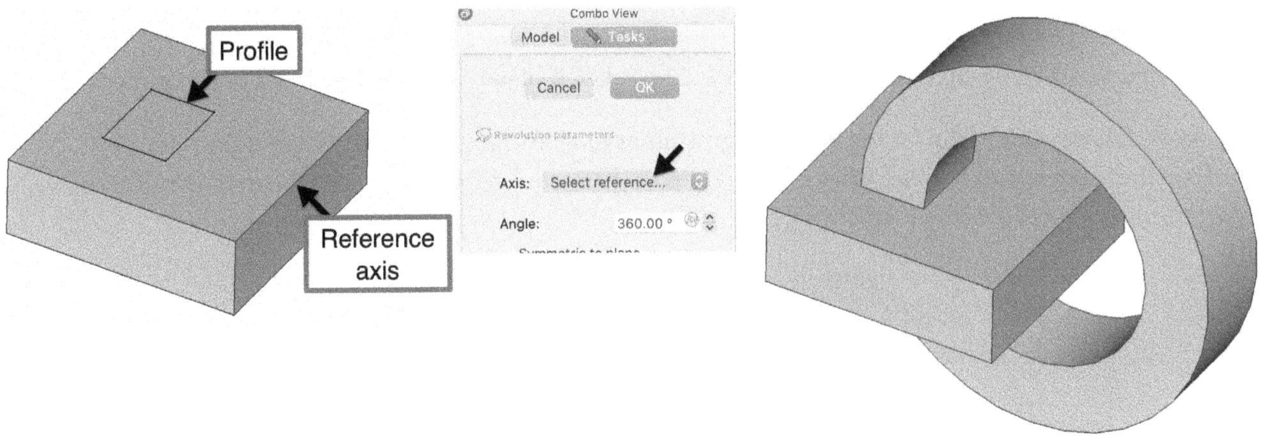

Next, you need to type-in a value in the **Angle** box. Check the Reversed option, if you want to reverse the direction of revolution.

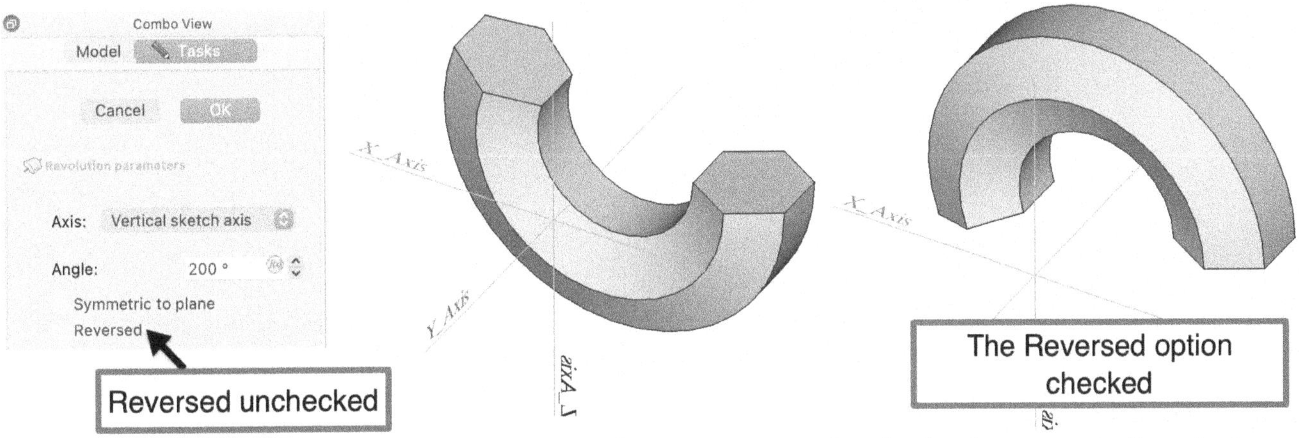

In addition to that, you can check the **Symmetric to plane** option to revolve the sketch symmetrically on both sides of the sketch plane.

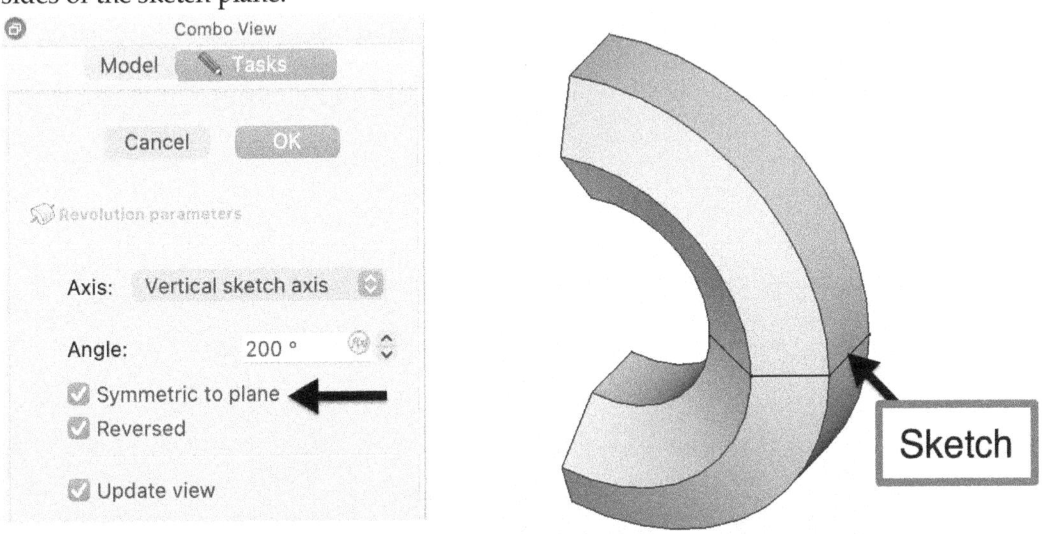

Click **OK** on the **Tasks** tab of the **Combo View** panel to complete the *Revolved* feature.

External Geometry

This command projects the edges of a 3D geometry onto a sketch plane. To do this, select a plane or model face and click the **Create sketch** icon on the **Part Design Helper** toolbar. Next, click the **External Geometry** icon on the **Sketcher geometries** toolbar. Click on the edges of the model geometry. Click **OK** on the **Project** dialog to project the edge on to the sketch plane.

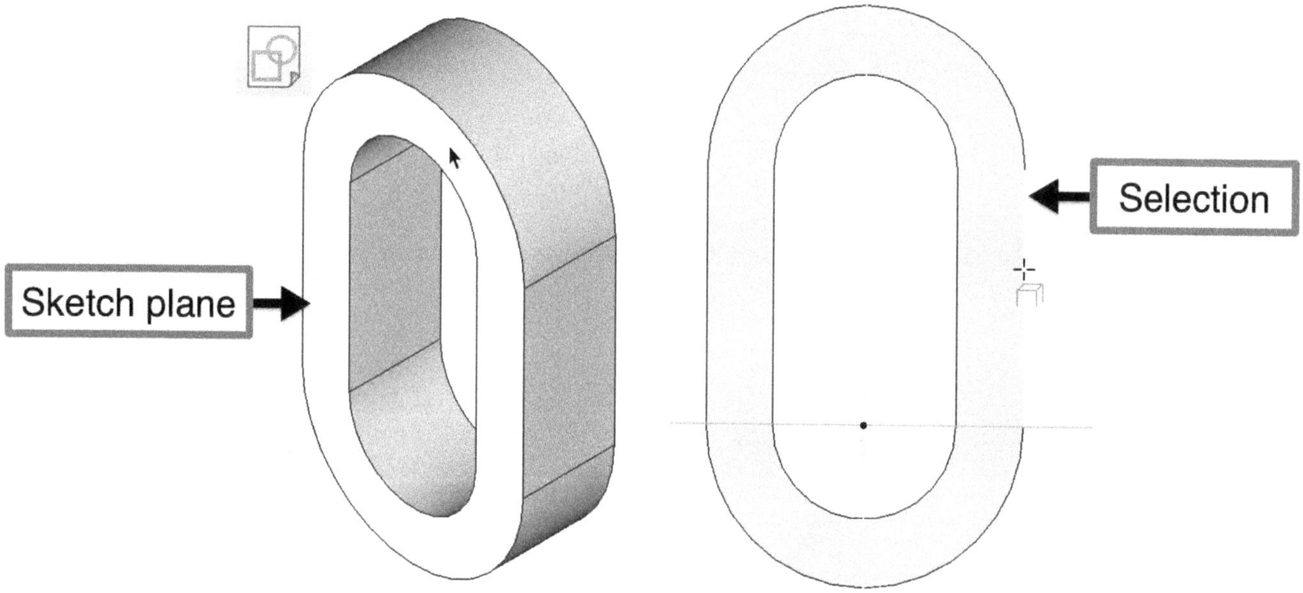

The projected element will be red in color and can be used to connect the sketch with the existing geometry. For example, create a rectangle and apply the Coincident constraint between the vertex of the rectangle and the endpoint of the projected element.

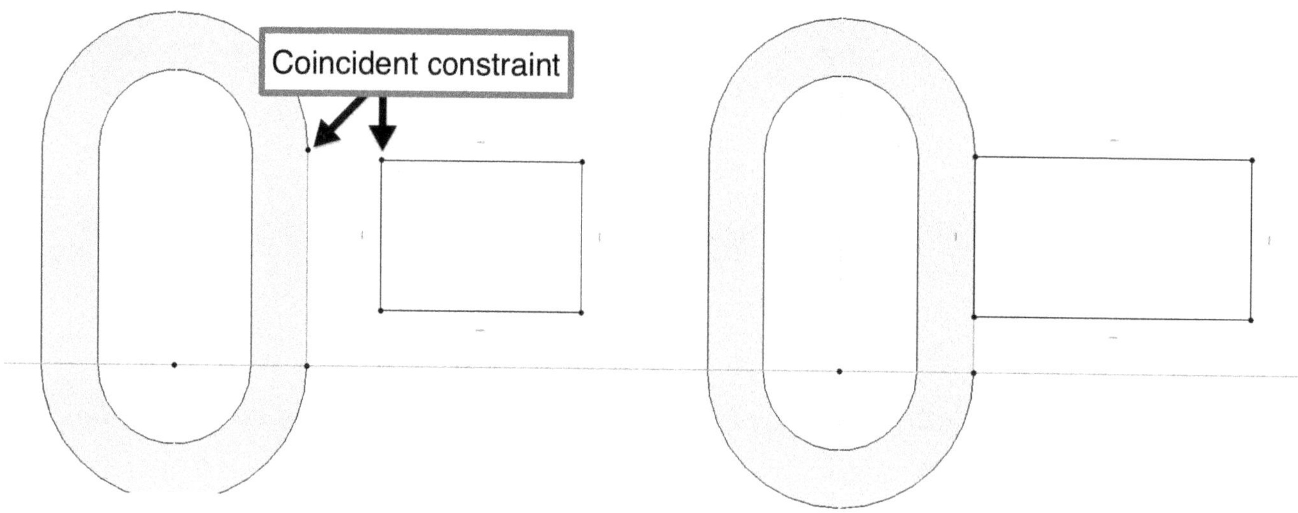

Pocket

This command removes material from the geometry by extruding a sketch. It functions on the same lines as the **Pad** command. Draw a sketch on a plane or a model face. Next, click the **Pocket** icon on the **Part Design Modeling** toolbar (or) click **Part Design > Create a subtractive feature > Pocket** on the Menu bar; the sketch is selected automatically. On the **Pocket parameters** section, type in a value in the **Length** box; the pocket's preview appears in the direction normal to the sketch. Click **OK** to create the pocket feature.

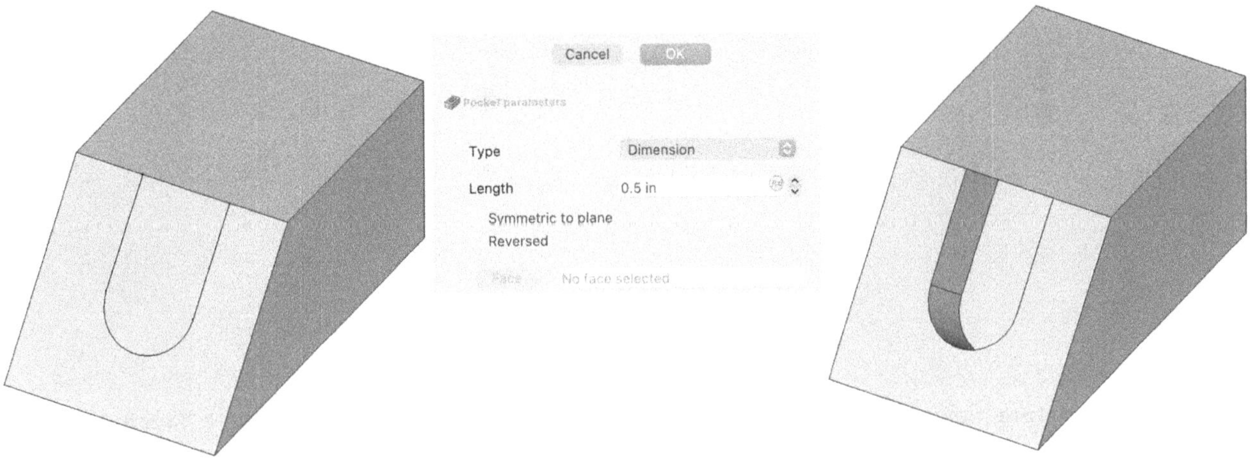

The Groove command

This command removes material from the geometry by revolving a sketch about an axis. It functions in a way similar to the **Revolution** command. Draw a sketch on a plane or a model face. Next, create an axis using the **Datum line** command. Click the **Groove** icon on the **Part Design Modeling** toolbar (or) click **Part Design > Create a subtractive feature > Revolution** on the Menu bar; the sketch is selected automatically. Select the **Construction Line 1** from the **Axis** drop-down. Next, specify the revolution angle and click **OK** to complete the groove feature.

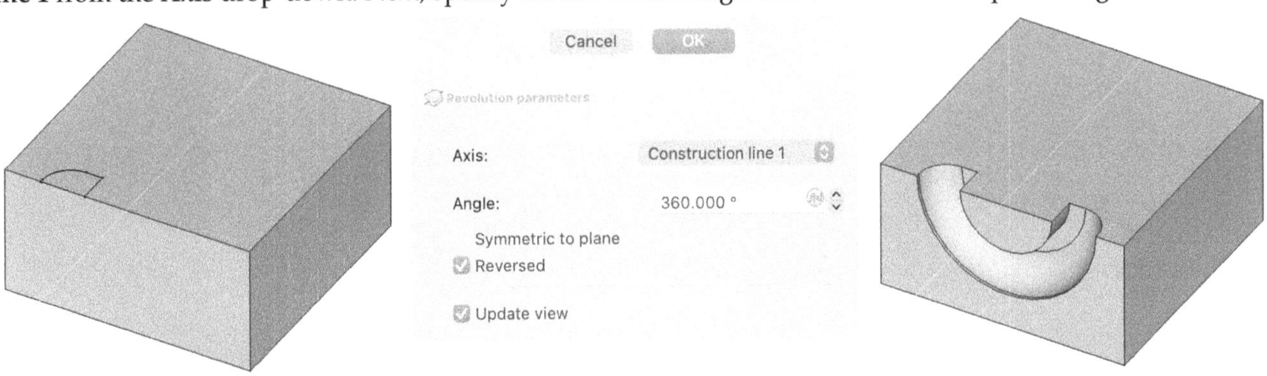

Planes

Planes are a specific type of element in FreeCAD 0.19, known as datum features. These features act as supports to your 3D geometry. Each time you start a new part file, FreeCAD 0.19 automatically creates default planes (XY, YZ, and XZ planes). Until now, you have learned to create sketches on any of the default planes. If you want to create sketches and geometry at locations other than default planes, you can create new planes manually. You can do this by using the **Create datum plane** command.

Creating a Parallel plane

Click the **Create datum plane** icon on the **Part Design Helper** toolbar (or) click **Part Design > Create a datum > Create a datum plane** on the menu bar. Click on a flat face of goemetry or a plane. Next, type a value in the **In z-direction** box on the **Tasks** tab of the **Combo View** panel. Check the **Flip sides** option located at the bottom of the **Combo View** panel to change the side on which the plane is created. Click **OK** on the **Tasks** tab of the **Combo View** panel to create a plane parallel to the selected face or another plane

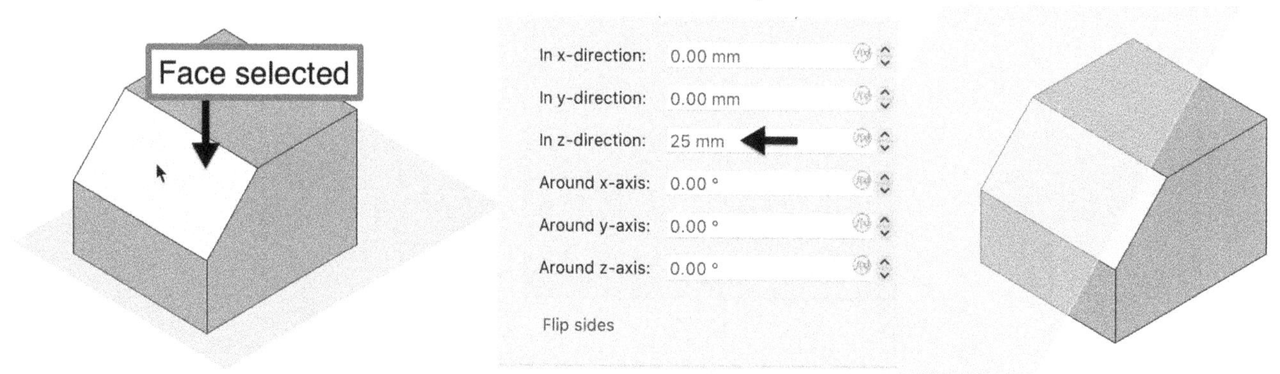

Inertia 2-3

Click the **Create datum plane** icon on the **Part Design Helper** toolbar (or) click **Part Design > Create a datum > Create a datum plane** on the menu bar. Select an edge from the model geometry and select the **Inertia 2-3** option from the **Attachment mode** section; the plane is created perpendicular to the selected edge and positioned at the midpoint.

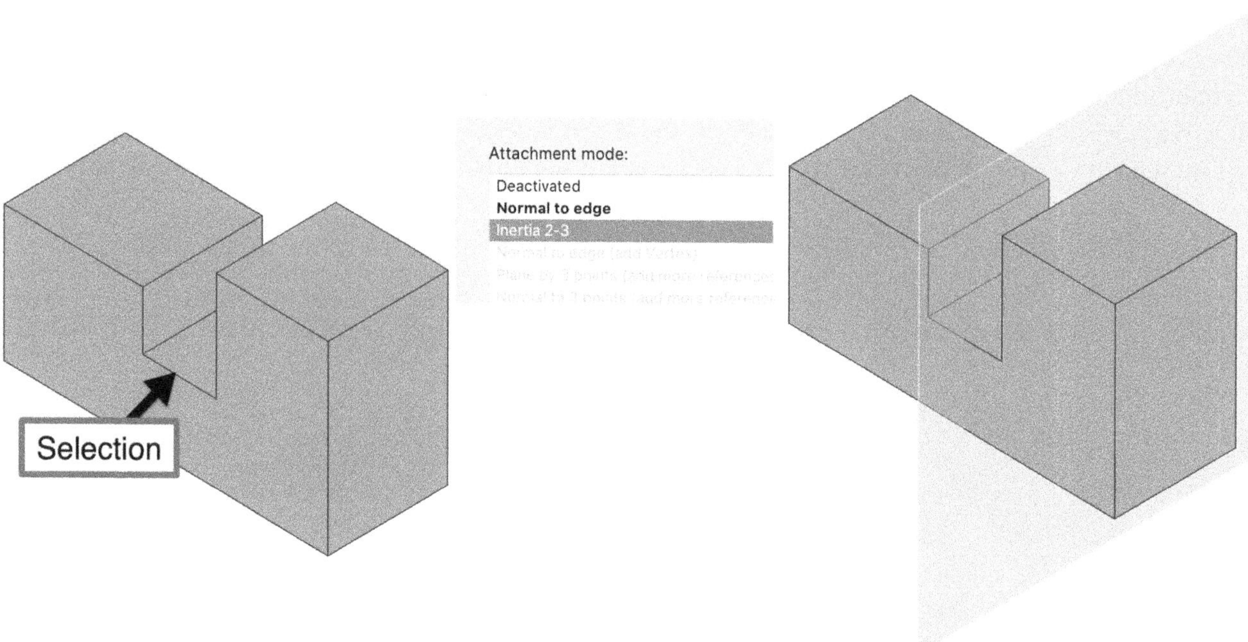

Plane at Angle

Click the **Create datum plane** icon on the **Part Design Helper** toolbar. Click on a face of the part geometry. Next, type in a value in the **Around x-axis** box; the plane is rotated at the specified angle around the X-axis of the selected face. You can also rotate the plane around the Y-axis of the selected face by entering a value in the **Around y-axis** box.

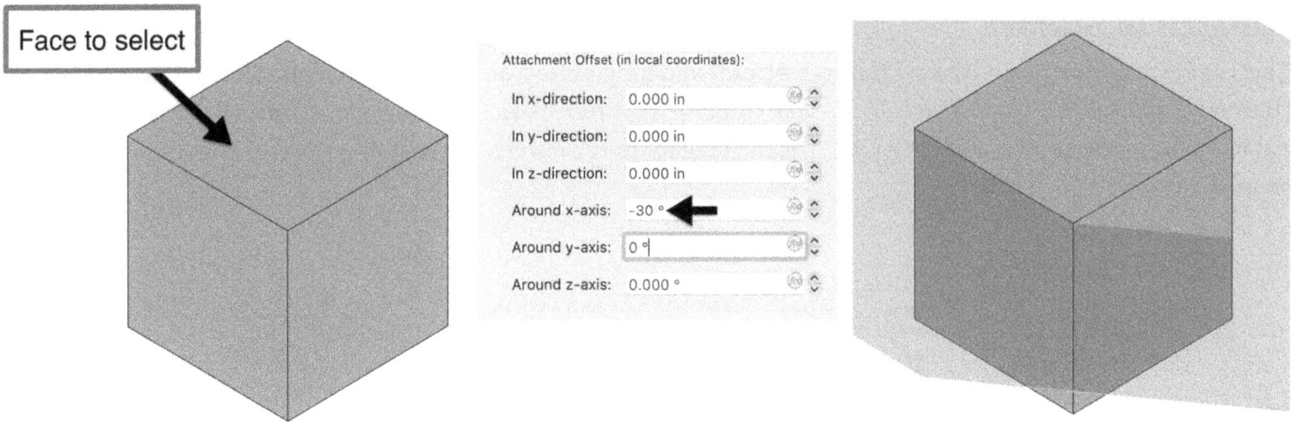

Plane by 3 Points

This option creates a plane passing through three points. Activate the **Create datum plane** command and select three points from the model geometry. Next, select the **Plane by 3 Points** option from the **Attachment mode** section and click **OK**. The plane will be created, passing through the selected points.

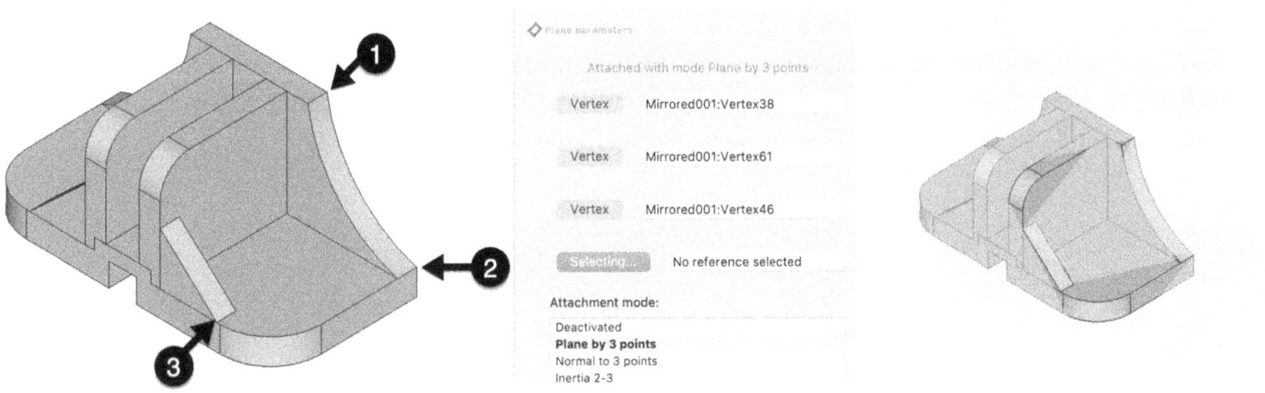

Normal to 3 Points

This option creates a plane normal to three points. Activate the **Create datum plane** command and select three points from the model geometry. Next, select the **Normal to 3 Points** option from the **Attachment mode** section and click **OK**. The plane will be created, normal to the selected points.

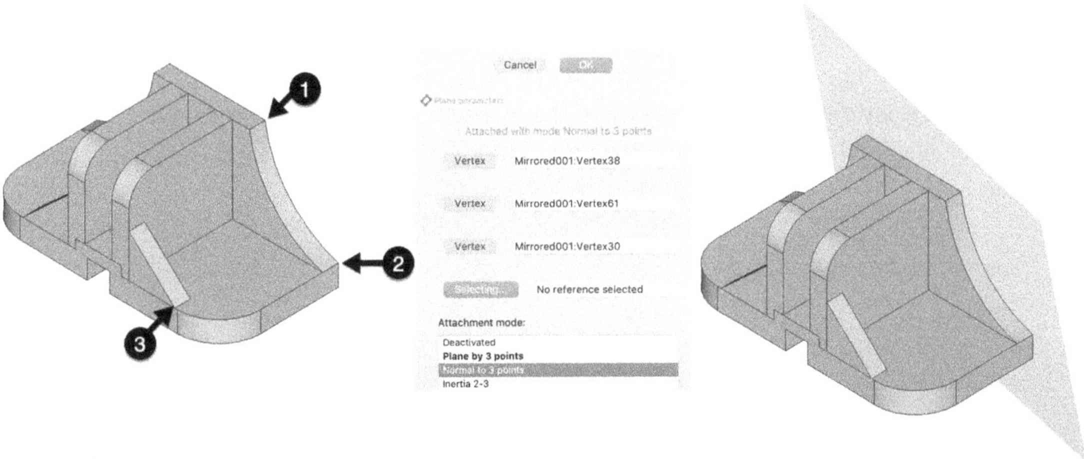

Tangent to Surface

This option creates a plane passing through a point and tangent to a curved face. Activate **Create datum plane** command and select a curved face. Next, select a point or vertex. Select the **Tangent to Surface** option from the **Attachment mode** section and click **OK**. A plane tangent to the curved face passing through a point will be created.

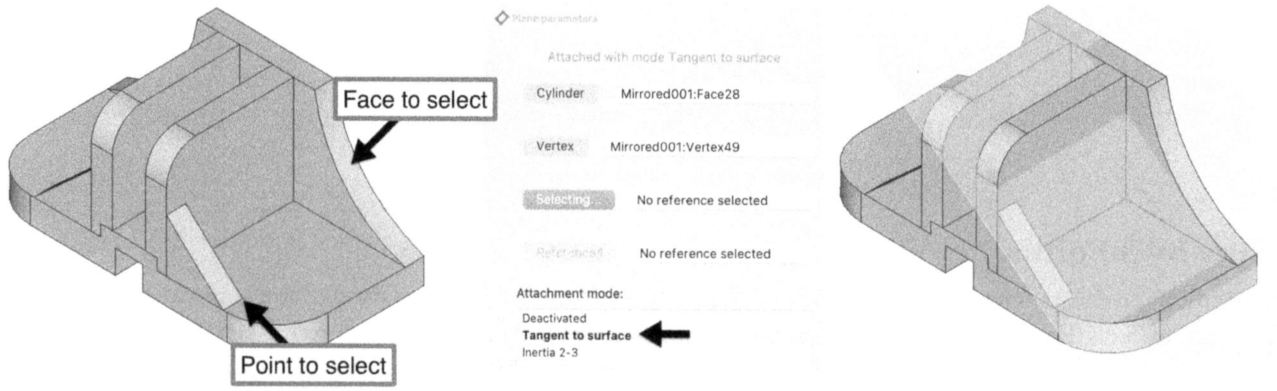

Normal to edge

This option creates a plane, which will be normal to a line, curve, or edge. Activate the **Create datum plane** command and select an edge, line, curve, arc, or circle. Next, you need to specify the location of the plane on the selected path. To do this, select a point from the graphics window. Next, select the **Normal to edge** option from the **Attachment mode** section and click **OK**. A plane normal to the edge will be created.

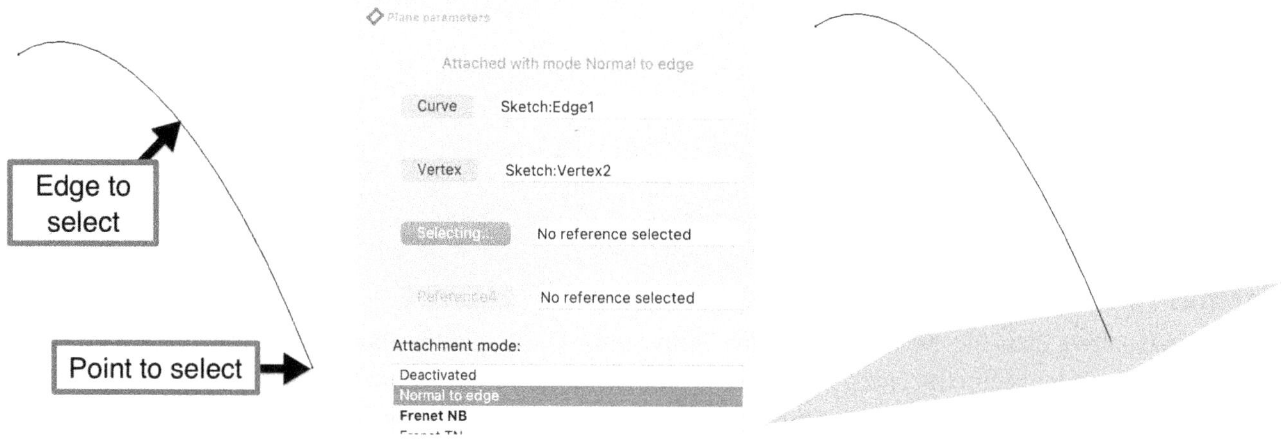

Revolution section

This option creates a plane, which will be perpendicular to a circular edge. Activate the **Create datum plane** command and select circular edge. Next, specify the location of the plane by selecting a point from the graphics window. Select the **Revolution section** option from the **Attachment mode** section and click **OK**. A plane normal to the edge will be created.

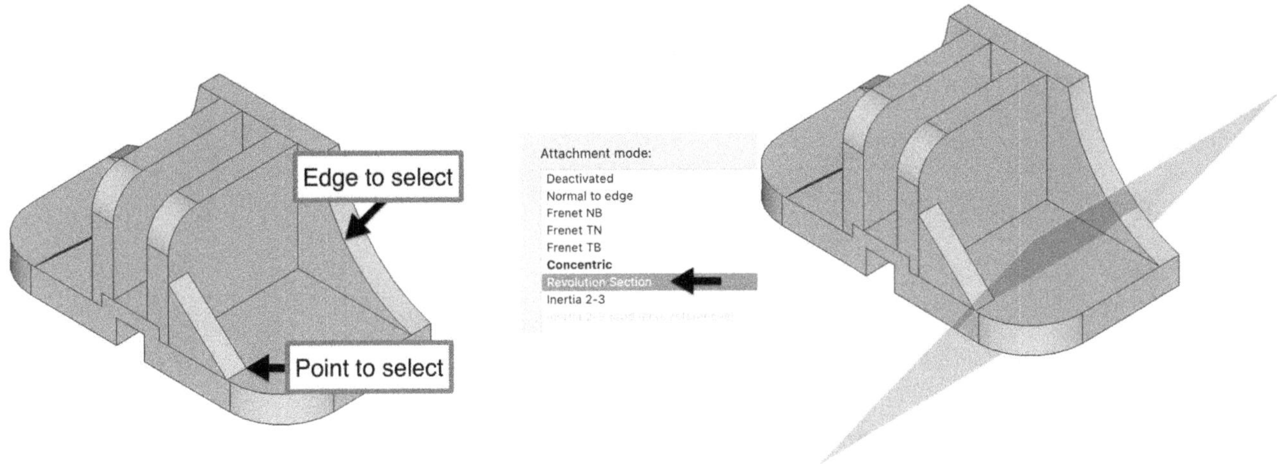

Datum Line

A datum line is another type of construction element that helps you in building a 3D model. For example, you can use a datum line to create a revolved feature. You can create various types of axes in FreeCAD 0.19. They are discussed next.

1st principal axis

This option creates an axis through a cylinder or cone or torus. Click the **Create a new datum line** icon on the **Part Design Helper** toolbar (or) click **Part Design > Create a datum > Create a datum line** on the menu bar. Next, select a round face, cone, or torus. Select the **1st principal axis** option from the **Attachment mode** section. You can also create a datum line perpendicular to flat face.

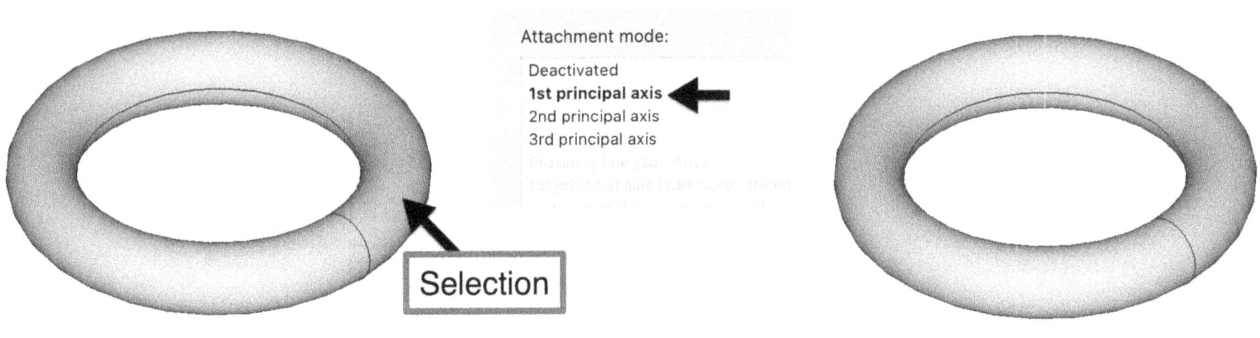

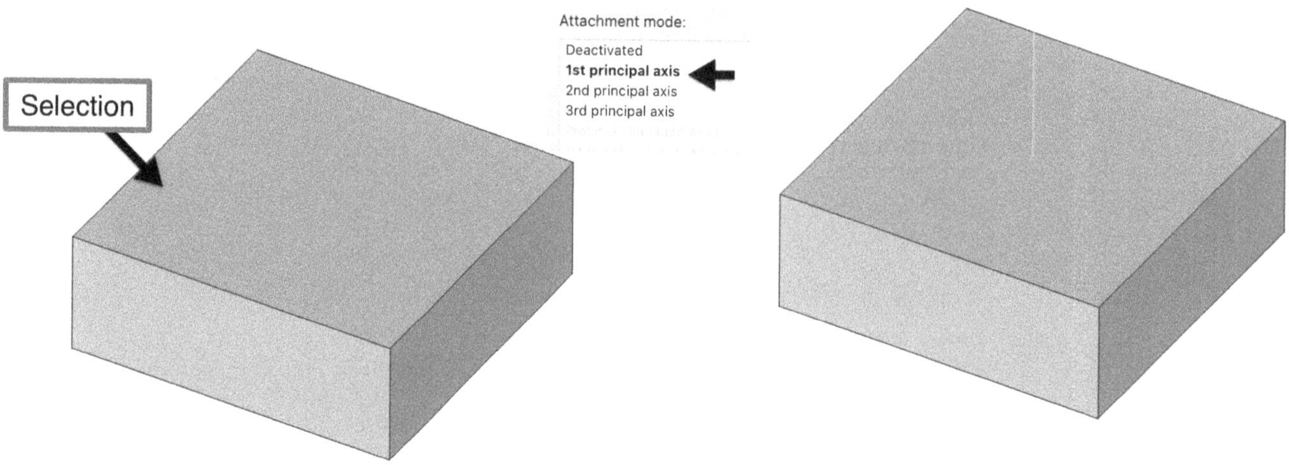

Through Two Points

This option creates a datum line passing through two selected points. Activate **Create a new datum line** command and select two points. Select two points or vertices, and then select the **Through two points** option from the **Attachment mode** section.

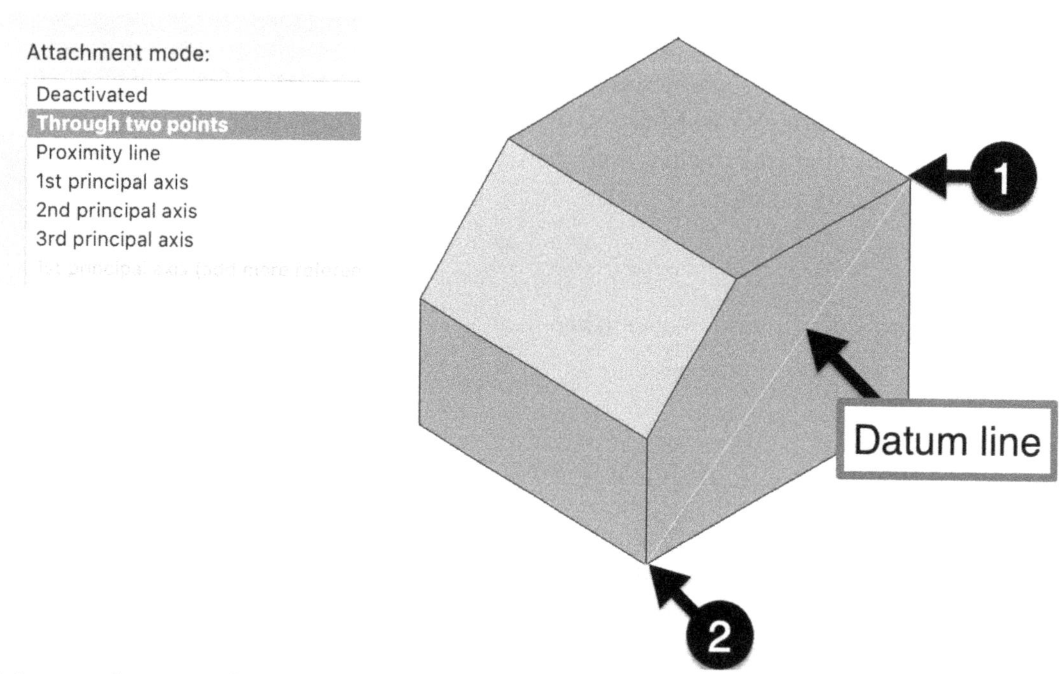

Normal to surface

This option creates an axis, which will be perpendicular to a face and point. Activate **Create a new datum line** command and select a face. Next, click on a point to define the location of the axis. Select the **Normal to surface** option from the **Attachment mode** section.

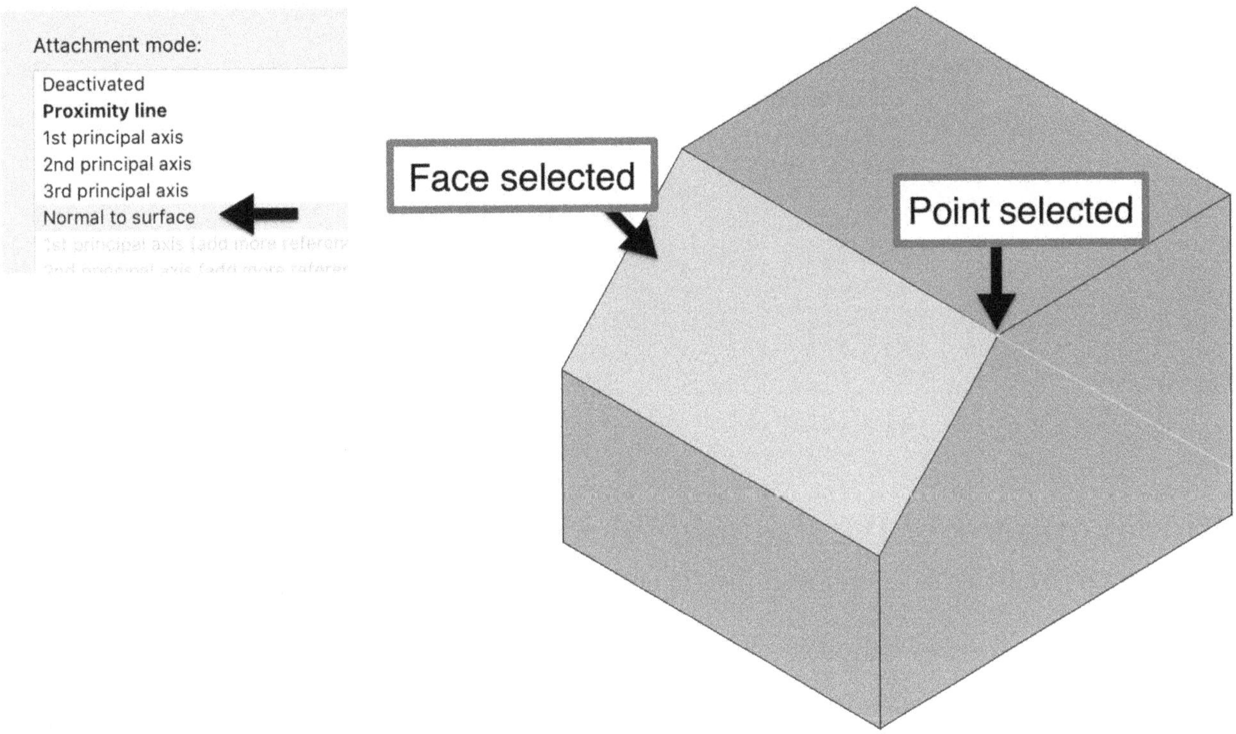

Datum Point

Points are another type of construction element in addition to planes and axes. The options to create construction points are explained next.

Center of mass

This option creates a datum point at the center of mass of a body. Click the **Create a new datum point** icon on the **Part Design Helper** toolbar (or) click **Part Design > Create a datum > Create a datum point** on the menu bar. Next, select a solid body and select the **Center of mass** option from the **Attachment mode** section.

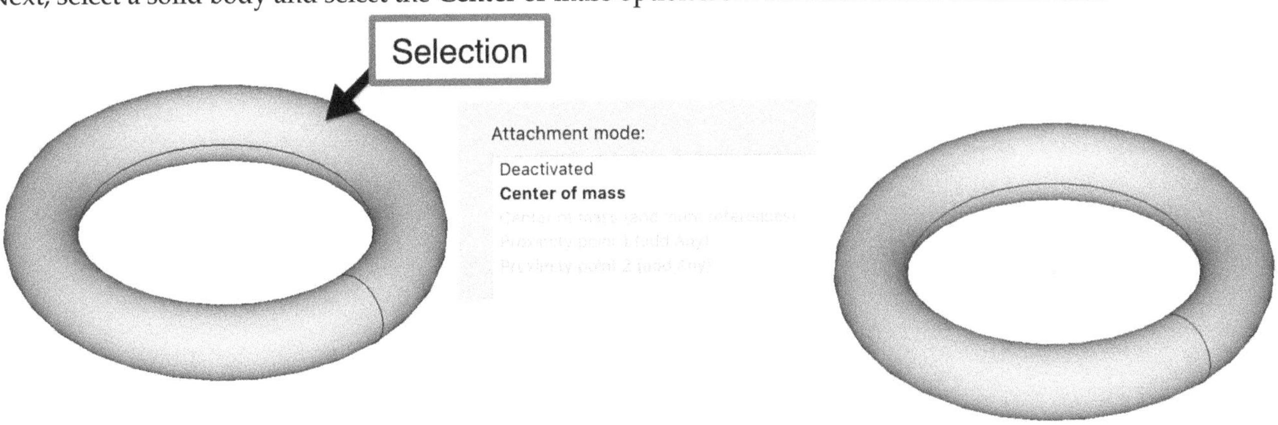

Vertex

This option creates a datum point at a selected point or vertex. Click the **Create a new datum point** icon on the **Part Design Helper** toolbar and select a point or vertex of the model. Next, select the **Vertex** option from the **Attachment mode** section.

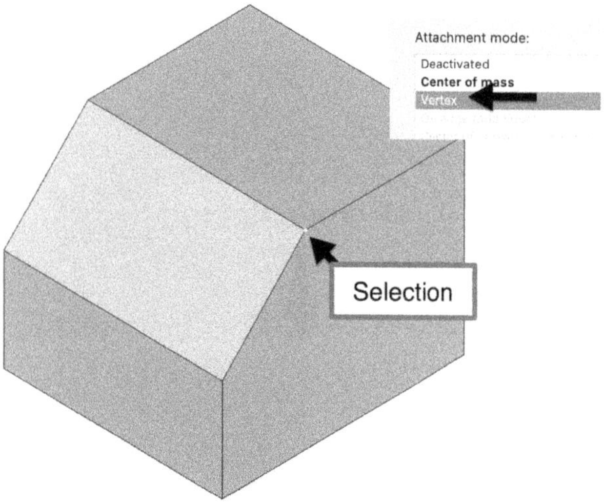

Additional options of the Pad and Pocket commands

The **Pad** and **Pocket** commands have some additional options to create complex features of a 3D geometry.

Extent Type

The **Type** drop-down has five options to define the end limit of the *Pad* or *Pocket* feature. These options are **Dimension**, **To Last**, **To first**, **Up to face**, and **Two Dimensions**.

The **Dimension** option pads the sketch up to the specified distance. On the **Tasks** tab of the **Combo View** panel, select **Dimension** from the **Type** drop-down. Next, specify the distance in the **Length** box.

The **Two dimensions** option pads the sketch in both the directions normal to the sketch plane. Select this option from the **Type** drop-down. Next, enter values in the **Length** and **2nd length** boxes.

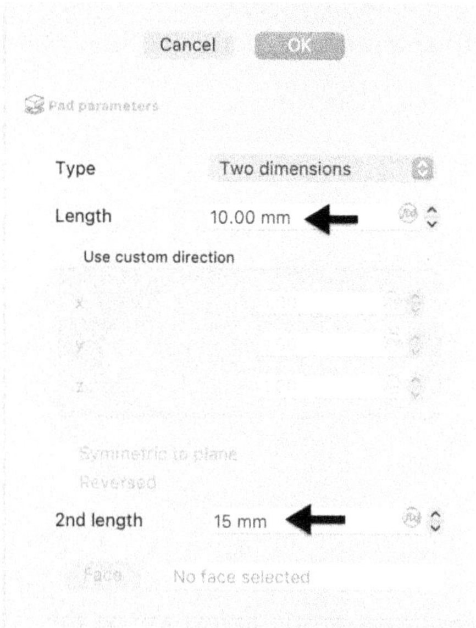

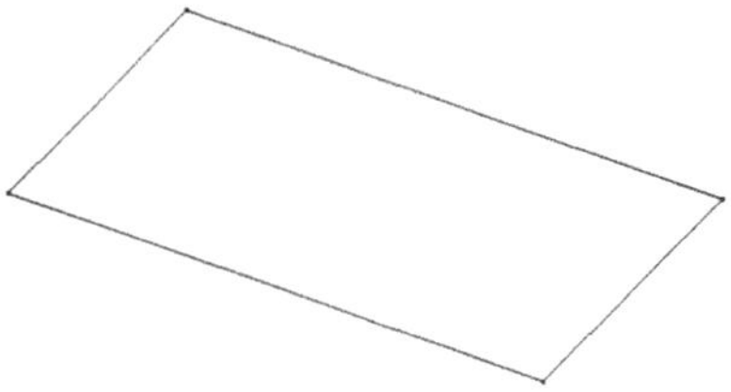

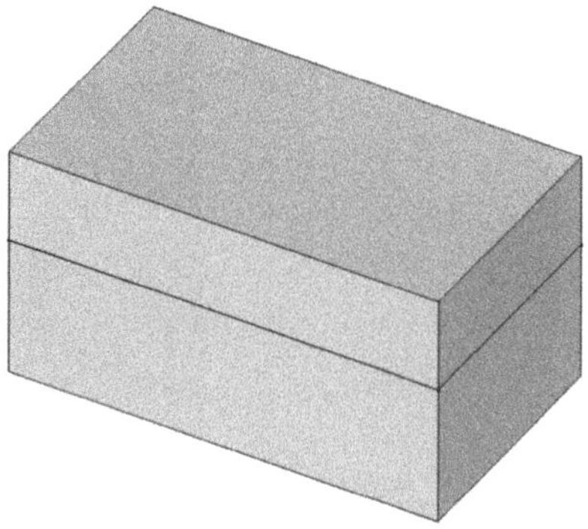

The **Up to face** option extrudes the sketch up to a selected face. On the **Tasks** tab of the **Combo View** panel, select **Up to face** from the **Type** drop-down (check the **Reversed** option if nothing is displayed in the graphics window). Next, select the face or plane; the sketch will be extruded up to the selected face or plane.

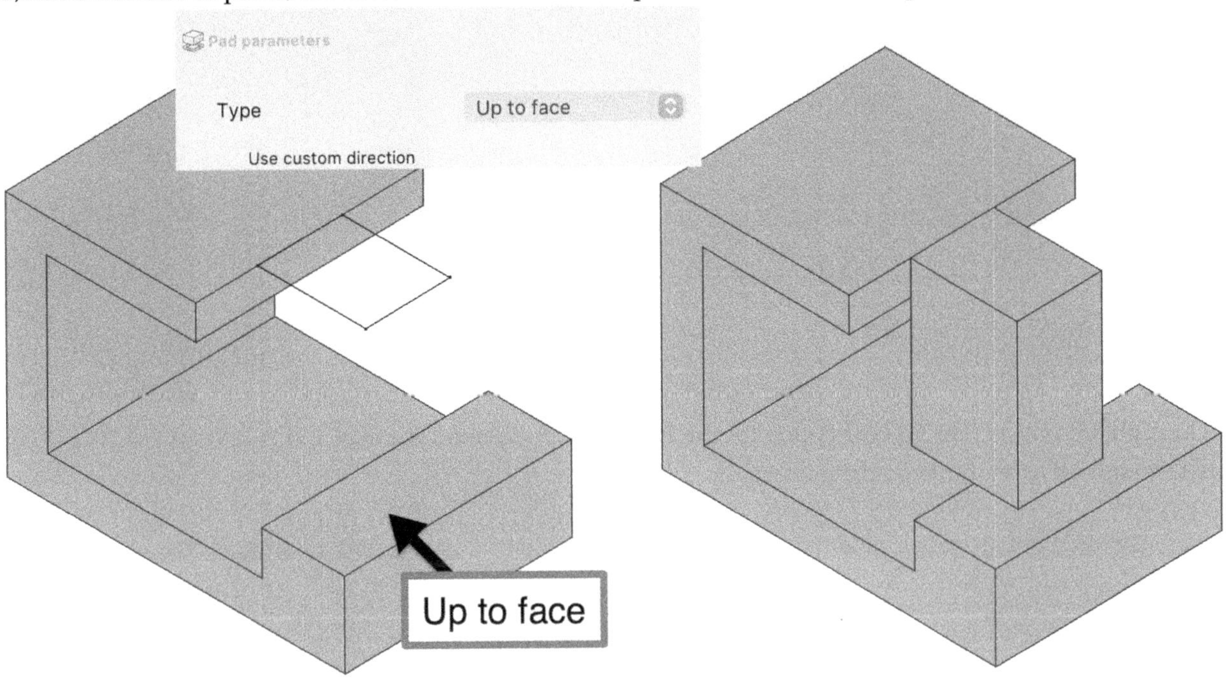

You can enter a value in the **Offset to face** box to offset the extrusion from the selected face.

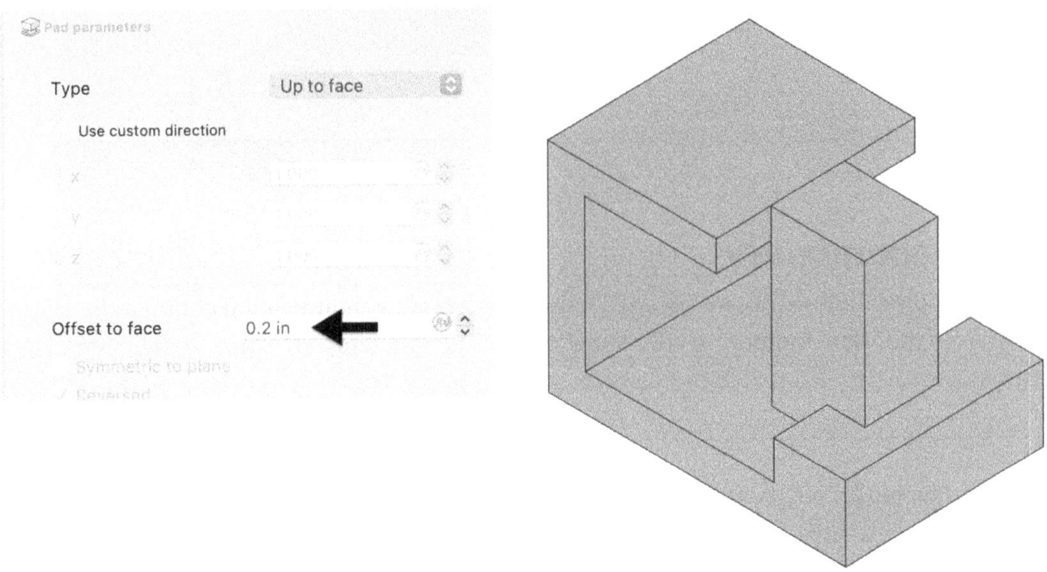

The **To first** option extrudes the sketch up to the next surface. Select **To first** from the **Type** drop-down. Check the **Reversed** option to reverse the direction.

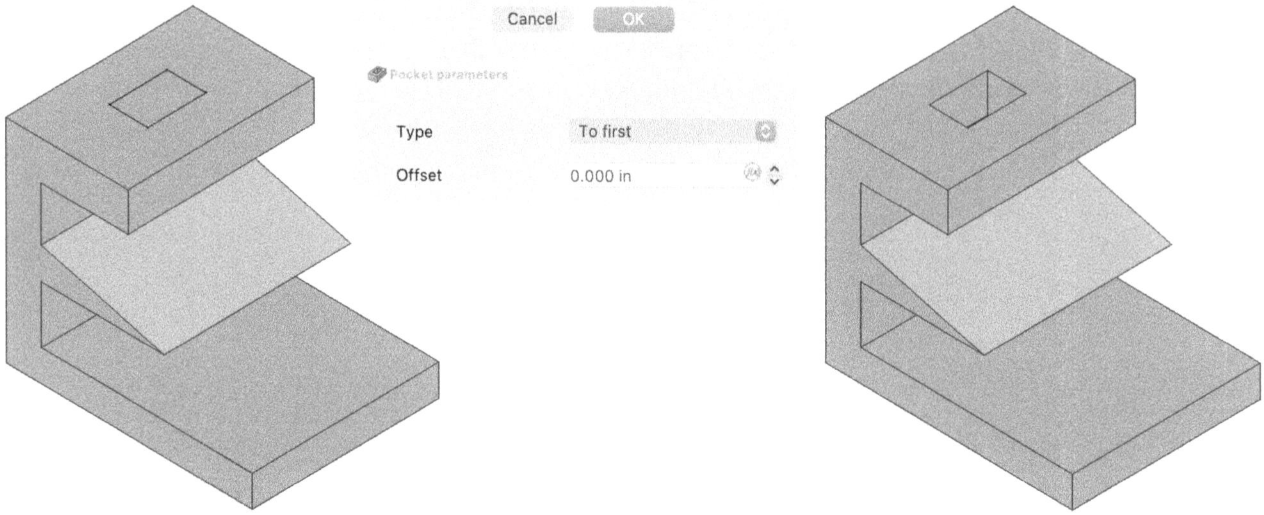

The **Through All** option (the **To Last** option for the **Pad** feature) extrudes the sketch throughout the 3D geometry. Select **Through all** from the **Type** drop-down.

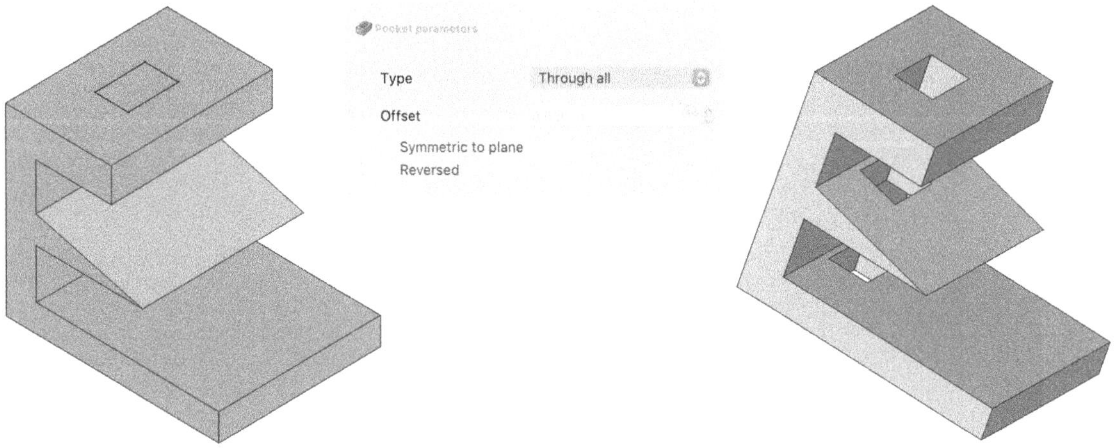

View Modification commands

The model displayed in the graphics window can be changed using various view modification commands. These commands can be accessed from the **View toolbar** (or) the **View** menu. The following are some of the main view modification commands:

	**Fit all**	The model will be fitted in the current size of the graphics window so that it will be visible completely. You can also fit the model in the graphics window by clicking the down arrow next to the Viewcube and selecting the **Zoom to fit** option.
	**Pan**	Press and hold the middle mouse button, and then drag the pointer to move the model view on the plane parallel to screen.
	**Free Orbit**	Press and hold the middle and right mouse button. Next, drag the pointer to rotate the model view.

	**Rotate Left**	On the menu bar, click **View > Standard Views > Rotate Left**; the model rotates to left.	
	**Rotate right**	On the menu bar, click **View > Standard Views > Rotate right**; the model rotates to right.	
	**Zoom In**	On the menu bar, click **View > Zoom > Zoom In** to zoom into the model.	
	**Zoom Out**	On the menu bar, click **View > Zoom > Zoom Out** to zoom out of the model.	
	**Box Zoom**	On the menu bar, click **View > Zoom > Box Zoom** and drag a rectangle. The contents inside the rectangle will be zoomed.	
Draw Style	**Flat Lines**	This represents the model with shades along with visible edges.	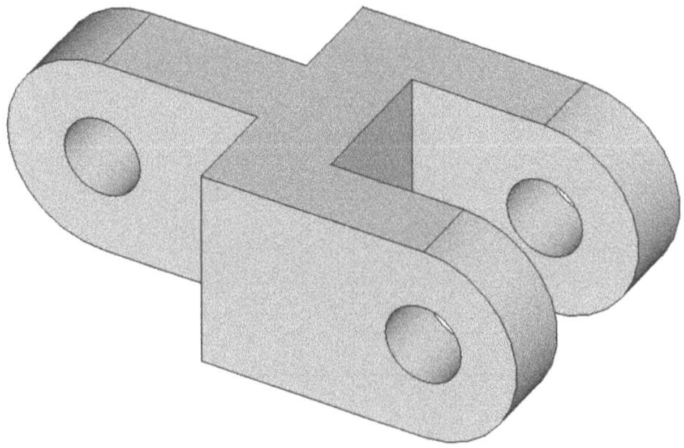
	Shaded	This represents the model with shades without visible edges.	

	Hidden line	This represents the model faces with triangular meshes.	
	Wireframe	This represents the model in wireframe along with the hidden edges	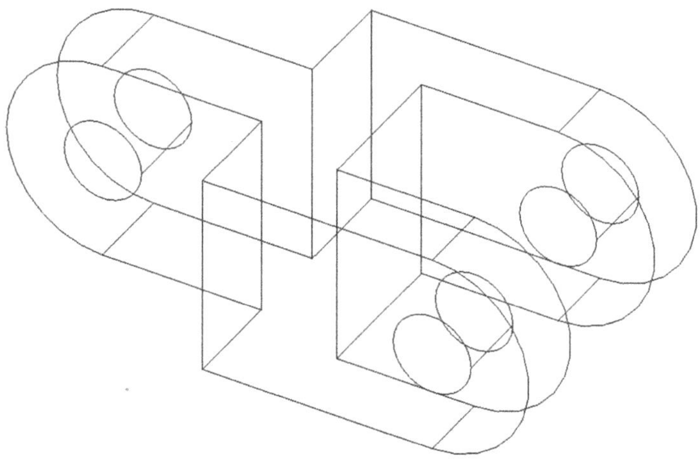
	Points	Only the vertices of the model are displayed	

	No Shading	This represents the model without any shading. The model edges, vertices and faces are displayed in solid color.	

Tutorial 1 (Millimeters)

In this example, you create the part shown below.

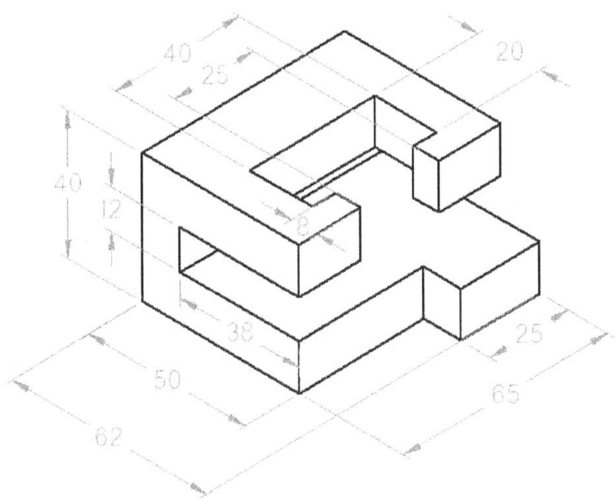

Creating a New Document

1. Click **FreeCAD** on the desktop.
2. On the menu bar, click **File > New**; it creates a new document.
3. On the **Workbench** toolbar, select **Workbench** drop-down **> Part Design**.
4. Click **Edit > Preferences** on the **Menu** bar; the **Preferences** dialog appears on the screen.
5. Click **Units** tab and select **Unit System > Standard (mm/kg/s/degree)**.
6. Select **Number of decimals > 2** and click **OK** on the **Preferences** dialog.

Creating a Sketch

1. Click the **Create sketch**  icon on the **Part Design Helper** toolbar, and then select the XZ Plane.
2. Click **OK** on the **Combo View** panel to start the sketch.
3. On the **Sketcher geometries** toolbar, click the **Create rectangle** 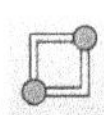 icon.
4. Click the origin point to define the first corner of the rectangle.
5. Move the pointer toward the top right and click to define the second corner – Press **Esc** to deactivate the tool.

6. On the **Sketcher Constraints** toolbar, click the **Constraint horizontal distance** icon. Next, select the horizontal line, type-in **50** in the **Length** box, and click **OK**.

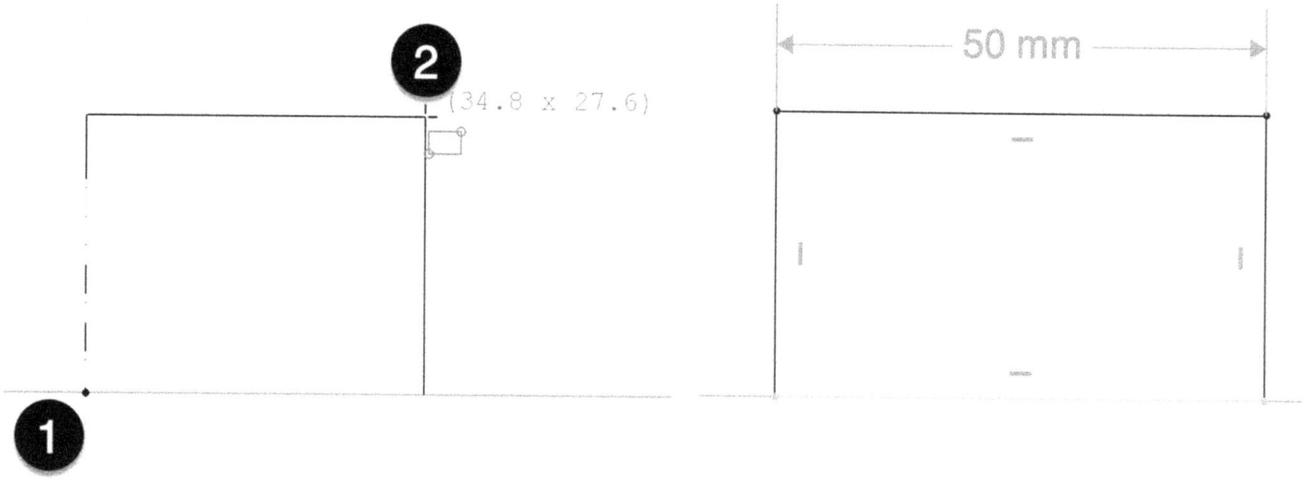

7. On the **Sketcher Constraints** toolbar, click the **Constraint vertical distance** icon. Next, select the vertical line, type-in **40** in the **Length** box, and click **OK**.

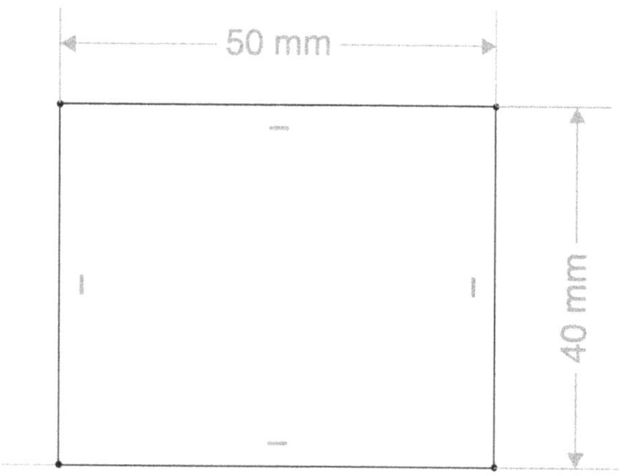

8. Click **Leave Sketch** on the **Part Design Helper** toolbar.

Creating a Pad Feature

1. Click the **Pad** icon on the **Part Design Modeling** toolbar.
2. Check the **Symmetric to plane** option on the **Pad parameters** panel.
3. Set the **Length** to **65** and click **OK** to create the pad feature.
4. Click the **Isometric** icon on the **View** toolbar.

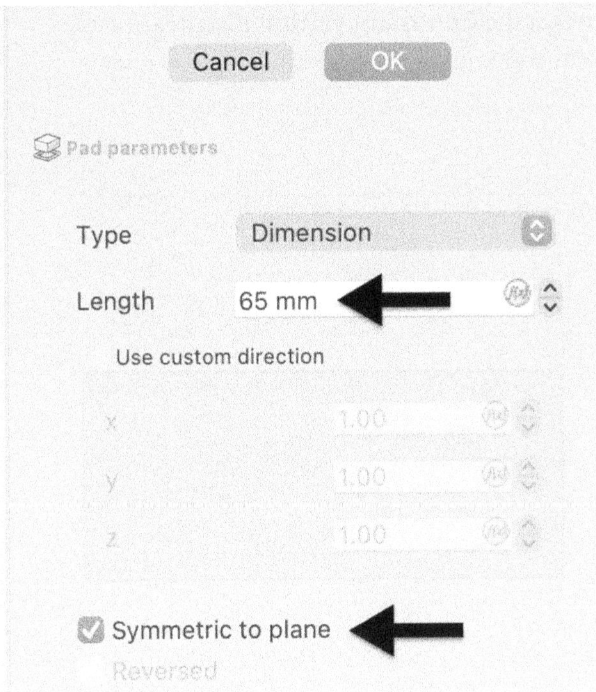

Creating a Padded Cut Feature

1. Click on the front face of the part geometry. Next, click the **Create Sketch** icon on the **Part Design Helper** toolbar.
2. Click **OK** on the **Combo View** panel.
3. Click the **External geometry** icon on the **Sketcher geometries** toolbar. Click on the right edge of the model.
4. On the **Sketcher geometries** toolbar, click the **Create rectangle** icon.
5. Click on the right edge of the model, move the pointer toward left, and then click.

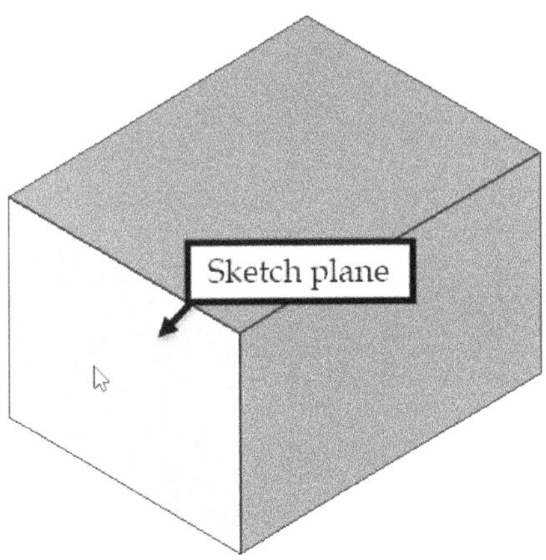

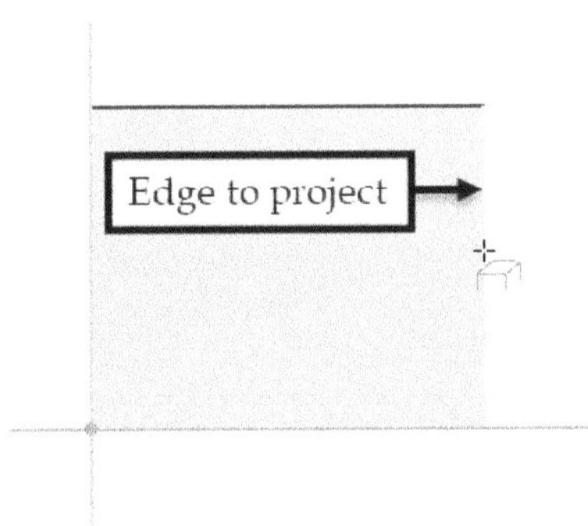

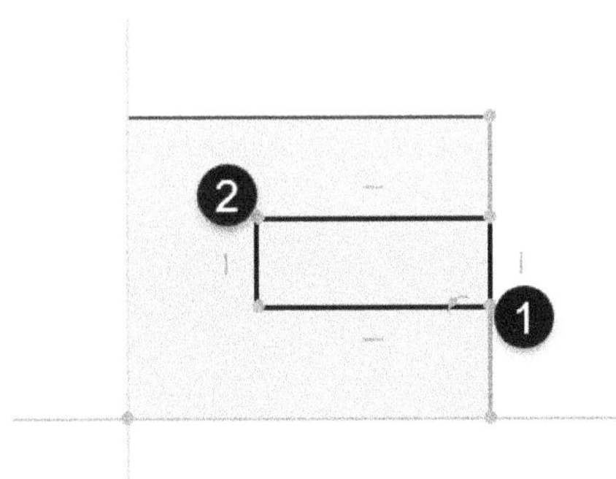

6. Activate the **Constraint horizontal distance** command and select the horizontal line of the rectangle. Next, type-in **38** in the **Length** box and click **OK** to add the dimension.
7. Click the **Constraint vertical distance** command and select the vertical line of the rectangle.

8. Type-in **12** in the **Length** box and then click **OK**. Make sure that the **Constraint vertical distance** is active.
9. Select the bottom right corner of the rectangle and the bottom endpoint of the projected edge, as shown.
10. Type-in **14** in the **Length** box and click **OK**.
11. Click **Leave Sketch** on the **Part Design Helper** toolbar.

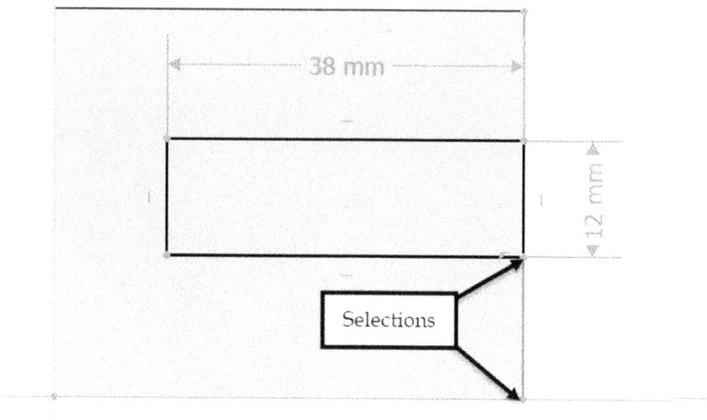

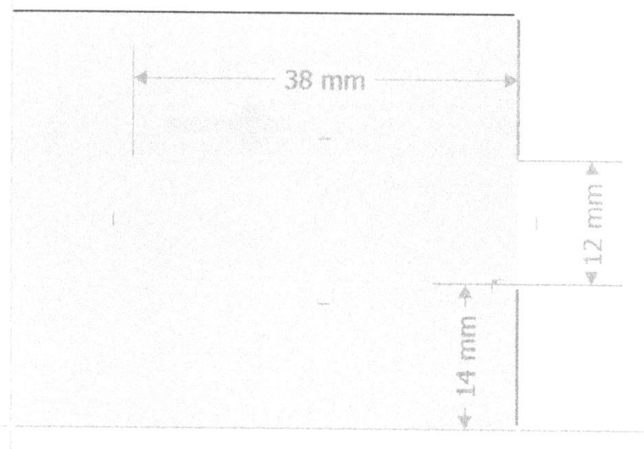

12. Click the **Pocket** icon on the **Part Design Modeling** toolbar.
13. On the **Combo View** panel, under the **Pocket Parameters** section, select **Type > Through All**.
14. Click **OK** on the **Combo View** panel to create the cut throughout the part design.

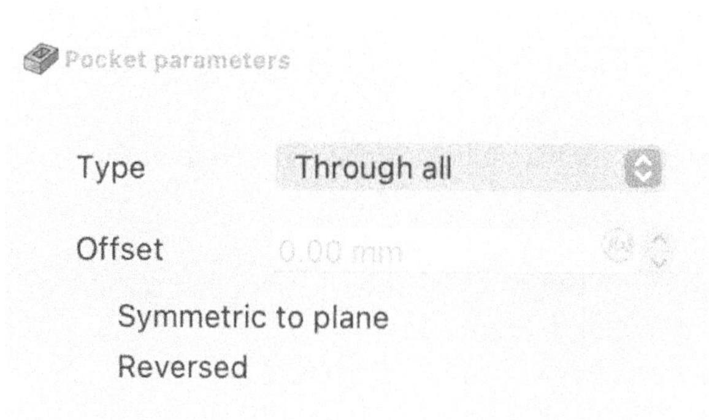

Extruding a sketch up to the Face next to the sketch plane

1. Click on the top face of the part design. Click the **Create Sketch** command on the **Part Design Helper** toolbar.

2. Click the **Set to top** icon on the **View** toolbar.
3. Click the **External Geometry** icon on the **Sketcher geometries** toolbar. Next, select the right vertical edge of the model; the selected edge is projected.
4. Activate the **Polyline** command (click the **Polyline** icon on the **Sketcher geometries** toolbar).
5. Create the sketch, as shown.
6. Select the inclined line and click the **Constrain vertical** icon on the **Sketcher constraints** toolbar, as shown.

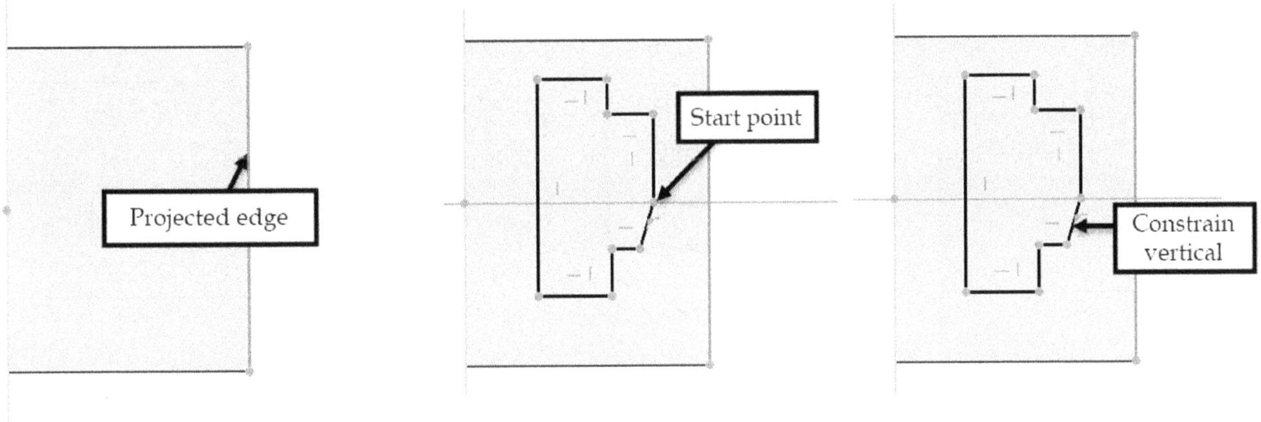

7. Click the **Constrain equal** icon on the **Sketcher constraints** toolbar and select the horizontal lines, as shown. Next, select the vertical lines, as shown.
8. Click the **Constrain equal** icon on the **Sketcher constraints** toolbar and select the vertical lines, as shown.

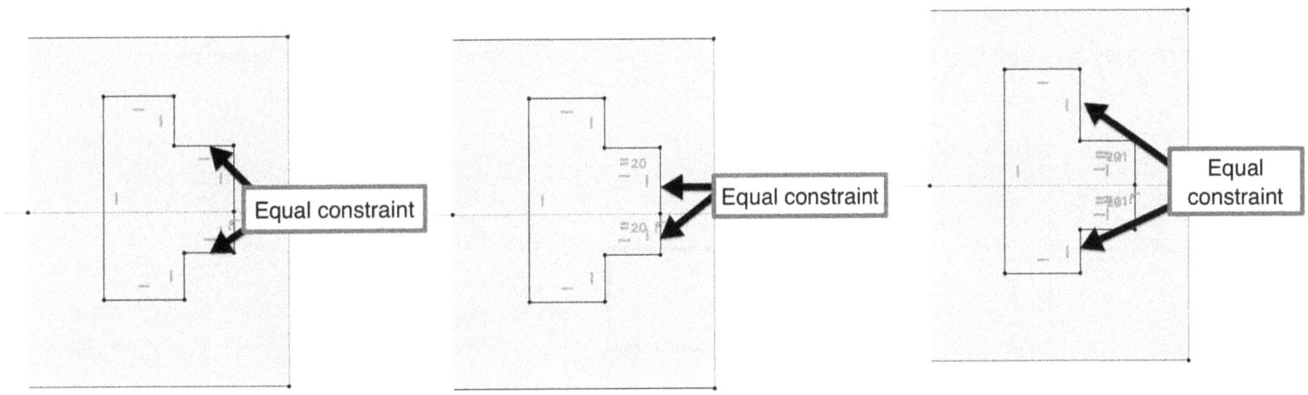

9. Click the **Constrain point onto an object** icon on the **Sketcher constraints** toolbar and select the projected edge and the start point of the sketch. The selected point is made coincident to the projected edge.

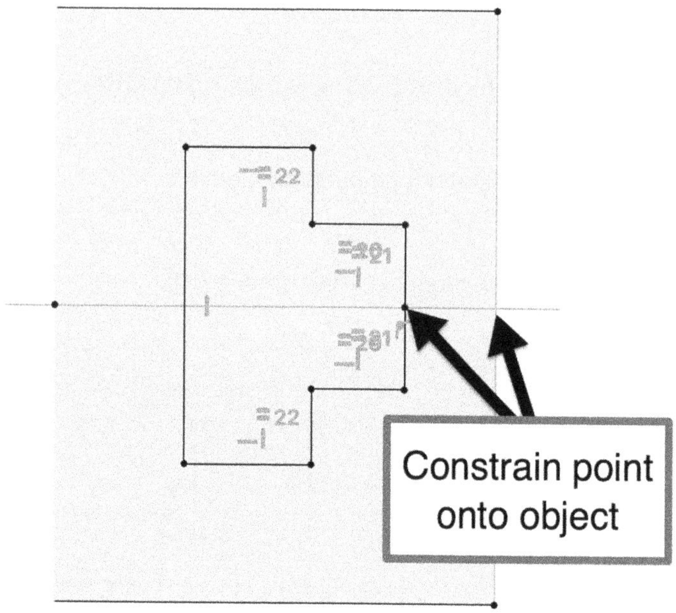

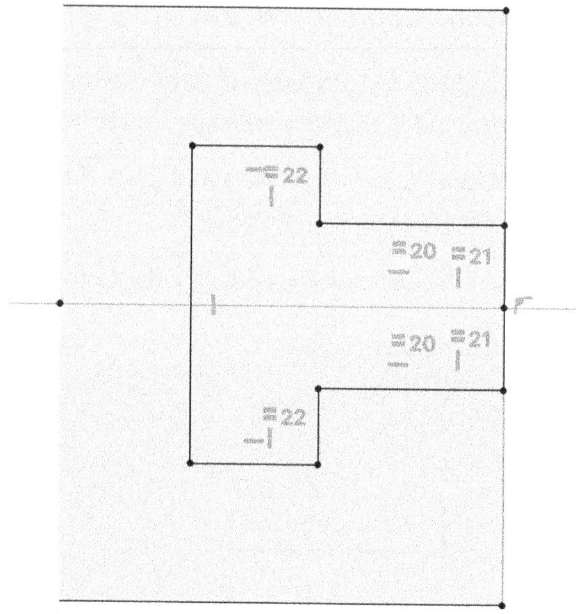

10. Click the **Constrain vertical distance** icon on the **Sketcher constraints** toolbar. Next, select the left vertical line. Type **40** in the **Length** box and click **OK**.
11. Select the endpoints of the right vertical line. Next, type **25** in the **Length** box, and then click **OK.**
12. Click the **Constrain horizontal distance** icon on the **Sketcher constraints** toolbar, and then select the left horizontal line. Next, type **20** in the **Length** box and click **OK**.
13. Select the right horizontal line and type **8** in the **Length** box, and then click **OK.**

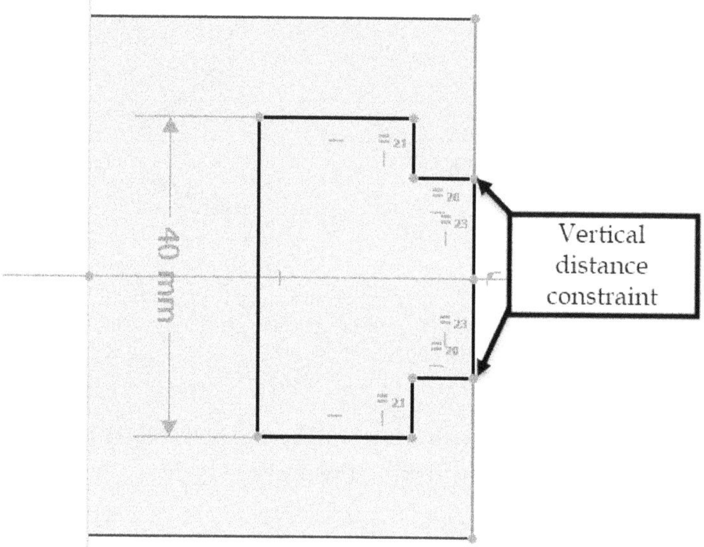

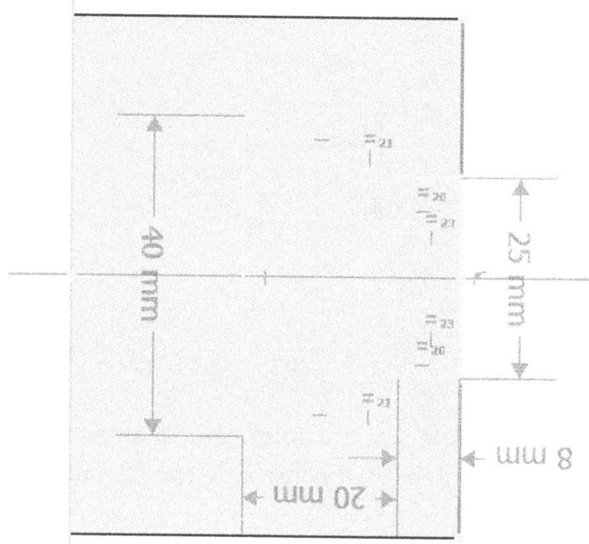

14. Click the **Close** button on the **Combo View** panel to close the sketch.
15. Activate the **Pocket** command (on the **Part Design Modeling** toolbar, click the **Pocket** icon)
16. On the **Pocket parameters** section, select **Type > To first.**

17. Click **OK** to remove the material up to the surface next to the sketch plane.

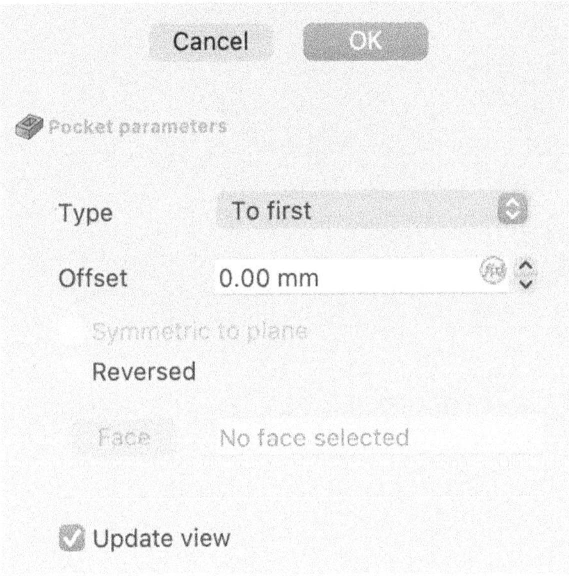

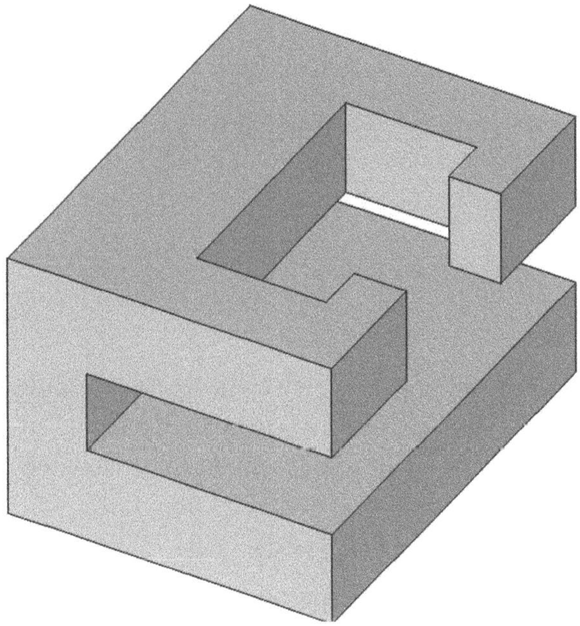

Extruding a sketch up to a selected face

1. Activate the **Create Sketch** command and select the XY plane. Next, click **OK** to start a sketch.
2. Click the **External Geometry** icon on the **Sketcher geometries** toolbar. Next, select the right vertical edge of the model; the selected edge is projected.
3. Draw a rectangle and apply constraints to it, as shown.
4. On the **Sketcher constraints** toolbar, click **Constrain Coincident**. Next, select the corner point of the rectangle and the model, as shown.
5. Click the **Close** button on the **Combo View** panel to close the sketch.

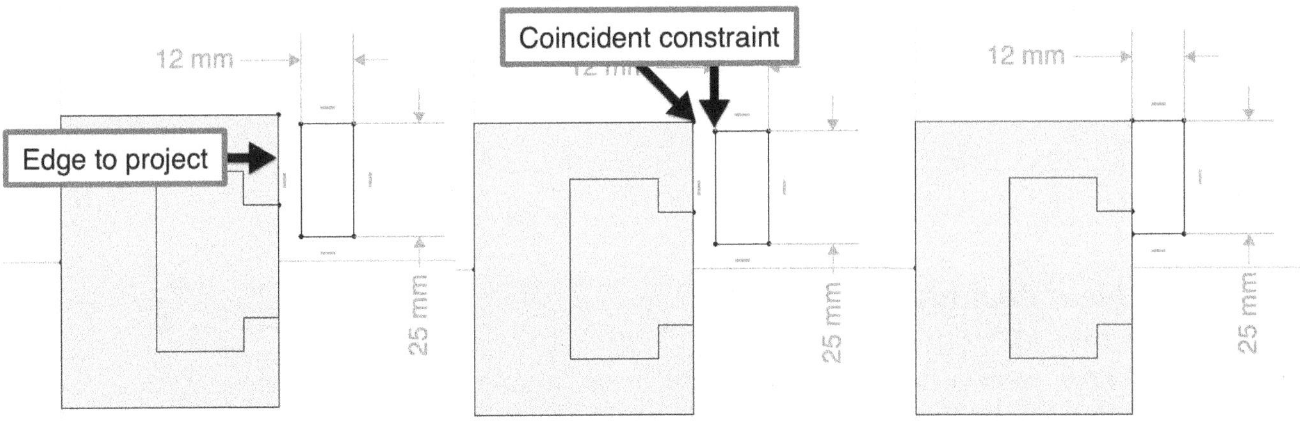

6. Activate the **Pad** command (on the **Part Design Modeling** toolbar, click the **Pad** icon)
7. Select **Type > Up to face** from the **Pad Parameters** section on the **Combo View** panel. Next, select the horizontal face of the part geometry, as shown.
8. Click **OK** on the **Pad parameters** section to complete the part.

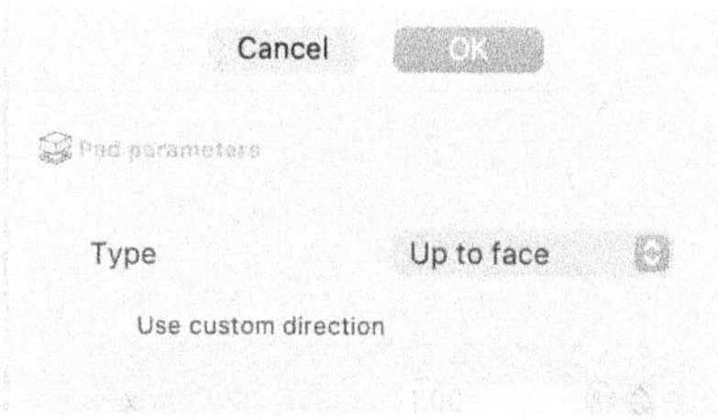

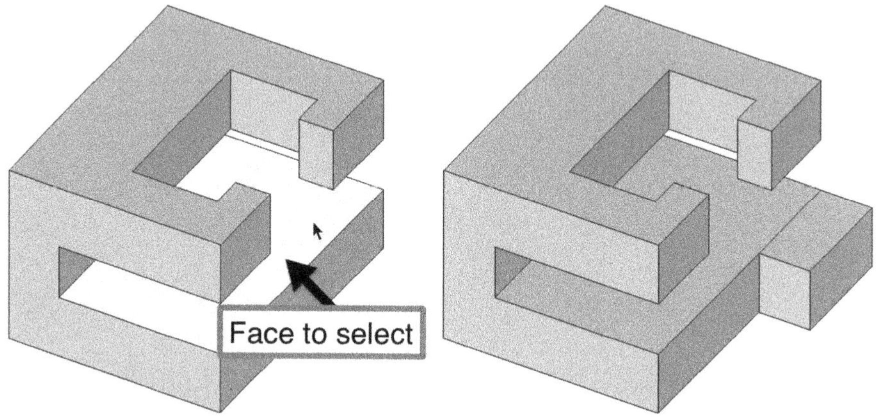

9. Click the **Save** icon on the **File** toolbar located at the top-left corner of the window. Next, type C3_exampl1 in the **File name** box, and then click **Save**.
10. Click **File > Close** on the menu bar to close the document.

Tutorial 2 (Inches)

In this example, you create the part shown below.

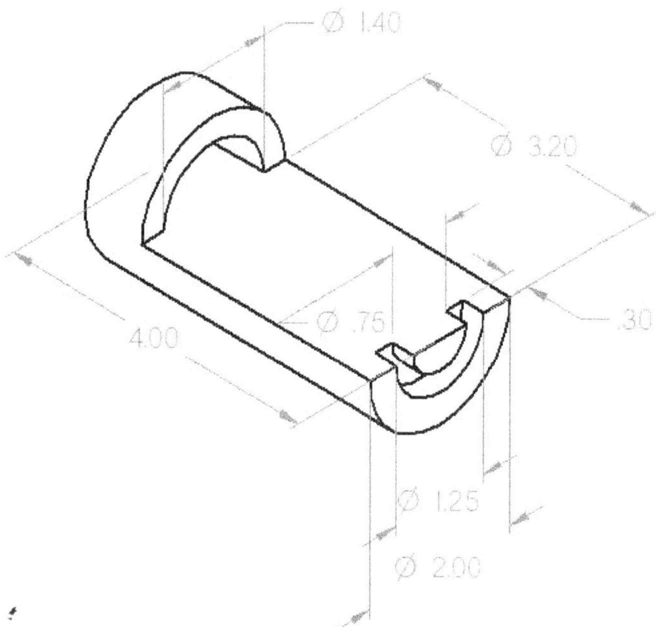

Creating a New document

1. Open a new **FreeCAD** file.
2. Click **Edit > Preferences** on the **Menu** bar; the **Preferences** dialog appears on the screen.
3. Click the **Units** tab on the **Preferences** dialog.
4. Select **Unit System > Imperial decimal** and type-in **3** in the **Number of decimals** box.
5. Click **OK** on the **Preferences** dialog.
6. Select the **Part Design** option from the **Workbenches** drop-down.

Creating a Revolved Feature

1. Click the **Create sketch** icon on the **Part Design Helper** toolbar, and then select the XY_Plane.
2. Click **OK** on the **Combo View** panel.
3. On the **Sketcher geometries** toolbar, click the **Create rectangle** icon.
4. Click on the origin point to define the first corner of the rectangle.
5. Move the pointer toward the top right corner and click to define the second corner.

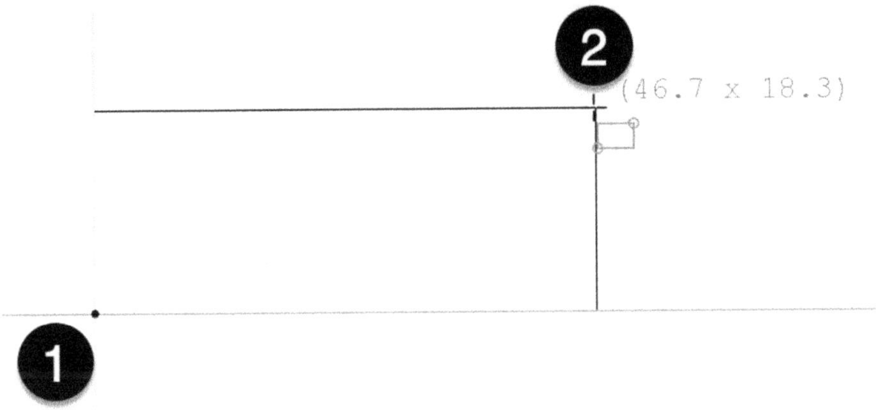

6. Click the **Constrain horizontal distance** tool on the **Sketcher constraints** toolbar.
7. Select the bottom horizontal line. Next, type-in **4** in the **Length** box and click **OK** on the **Insert Length** dialog.
8. Click the **Constrain vertical distance** tool on the **Sketcher constraints** toolbar. Next, select the vertical line.
9. Type-in **1** in the **Length** box and click **OK** on the **Insert Length** dialog.

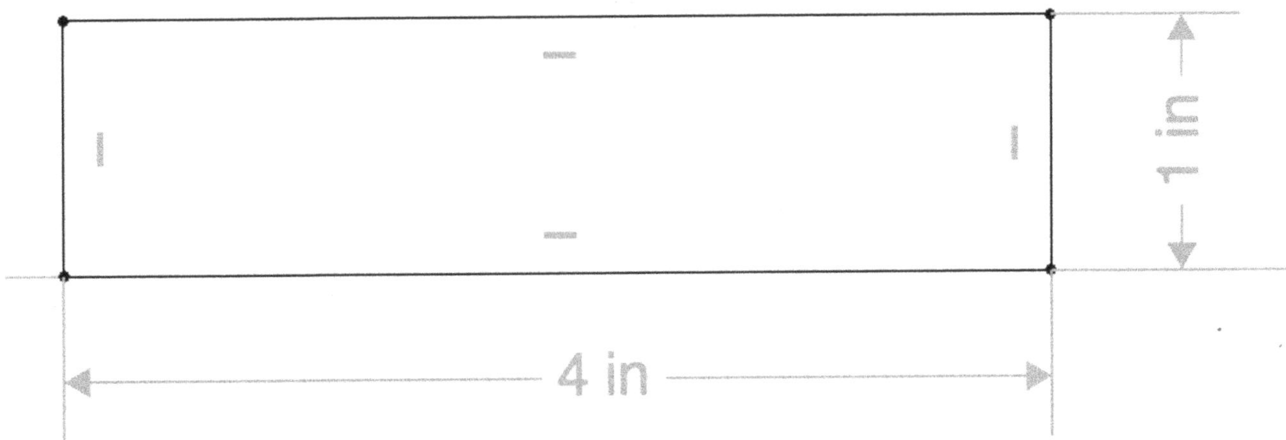

10. Click **Leave Sketch** on the **Part Design Helper** toolbar. Next, click the **Isometric** icon on the **View** toolbar.
11. Click the **Revolution** icon on the **Part Design Modeling** toolbar.
12. Select **Axis > Horizontal Sketch Axis** from the **Revolution parameters** section on the **Combo View** panel.
13. Type-in **180** in the **Angle** box and check the **Reversed** option. Next, click **OK** on the **Combo View** panel to create the *Revolution* feature.

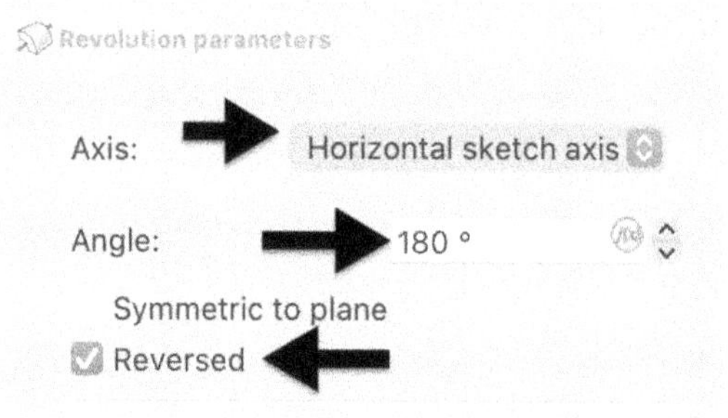

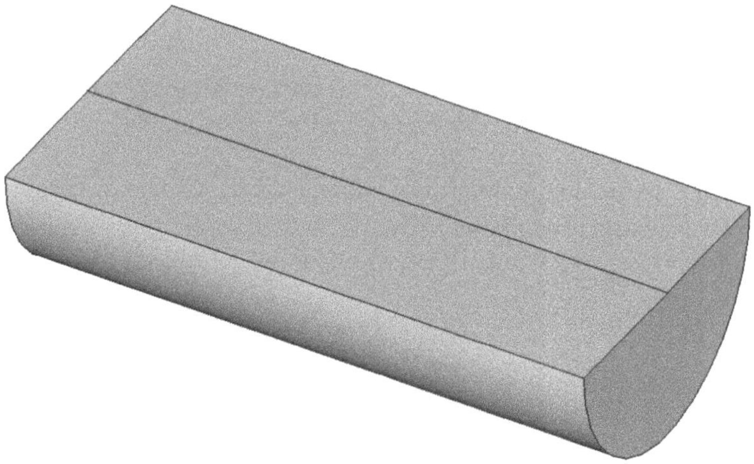

Creating a Groove feature

1. Select the top face of the part geometry and click the **Create Sketch** tool on the **Part Design Helper** toolbar.
2. Draw a rectangle and apply the dimensional constraints to it, as shown.
3. Click the **External geometry** icon on the **Sketcher geometries** toolbar and select the right vertical edge.
4. Click the **Constrain vertical distance** icon on the **Sketcher constraints** toolbar.
5. Select the top right corner of the rectangle and the top endpoint of the projected edge. Type **0.375** in the **Length** box and click **OK**.
6. Click the **Constrain point onto object** icon on the **Sketcher constraints** toolbar and select the top right corner of the rectangle. Next, select the projected edge. Click the **Close** button on the **Combo View** panel.

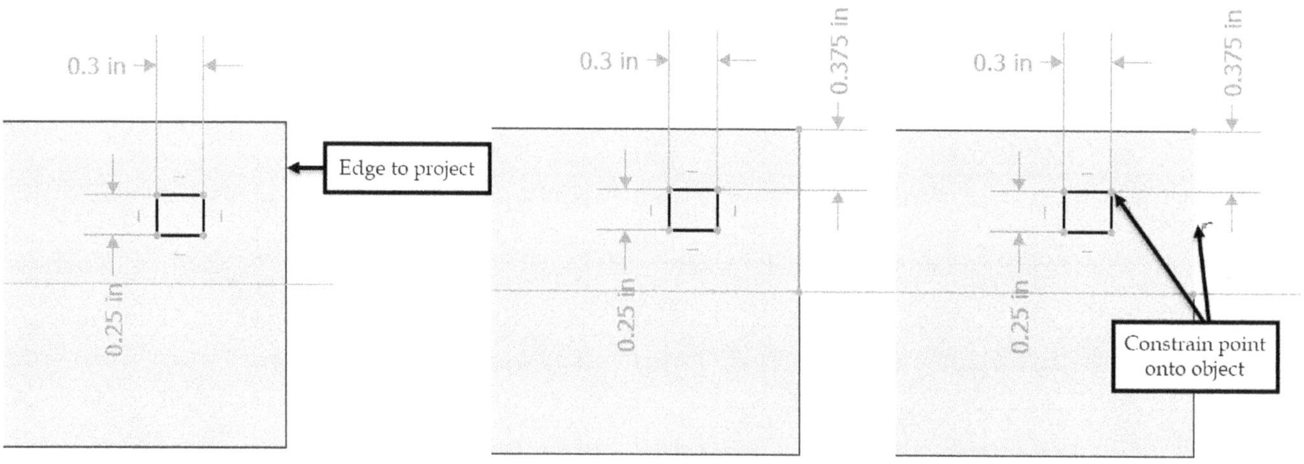

7. On the **Part Design Modeling** toolbar, click the **Groove** icon.
8. On the **Groove parameters** section, select **Axis > Horizontal sketch axis**. Next, type **180** in the **Angle** box and click **OK** to create the groove feature.

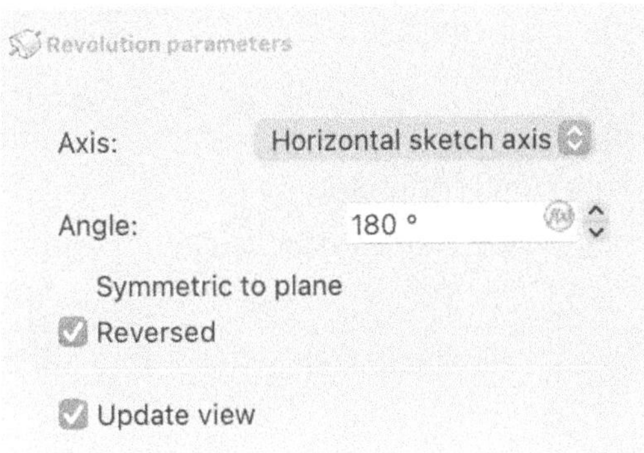

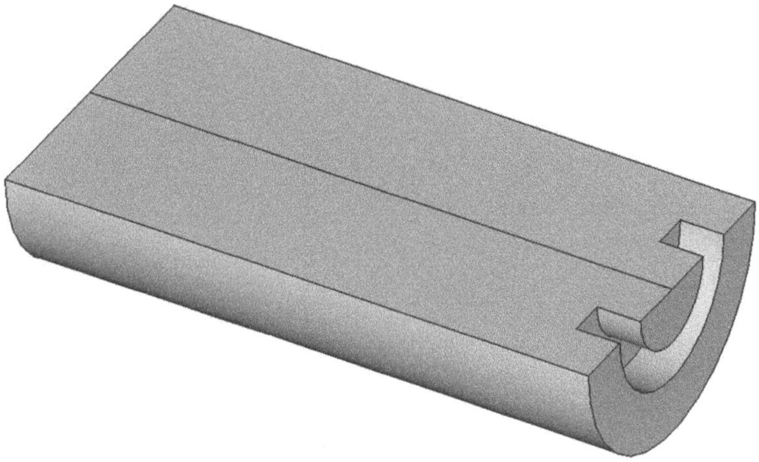

Adding a Revolution Feature to the Model

1. Select the top face of the part geometry and click the **Create Sketch**  tool on the **Part Design Helper** toolbar.
2. Draw a rectangle and add dimensional constraints to it, as shown.
3. Click the **External geometry**  icon on the **Sketcher geometries** toolbar and select the bottom horizontal edge.
4. Click the **Constrain coincident** icon on the **Sketcher constraints** toolbar. Next, select the bottom left corner point of the rectangle and the left endpoint of the projected edge.
5. Click the **Close** button on the **Combo View** panel.

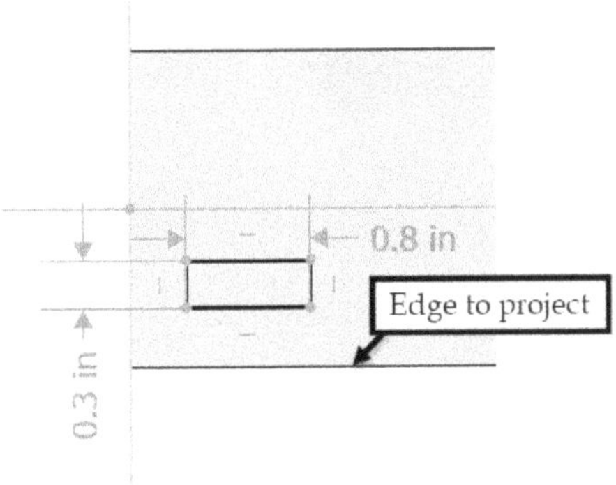

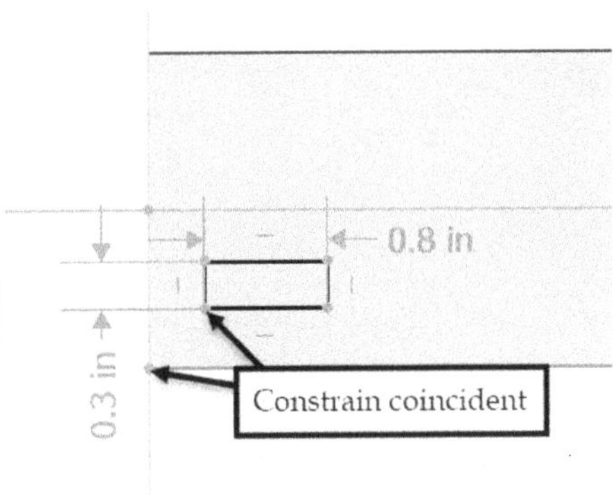

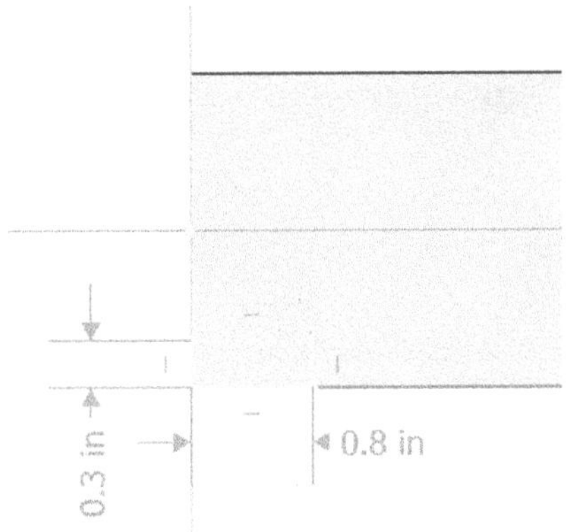

6. Click the **Revolution**  icon on the **Part Design Modeling** toolbar.
7. On the **Revolution parameters** section, select **Axis > Horizontal sketch axis.**
8. Type 180 in the **Angle** box and click **OK.**

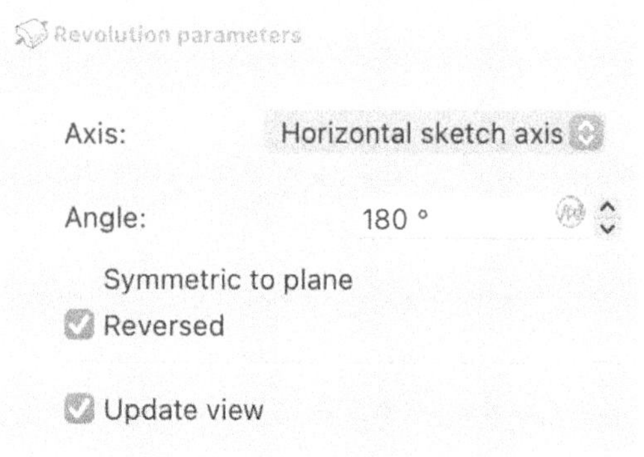

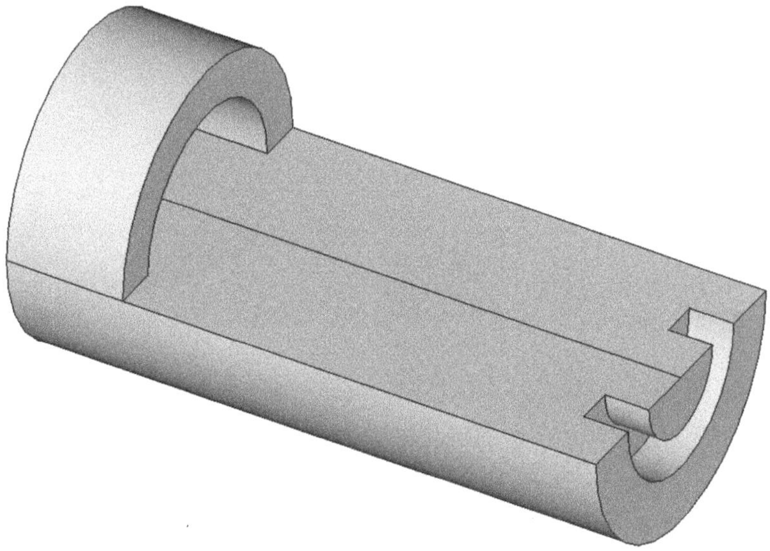

9. Click **File > Save** on the menu bar. Next, type **C3_example2** in the **File name** box, and then click **Save**.
10. Click **Close Tab** on the bottom left corner of the window.

Questions

1. How to create parallel planes in FreeCAD?
2. What are the **Type** options available on the **Pad Parameters** section?
3. How to create angled planes in FreeCAD?

Exercises

Exercise 1

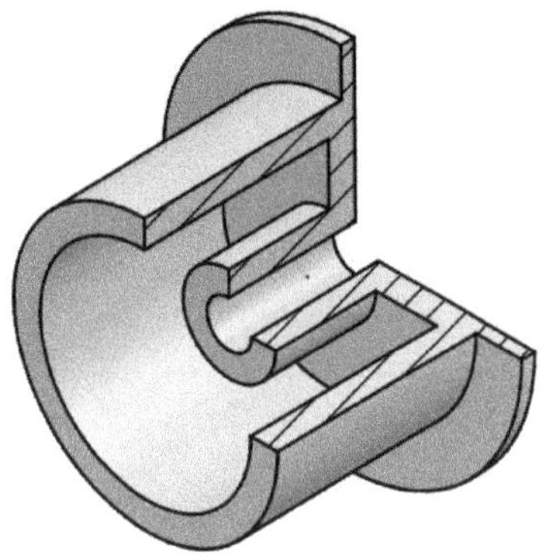

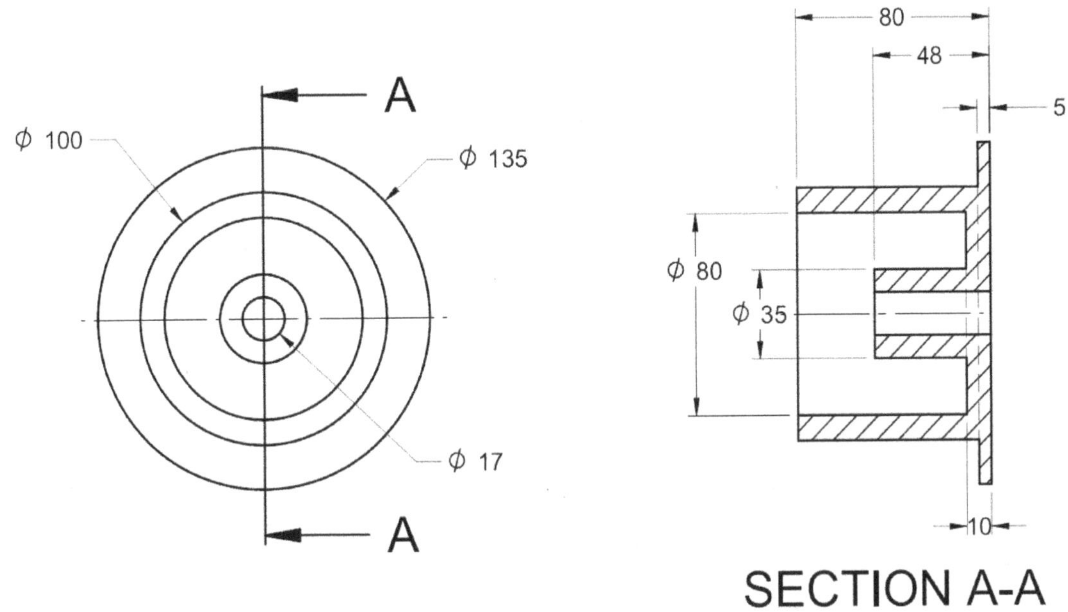

Exercise 2

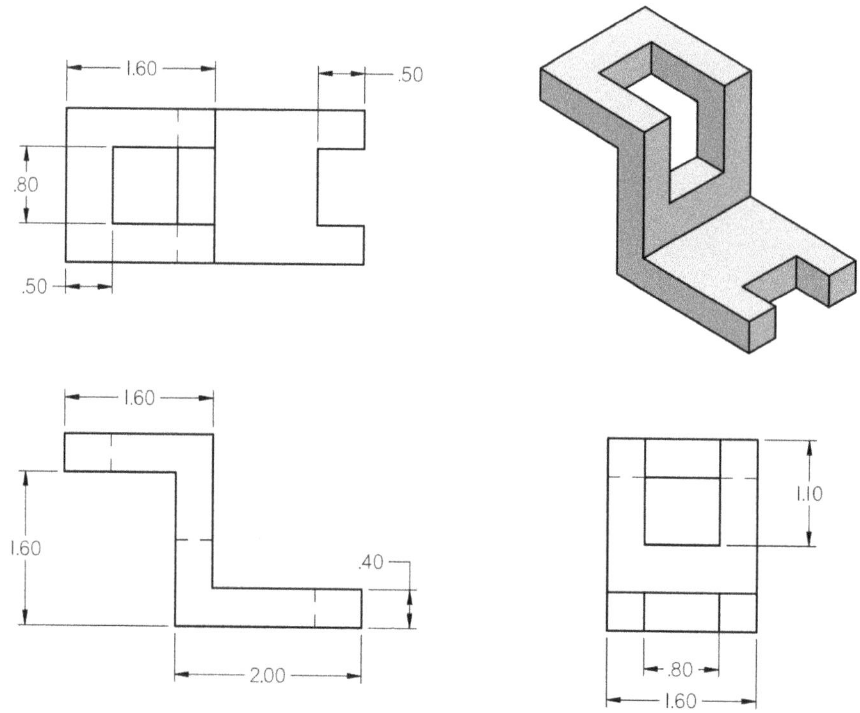

Exercise 3

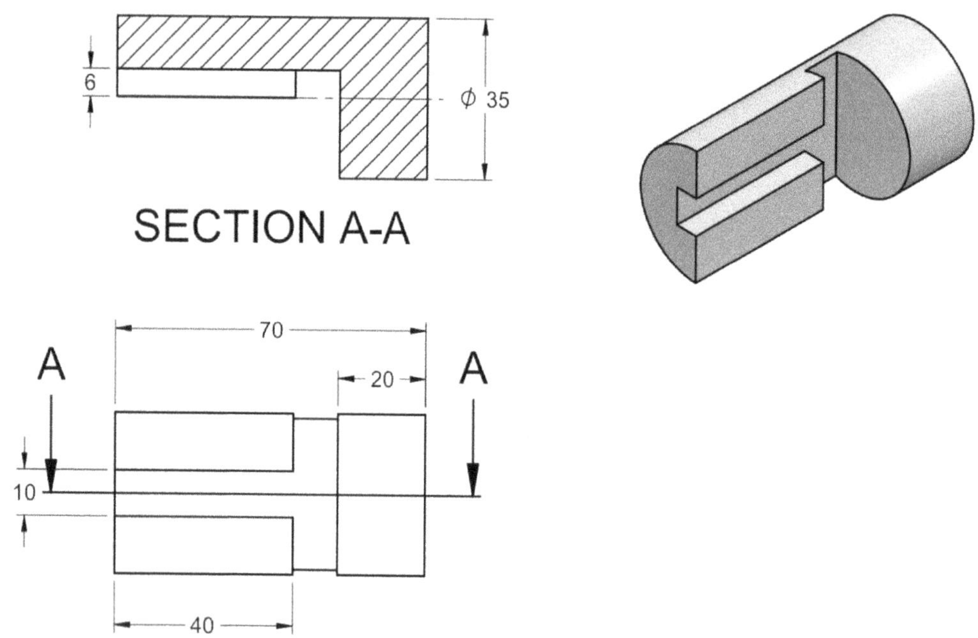

Chapter 4: Dress-up Features

So far, all of the features that were covered in the previous chapter were based on two-dimensional sketches. However, there are certain features in FreeCAD 0.19 that do not require a sketch at all. Features that do not require a sketch are called dress-up features. You can simply place them on your models. However, to do so, you must have some existing geometry. Unlike a sketch-based feature, you cannot use a dress-up feature for the first feature of a model. For example, to create a *Fillet* feature, you must have an already existing edge. In this chapter, you will learn how to add dress-up features to your design.

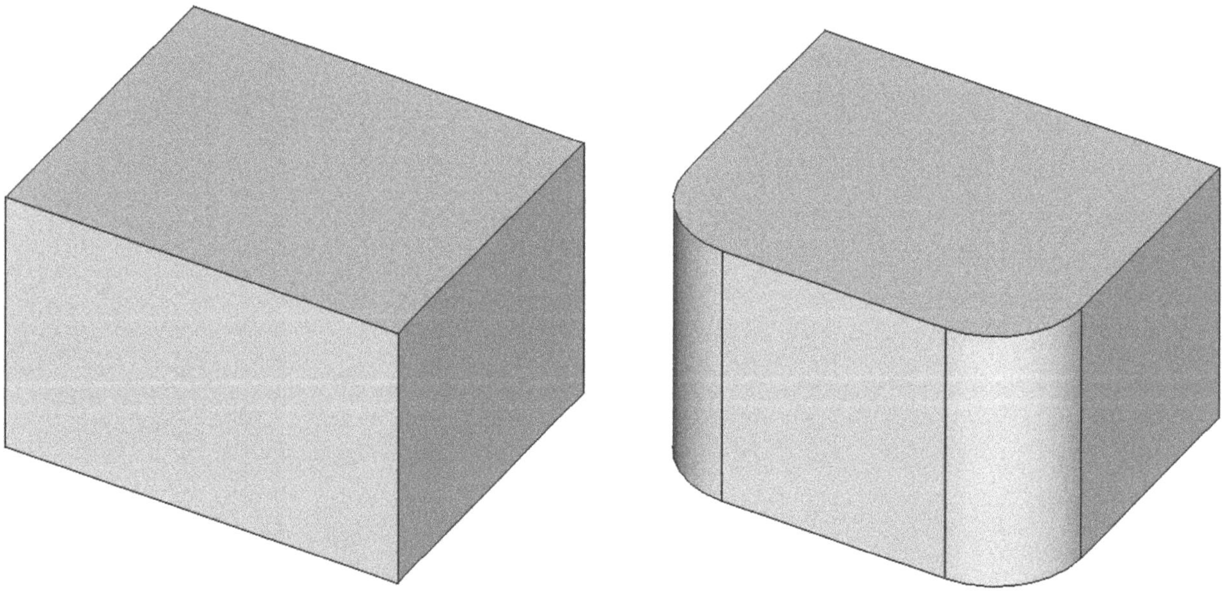

The topics covered in this chapter are:

- *Holes*
- *Threads*
- *Fillets*
- *Chamfers*
- *Drafts*
- *Shells*

 Hole

As you know, it is possible to use the *Pocket* command to create cuts and remove material. But, if you want to drill holes that are of standard sizes, the **Hole** command is a better way to do this. The reason for this is it has many hole types already predefined for you. All you have to do is choose the correct hole type and size. The other benefit is when you are going to create a 2D drawing, FreeCAD can place the correct hole annotation automatically. To create a hole first, click on the placement face and then click **Create a new Sketch** on the **Part Design Helper** toolbar. Next, create one or more circles, and then add dimensions and constraints to them.

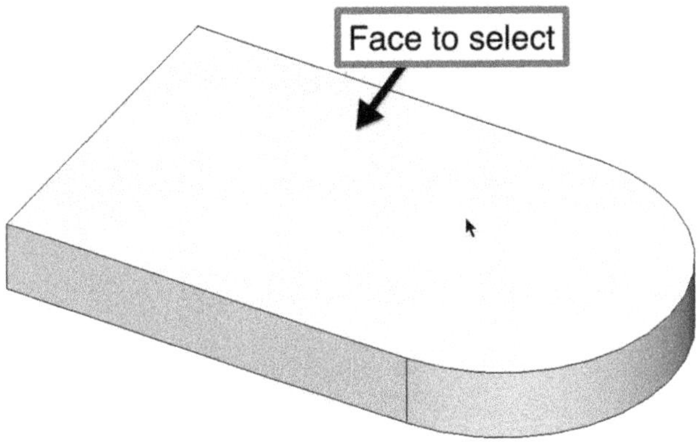

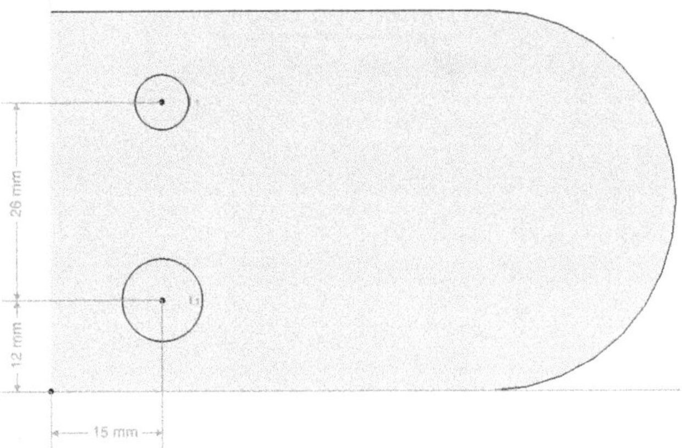

If you want to create a hole concentric to a circular edge, then click the **External geometry** icon on the **Sketcher geometries** toolbar. Next, click on the circular edge. Activate the **Create circle** command and select the centerpoint of the external geometry. Move the pointer outward and click to create the circle. Click **Leave Sketch** on the **Sketcher** toolbar.

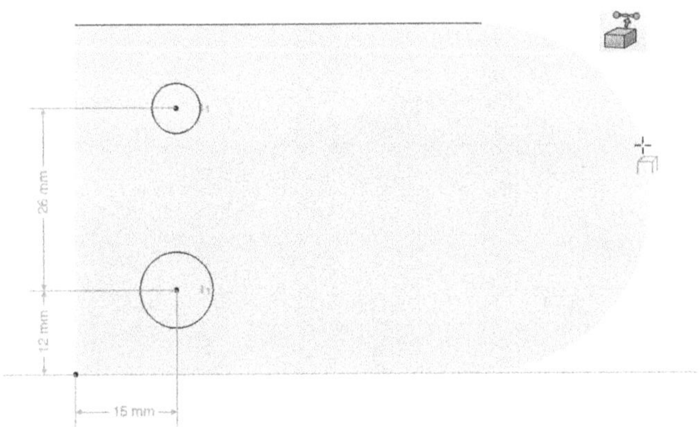

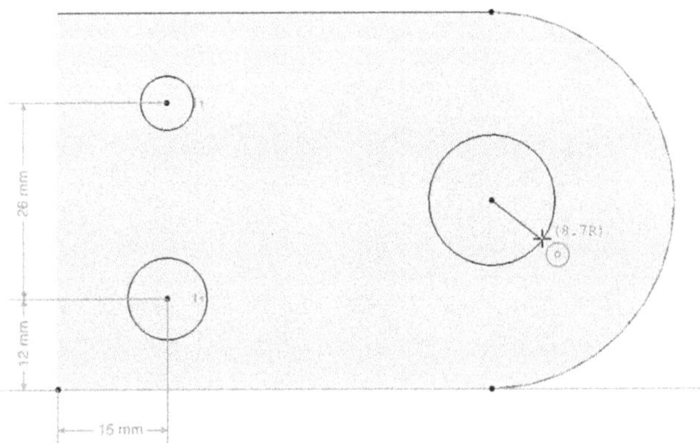

Activate the **Hole** command (Click the **Hole** icon on the **Part Design Modeling** toolbar (or) click **Part Design > Create a subtractive feature > Hole** on the menu bar); you will notice that the **Hole Parameters** panel appears on the left-side. There are options on this panel that make it easy to create different types of holes.

Simple Hole

To create a simple hole feature, select **Type > None** under the **Hole cut** section of the **Hole Parameters** panel. Next, select the **Depth** type. If you want a through-hole, select **Depth > Through all**. If you want the hole only up to some depth, then select **Depth > Dimension**, and then type-in a value in the **Depth** box.

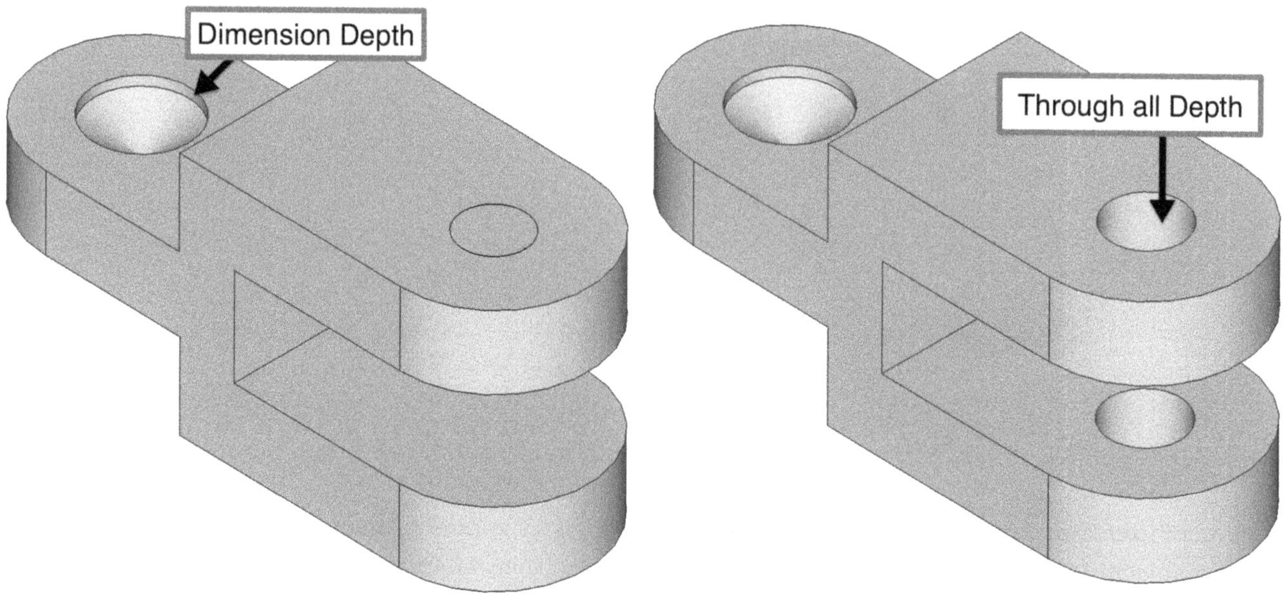

The **Drill Point** section has two options to define the end condition of the hole: **Flat** and **Angled**. The **Flat** option creates a hole with a flat bottom. The **Angle** option creates a hole with an angled bottom. The **Angled** box defines the angle of the cone tip at the bottom.

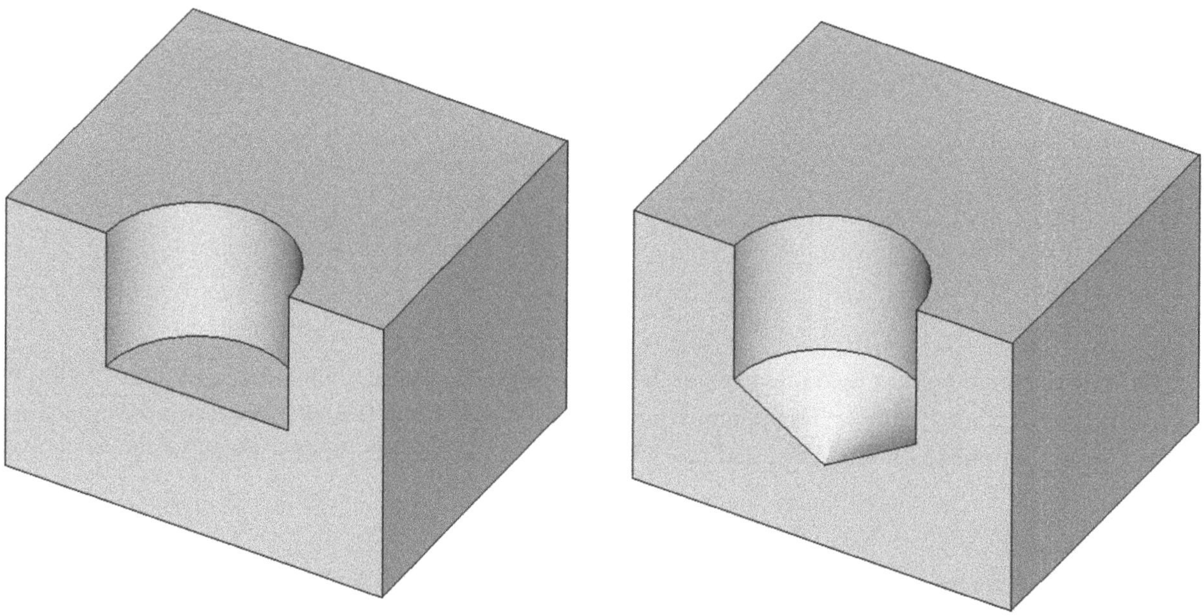

The **Take into account for depth** checkbox includes the tip of the angled bottom in the hole depth. If you uncheck this option, the angled bottom is excluded from the hole depth.

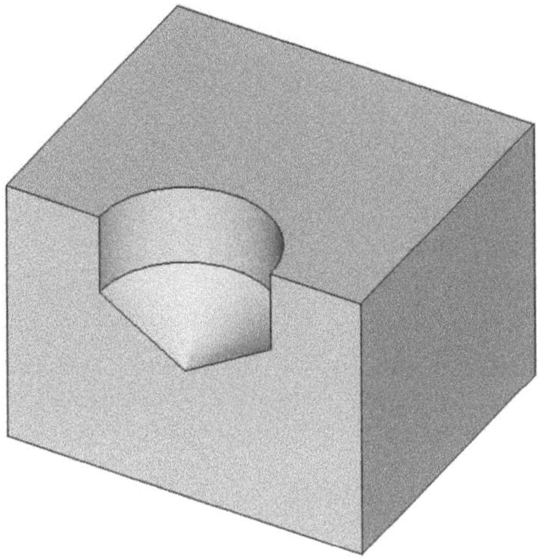

Take into account for depth checked

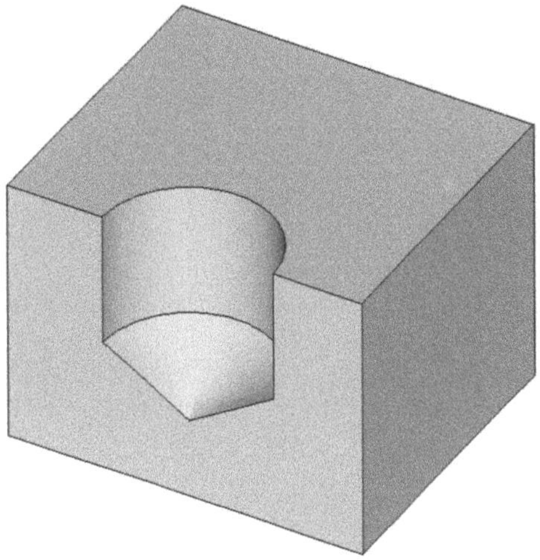

Take into account for depth unchecked

Next, specify the **Diameter** of the hole and click **OK** to complete the hole feature.

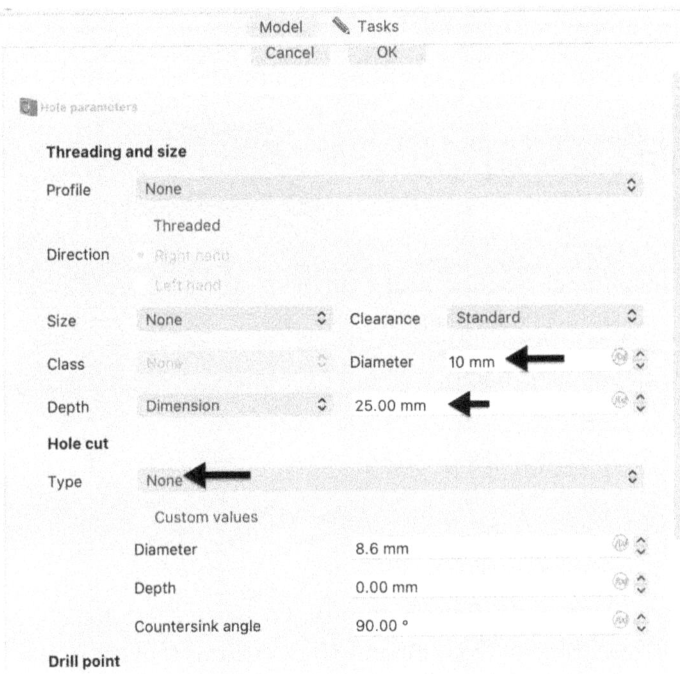

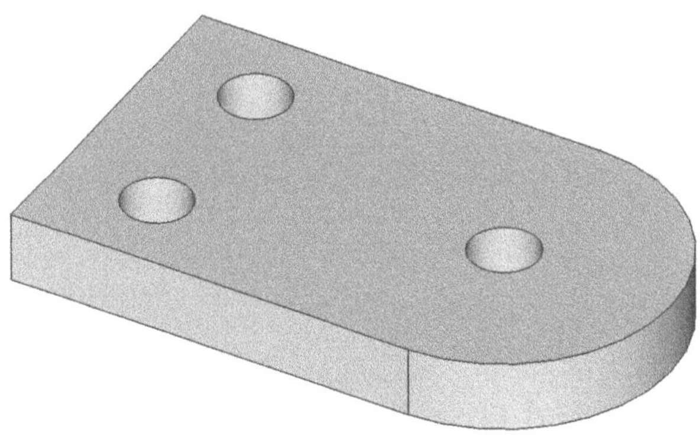

Counterbored Hole

A counterbore hole is a large diameter hole added at the opening of another hole. It is used to accommodate a fastener below the level of the workpiece surface. To create a counterbore hole, select **Type > Counterbore from** the **Hole cut** section. Next, specify the Hole Diameter, Hole Depth, Counterbore Diameter, and Counterbore Depth. Next, specify the desired **Drill Point** type (**Flat** or **Angled**). If you click the **Angled** option, then specify the **Tip Angle** value.

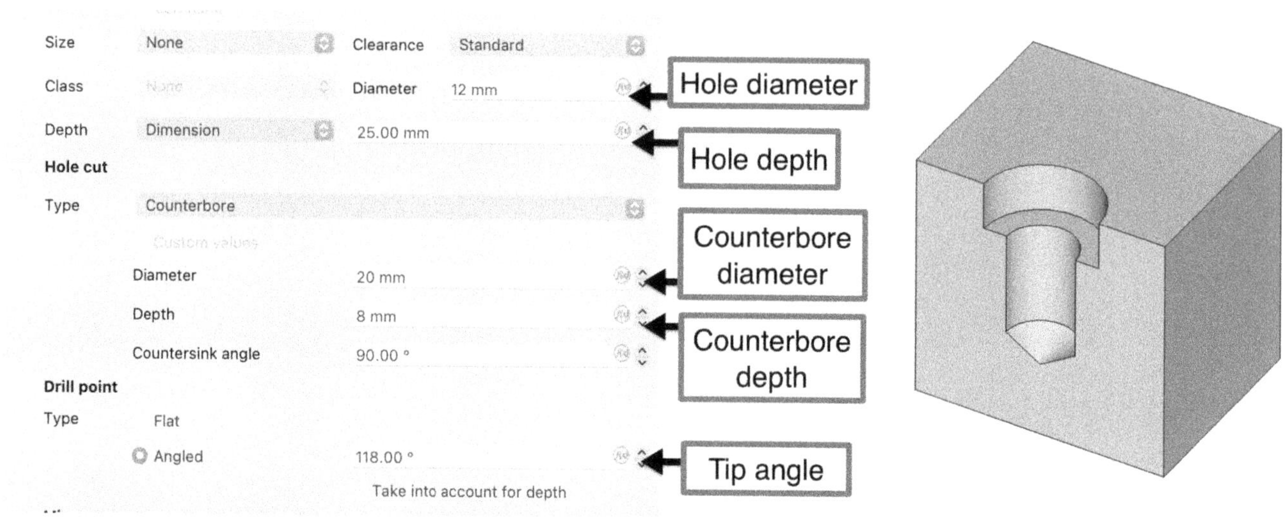

Countersink Hole

A countersunk hole has an enlarged V-shaped opening to accommodate the fastener below the level of the workpiece surface. To create a countersink hole, select **Type > Countersink** from the **Hole cut** section. Next, specify the Hole Diameter, Countersink Diameter, and Countersink angle. Set the hole diameter and drill point.

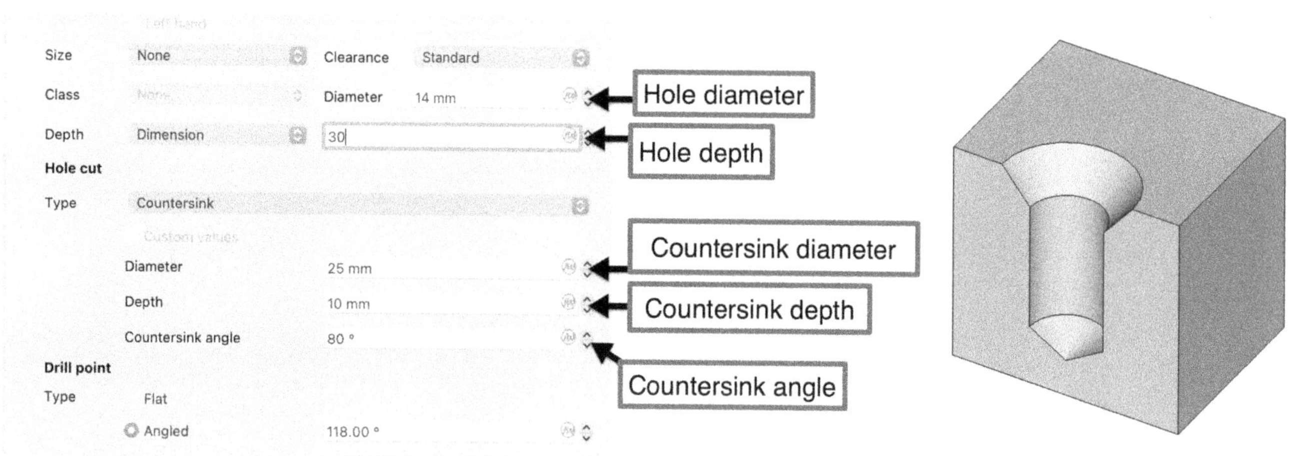

Threaded Hole

To create a threaded hole feature, first select the thread profile from the **Profile** drop-down. Next, check the **Threaded** option. Next, specify the **Direction**, **Size**, and **Class**. Specify the remaining hole options that are similar to the simple hole feature. Click **OK** to create the threaded hole.

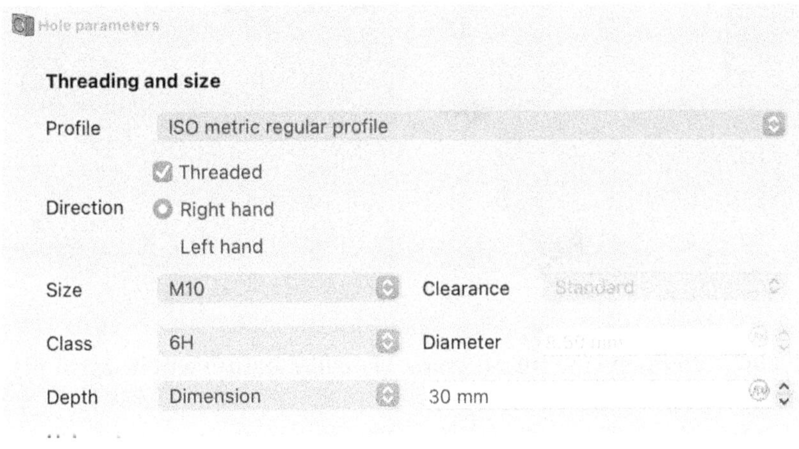

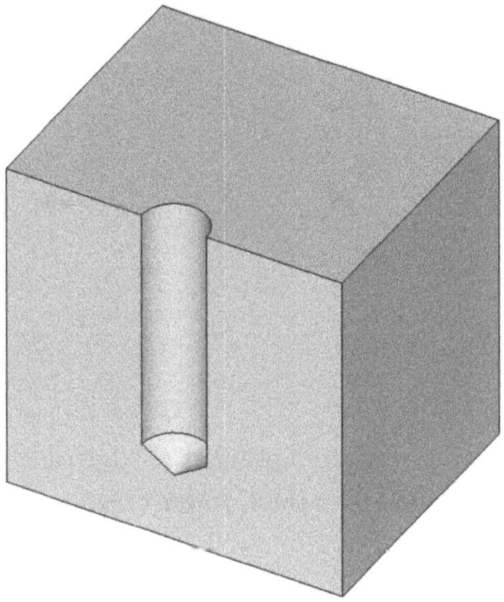

Fillet

This command breaks the sharp edges of a model and blends them. You do not need a sketch to create a fillet. All you need to have is model edges. Press and hold the CTRL key (COMMAND key for Mac users) and select the edges to fillet. Next, click **Part Design** > **Apply a dress-up feature** > **Fillet** on the menu bar (or) click the **Fillet** icon on the **Part Design Modeling** toolbar.

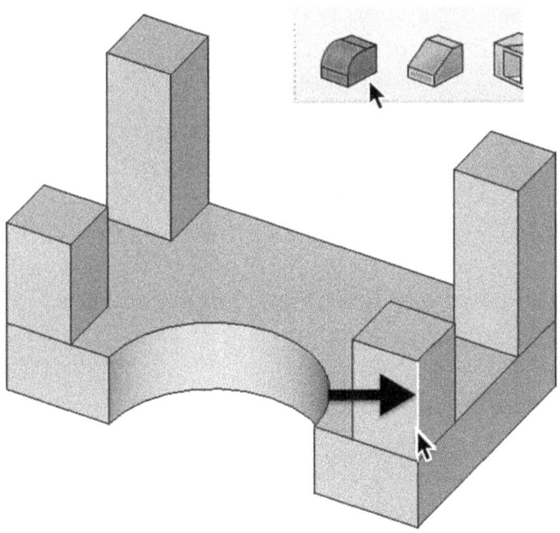

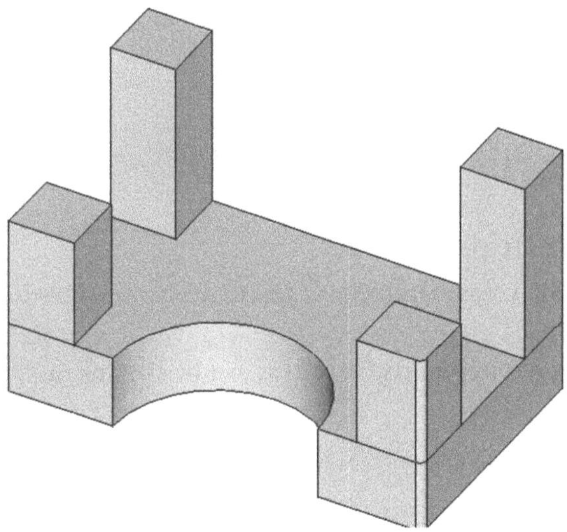

If you want to select more edges to be filleted, then click the **Add** button on the **Fillet parameters** panel and select the edges to be filleted. By mistake, if you have selected the wrong edge, you can deselect it. To do this, click the **Remove** button on the **Fillet parameters** panel and select the edge(s) to be removed. You can change the radius by typing a value in the **Radius** box available on the **Fillet parameters** panel. As you change the radius, all the selected edges will be updated. This is because they are all part of one instance. If you want the edges to have different radii, you must create fillets in separate instances.

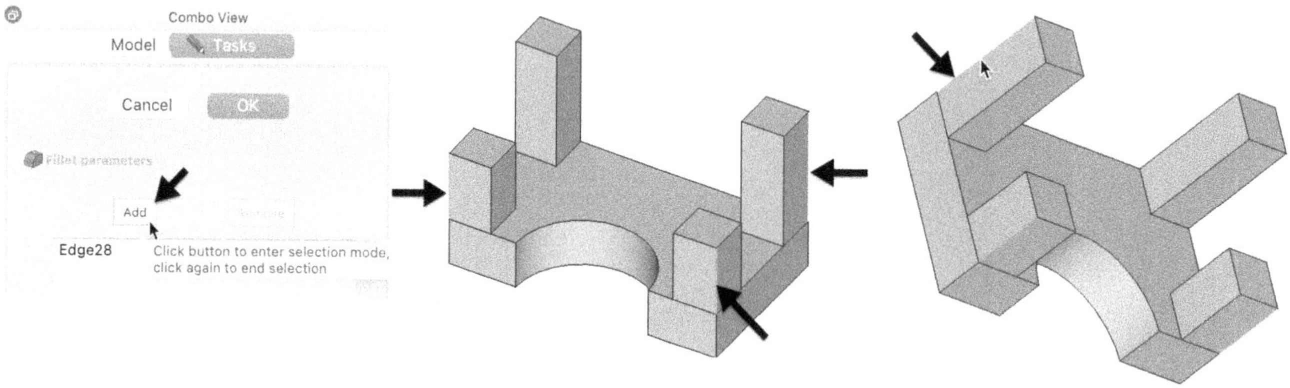

Select the required number of edges and click **OK** to complete the fillet feature. The *Fillet* feature will be listed in the **Model** tab of the **Combo View** panel.

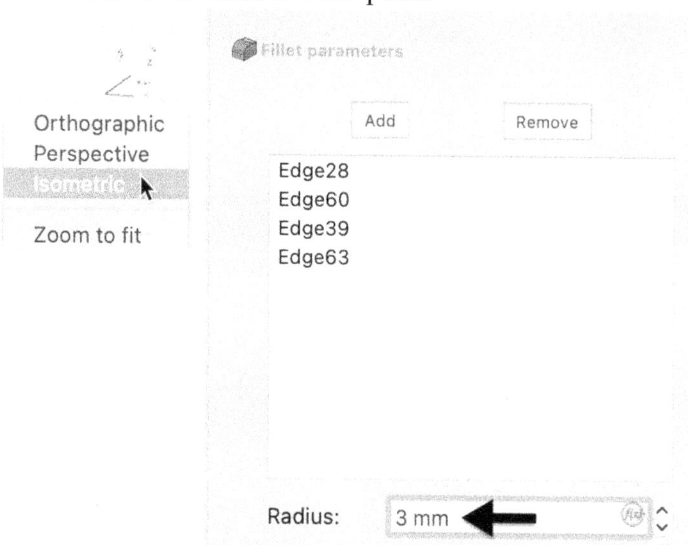

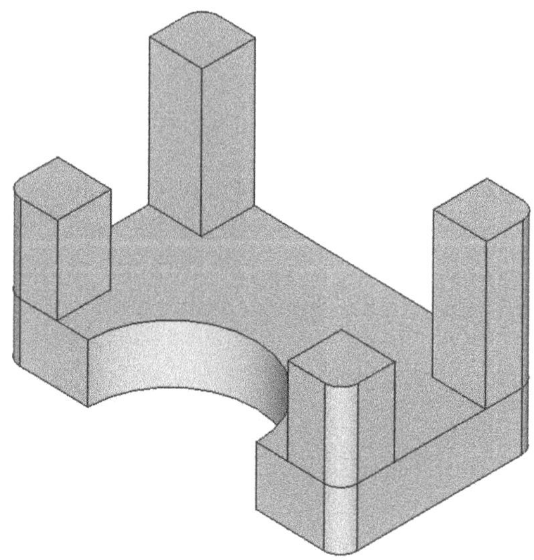

Selection Modes

The **Fillet** command allows you to select the edges to be filleted using two selection modes: **Edges** and **Faces**.

You can select individual edges just by clicking on them.

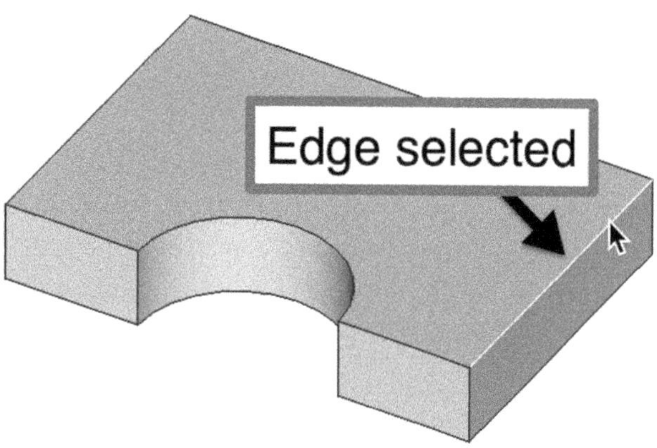

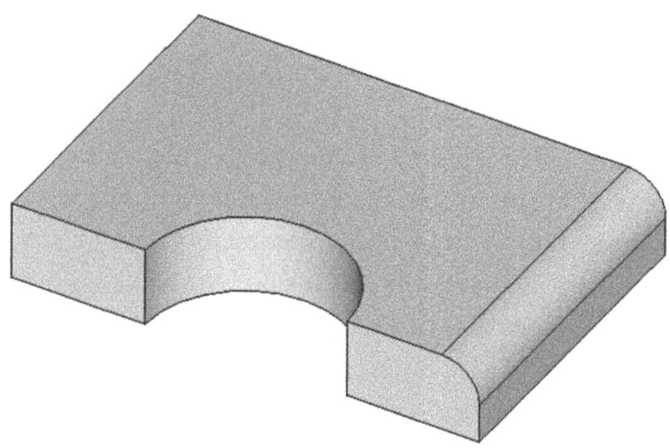

You can select all the edges of a face by merely clicking on it.

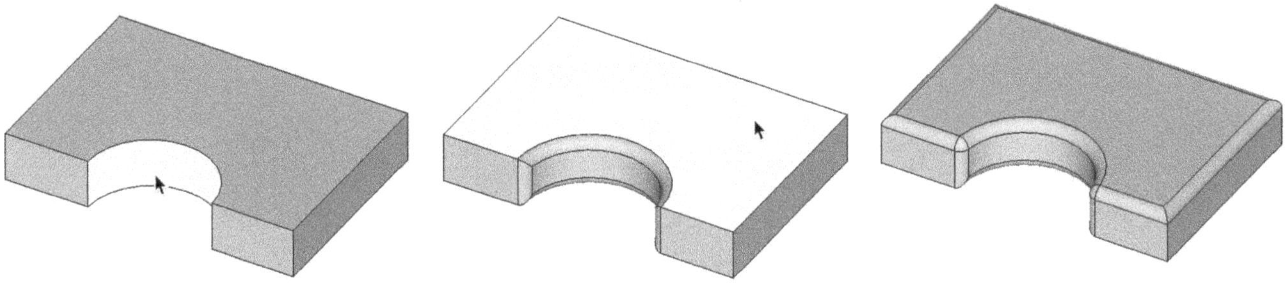

 Chamfer

The **Chamfer** and **Fillet** commands are commonly used to break sharp edges. The difference is that the **Chamfer** command adds a beveled face to the model. A chamfer is also a dress-up feature. There are three different chamfer types: *Equal Distance, Two Distances* and *Distance and Angle*. The default chamfer **Type** is Equal distance.

Equal Distance chamfer

This option is used to create a chamfer with equal distance on both sides of the edge. Press and hold the CTRL key (COMMAND key for Mac users) and select the edges to fillet. Next, click **Part Design** > **Apply a dress-up feature** > **Chamfer** on the menu bar (or) click the **Chamfer** icon on the **Part Design Modeling** toolbar. On the **Chamfer parameters** panel, select **Type > Equal Distance**. Next, type-in a value in the **Size** box. Click **OK** to complete the chamfer.

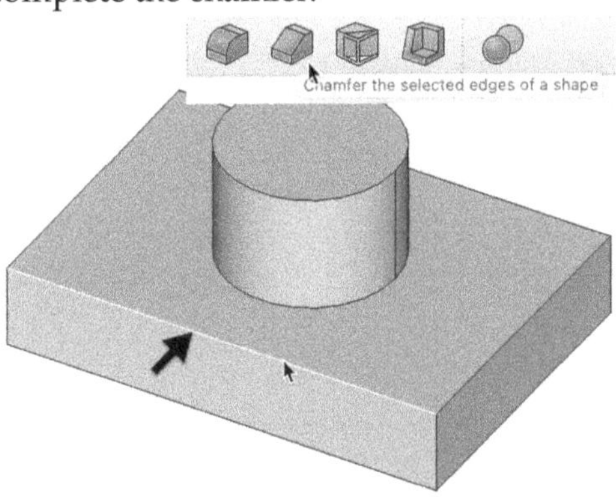

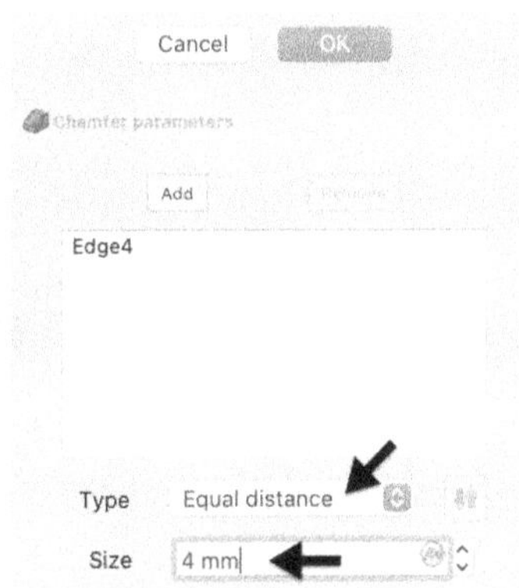

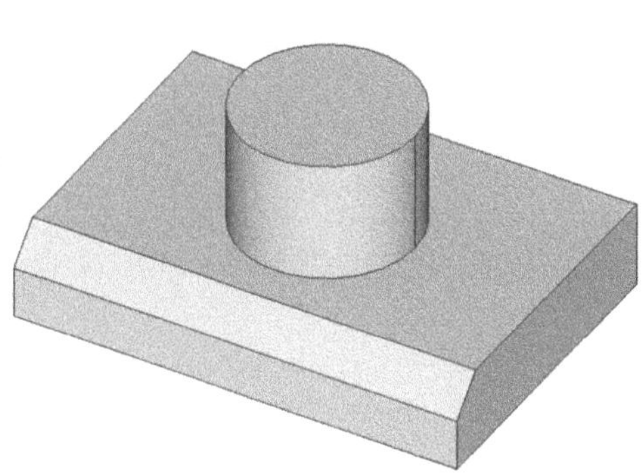

Distance and Angle chamfer

This option lets you create a chamfer by defining its distance and angle values. Select the edge to chamfer and click the **Chamfer** icon on the **Part Design Modeling** toolbar. On the **Chamfer parameters** panel, select **Type > Distance and angle.** Type-in values in the **Size** and **Angle** boxes; the distance and angle values are measured from the vertical and horizontal faces, respectively. Click **OK** to complete the feature.

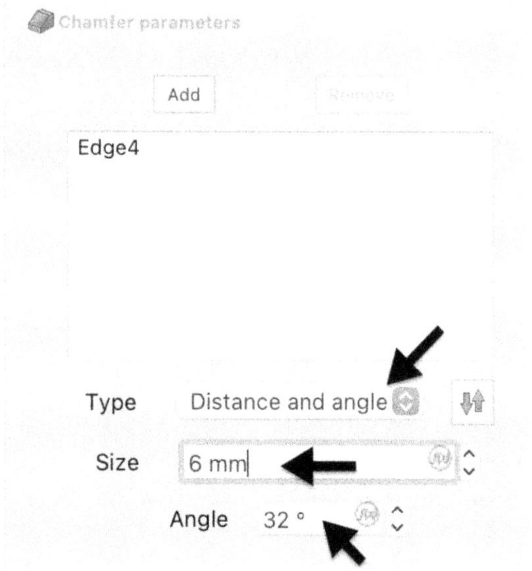

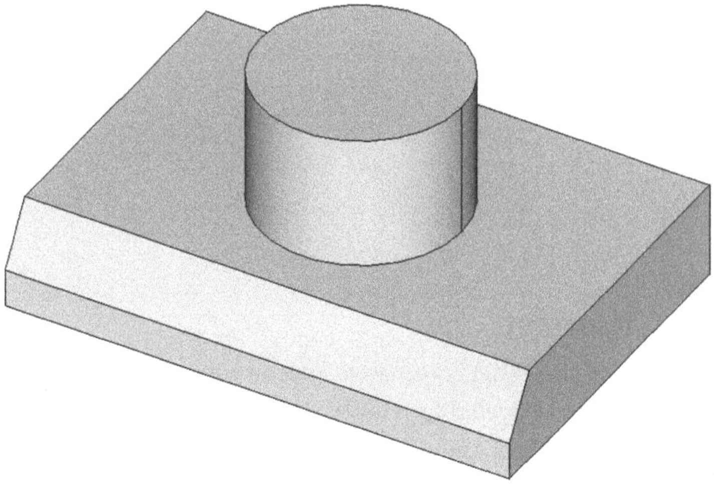

Two Distance chamfer

If you want a chamfer to have different setbacks on both sides of the edge, then select **Type > Two distances** option on the **Chamfer parameters** panel. Type-in values in the first **Distance** and the second **Distance** boxes on the **Chamfer parameters** panel. Click the **Flip Direction** icon to reverse the sides. Click **OK** to complete the feature.

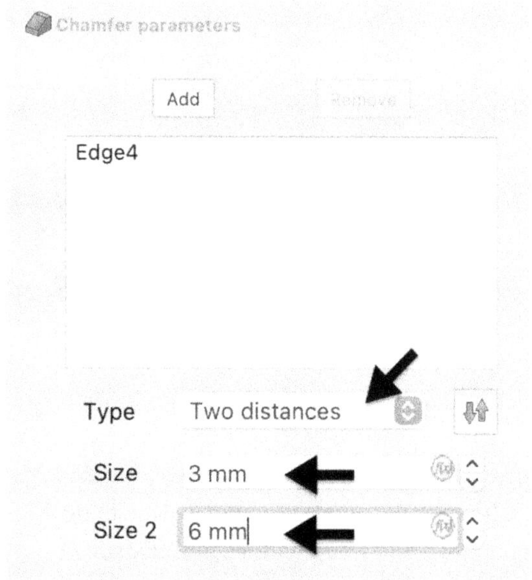

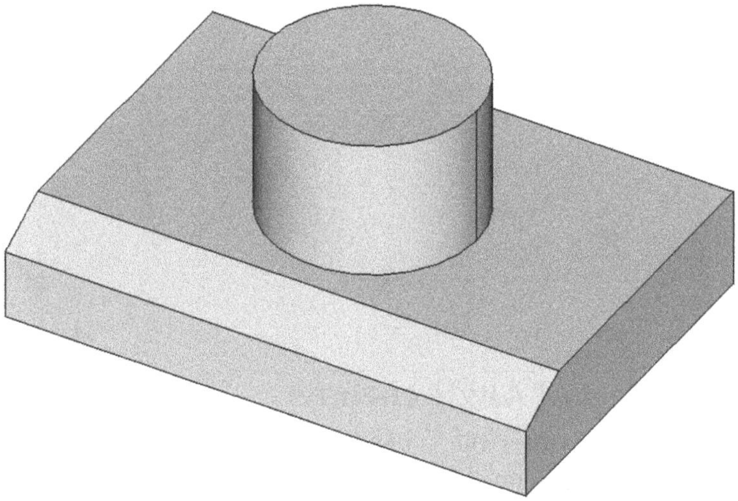

The Draft command

When creating cast or plastic parts, you are often required to add a draft on them so that they can be molded easily. A draft is an angle or taper applied to the faces of components so that they can be removed from the mold easily. The following illustration shows a molded part with and without a draft.

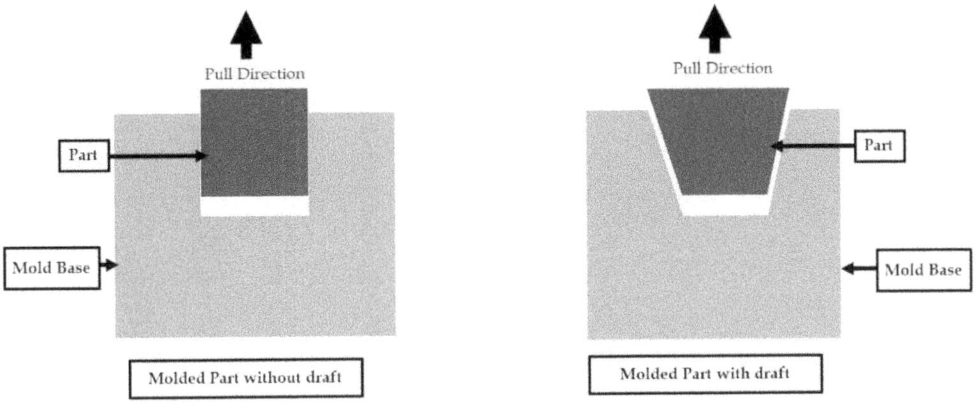

The **Draft** command will help you to apply a draft to the model geometry. To do this, first select the face to be drafted. Next, click **Part Design > Apply a dress-up feature > Draft** on the menu bar (or) click the **Draft** icon on the **Part Design Modeling** toolbar. On the **Draft parameters** panel, click the **Neutral plane** button and select the plane or face to define the neutral plane. Next, click the **Pull direction** button and select an edge to define the pull direction. Type-in a value in the **Draft angle** box to specify the angle.

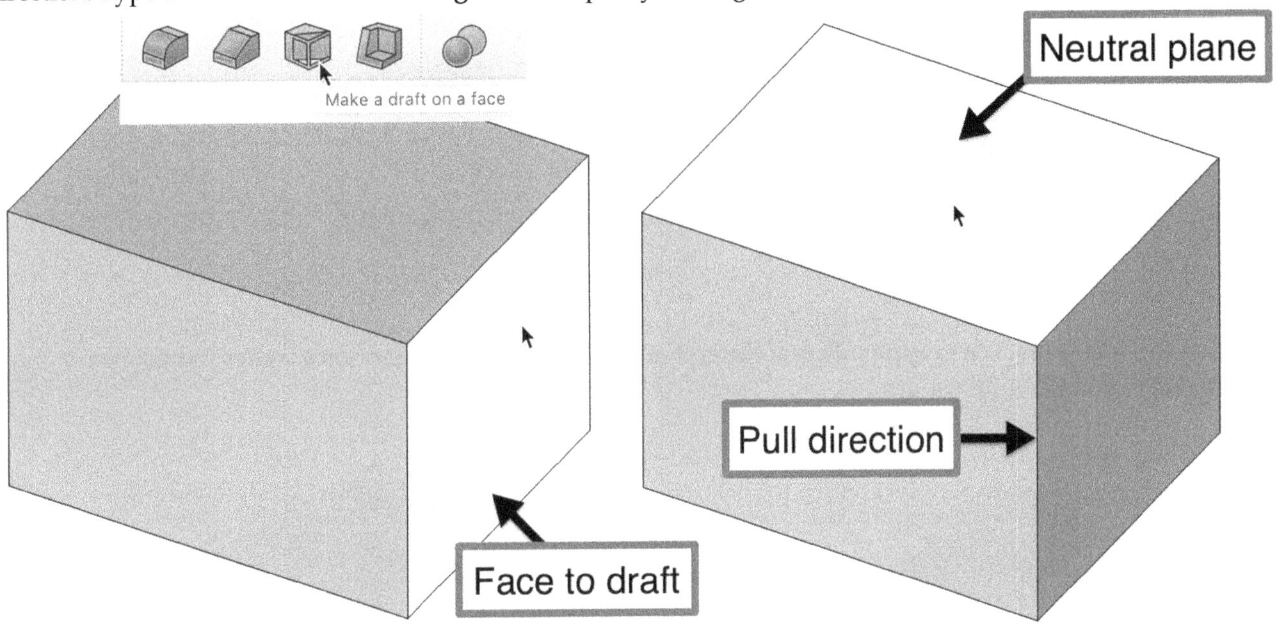

Check the **Reverse pull direction** option to reverse the direction of the draft. Click **OK** to create the draft feature.

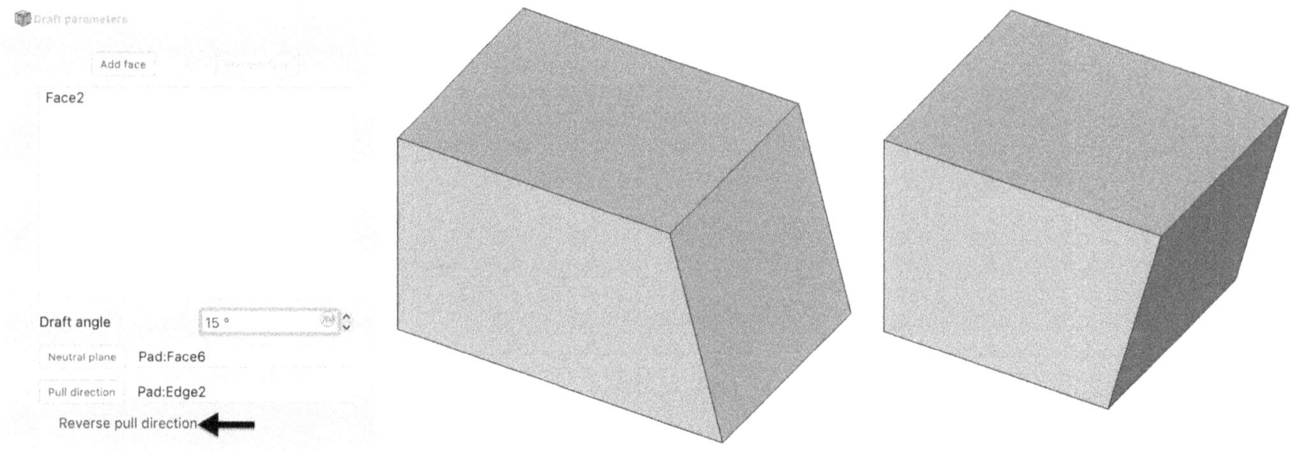

You can select tangentially connected faces by clicking anyone of them.

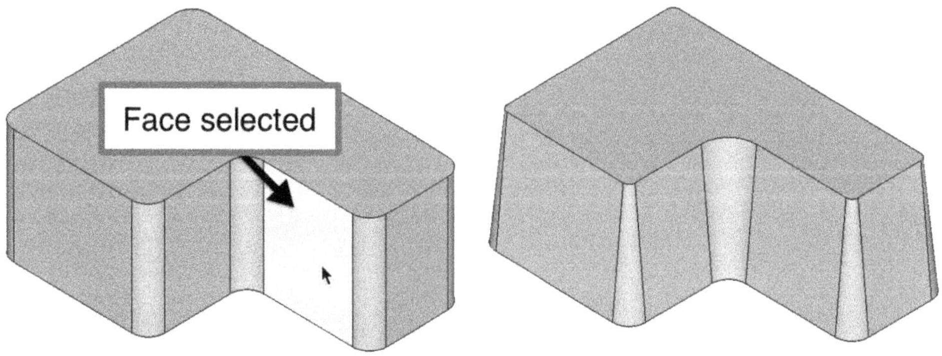

Note that it is a best practice to apply fillets at the end of the model. You may get some undesired geometry if you apply fillets before the draft feature.

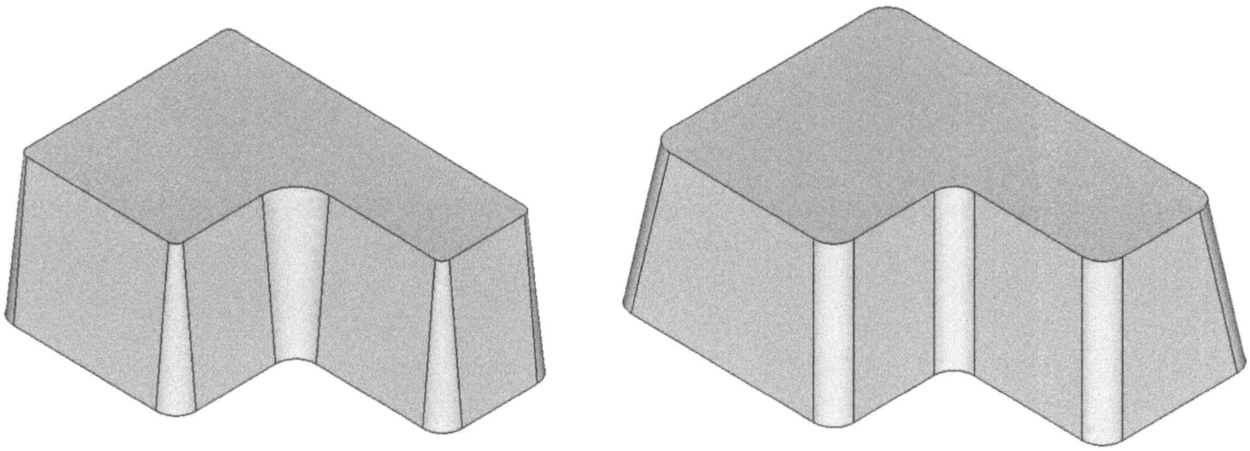

Fillets created before the draft

Fillets created after the draft

Thickness

The **Thickness** command is another useful command that can be applied directly to a solid model. It allows you to take a solid geometry and make it hollow. This can be a powerful and timesaving technique when designing parts that call for thin walls such as bottles, tanks, and containers. This command is easy to use. You should have a solid part to use this command. Select a face to remove and click **Part Design** > **Apply a dress-up feature** > **Thickness** on the menu bar (or) click the **Thickness** icon on the **Part Design Modeling** toolbar. Type-in the wall thickness in the **Thickness** box. The thickness is added outside the model. Check the **Make thickness inside** option, if you want to add thickness inside the model.

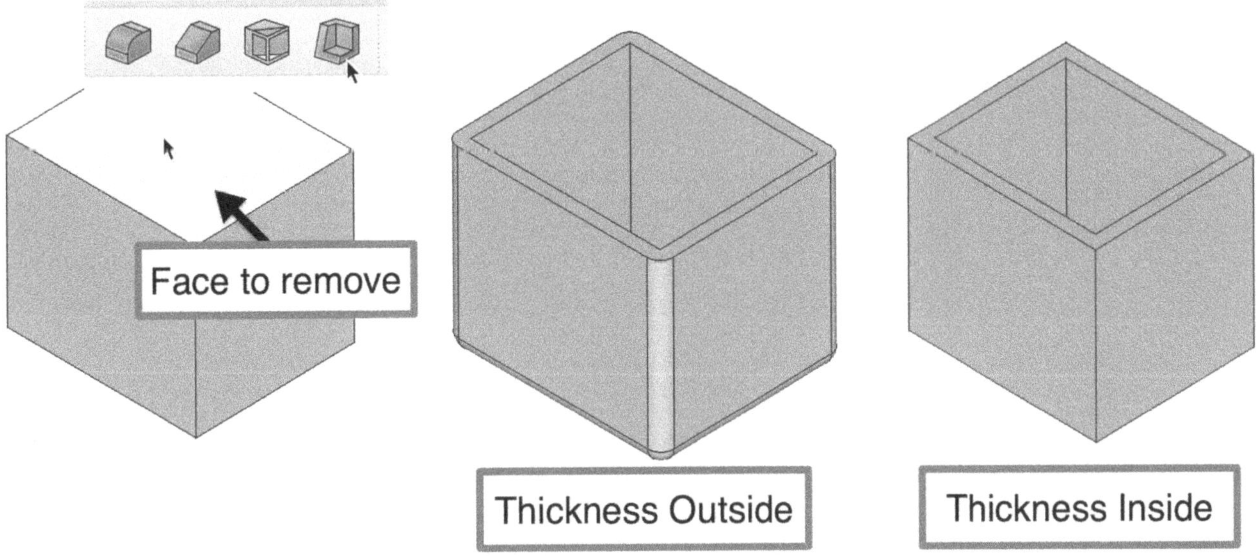

Next, specify the **Join Type**: **Arc** or **Intersection**. The **Arc** option creates an arc at the intersection of two faces. The **Intersection** option leaves sharp edges at the intersection of two faces. Click **OK** to create the *Thickness* feature.

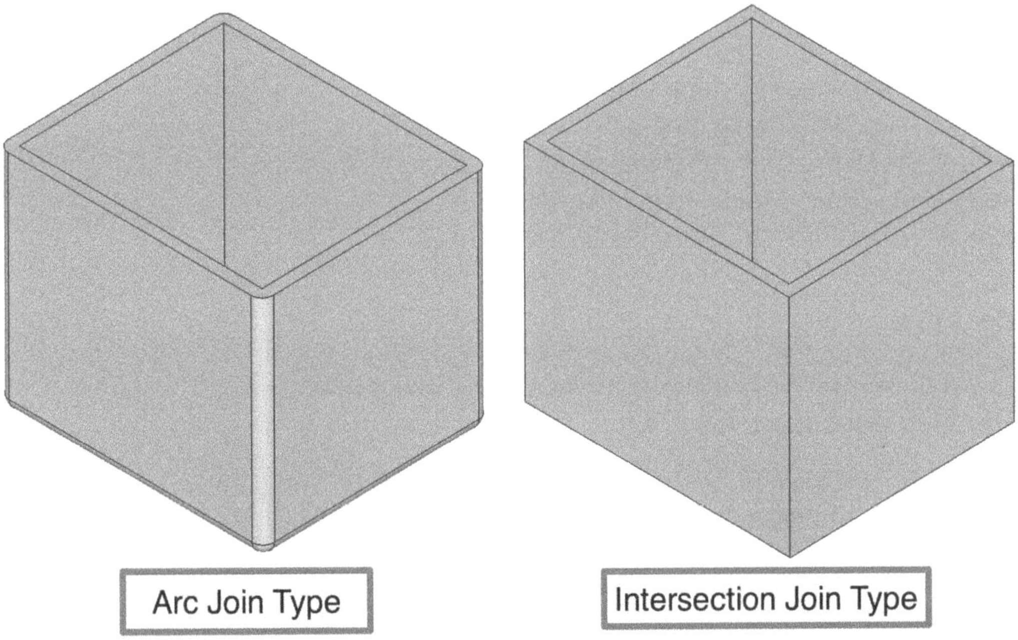

Tutorial 1 (Millimetres)

In this example, you create the part shown below.

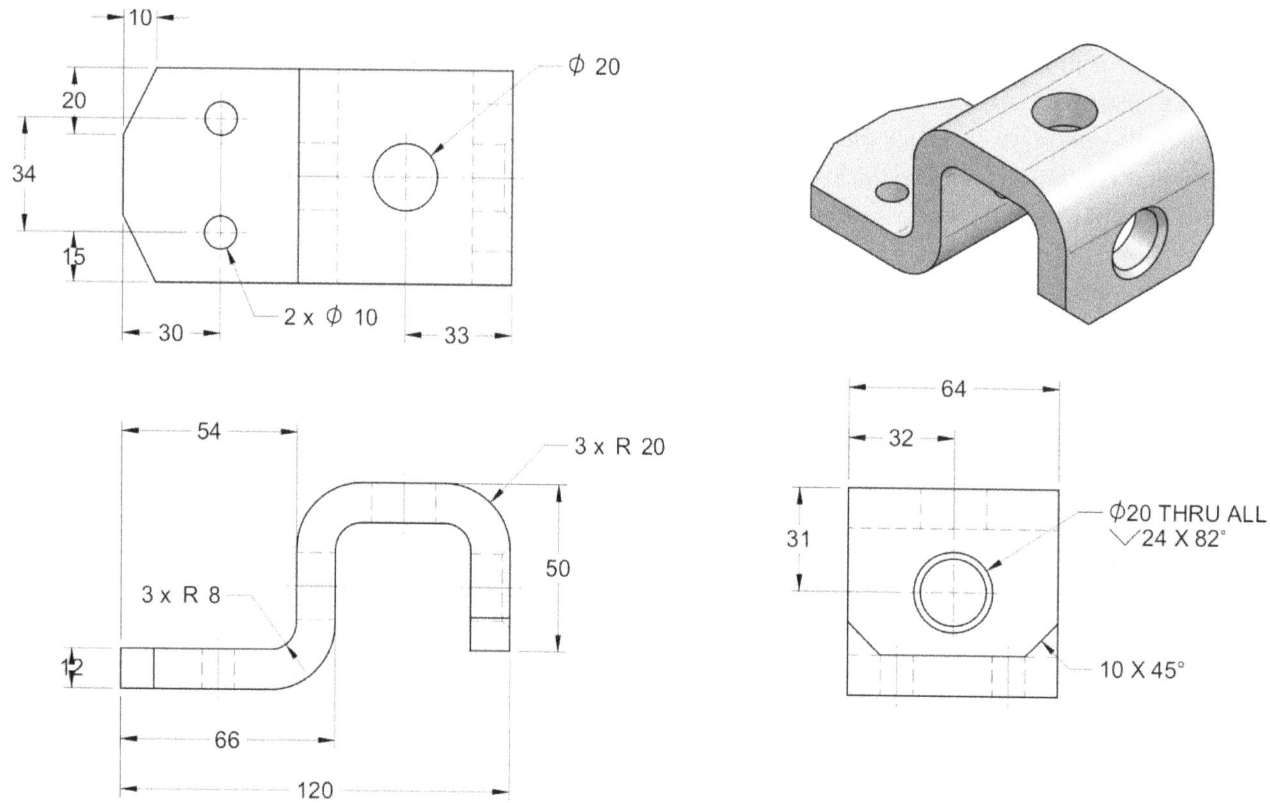

Creating a New document

1. Click **FreeCAD** on the desktop.
2. On the menu bar, click **File > New**; a new document opens.
3. On the **Workbench** toolbar, select **Workbench** drop-down **> Part Design.**
4. Click **Edit > Preferences** on the **Menu** bar; the **Preferences** dialog appears on the screen.
5. Click **Units** tab and select **Unit system > Standard (mm/kg/s/degree).**
6. Select **Number of decimals > 2** and click **OK** on the **Preferences** dialog.

Creating the Padded Feature

1. Click **Create Sketch** command on the **Part Design Helper** toolbar and then select the **XZ** plane.
2. Click **OK** on the **Combo View** panel to start the sketch.
3. Click the **Create line** icon on the **Sketcher geometries** toolbar. Next, select the origin point of the sketch.
4. Move the pointer horizontally toward the right and click to create the horizontal line.
5. Click the **Constrain horizontal distance** icon on the **Sketcher constraints** toolbar. Next, select the horizontal line. Type **66** and click **OK.**

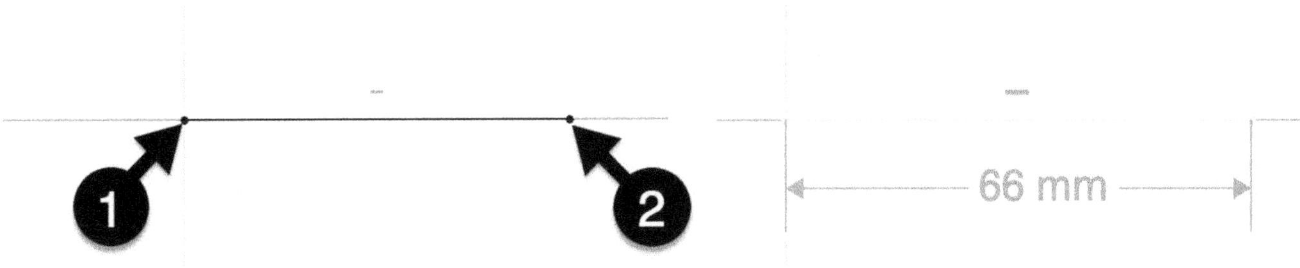

6. Click the **Create polyline** icon on the **Sketcher geometries** toolbar. Select the endpoint of the horizontal line.
7. Create a closed sketch, as shown below. Use the **Constrain Coincident** command and the endpoints, if they are not connected properly.

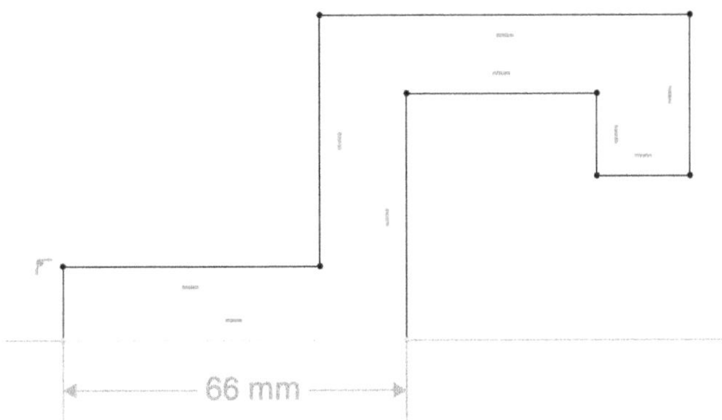

8. Click the **Constrain equal** icon on the **Sketcher constraints** toolbar. Next, select the small vertical and horizontal lines, as shown.

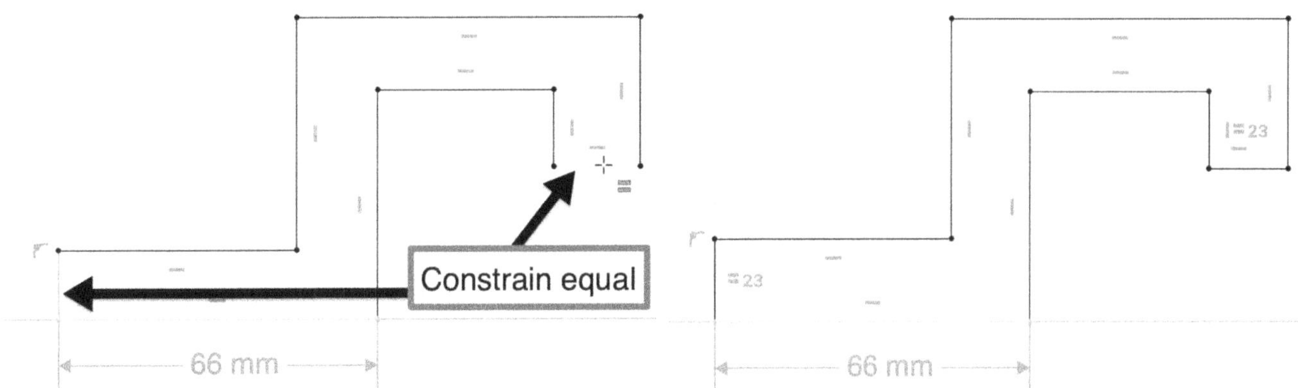

9. Click the **Constrain horizontal distance** icon on the **Sketcher constraints** toolbar. Next, select the points of the vertical lines, as shown. Type **120** and click **OK**.

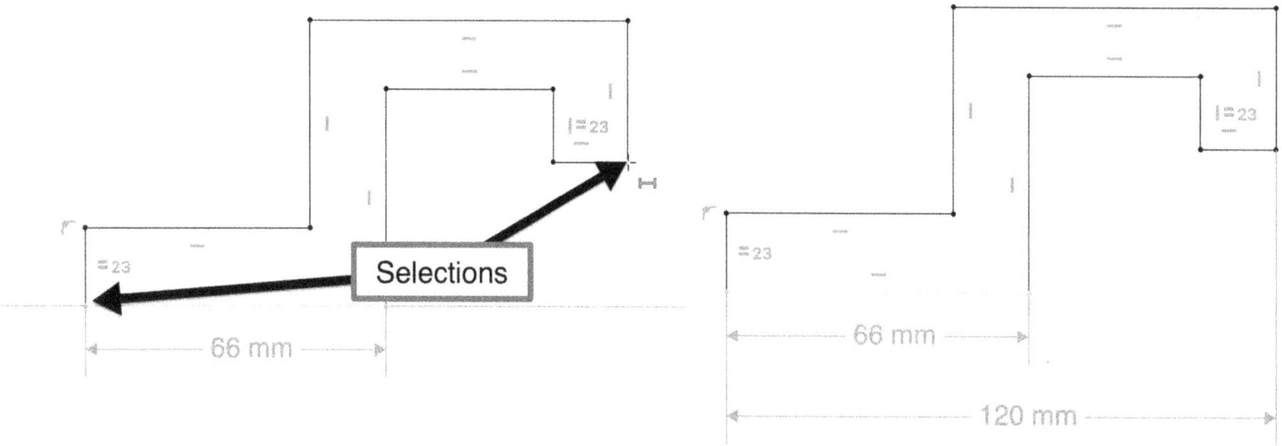

10. Click the **Constrain vertical distance** icon on the **Sketcher constraints** toolbar. Select the corner point of the sketch, as shown. Type **62** and click **OK**.

11. Add the remaining vertical and horizontal distance constraints, as shown.

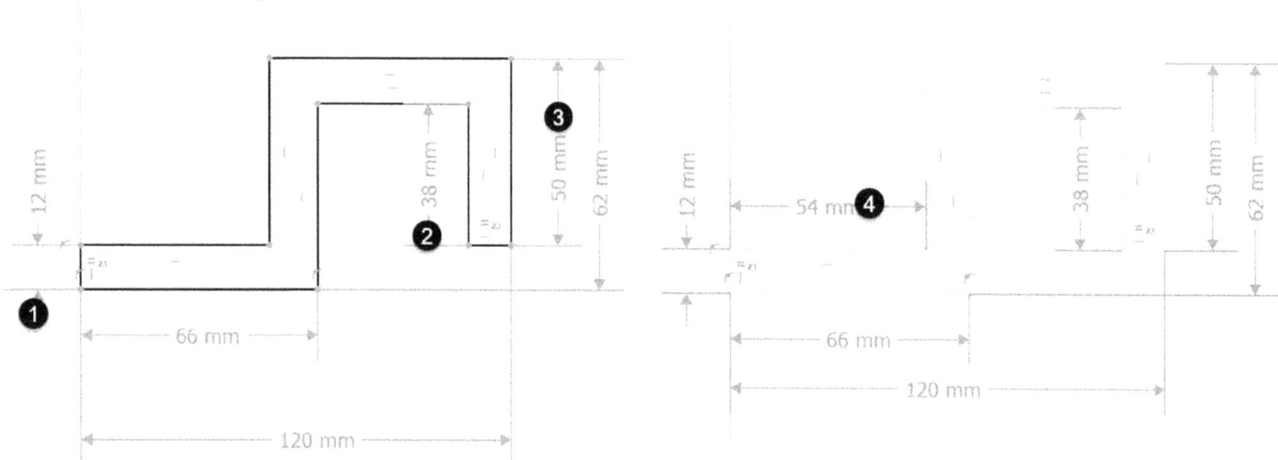

12. Click the **Leave Sketch** on the **Part Designer Helper** toolbar.
13. Click the **Pad** icon on the **Part Design Modeling** toolbar.
14. Type **64** in the **Length** box under the **Pad Parameters** section of the **Combo View** panel.
15. Select the **Symmetric to Plane** option and click **OK** to create the *Pad* feature.

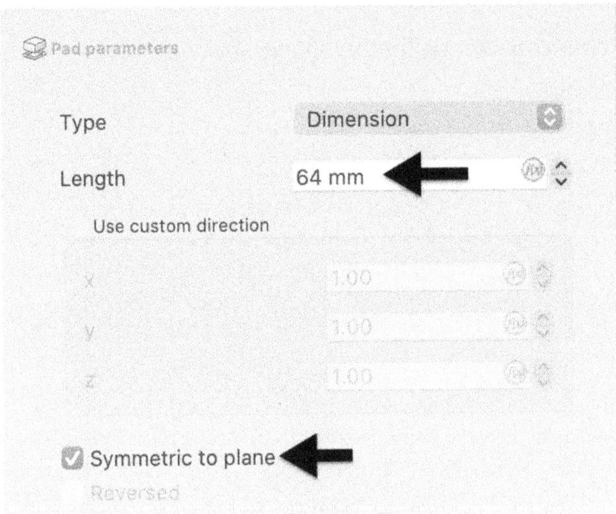

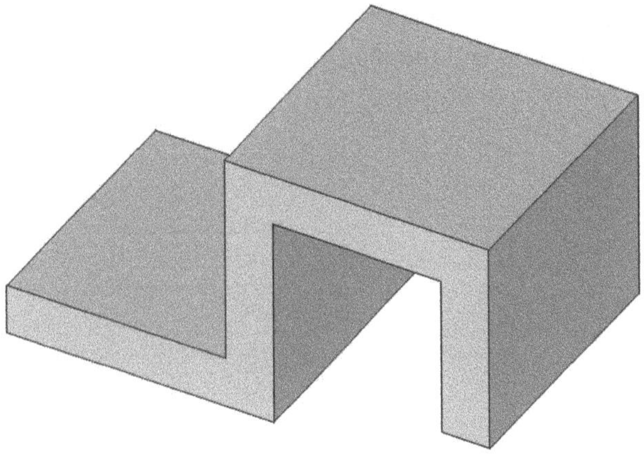

Creating the Hole Features

1. Select the right-side face of the model and click the **Create Sketch**  icon on the **Part Design Helper** toolbar.

2. Click the **Set to right view**  icon on the **View** toolbar.

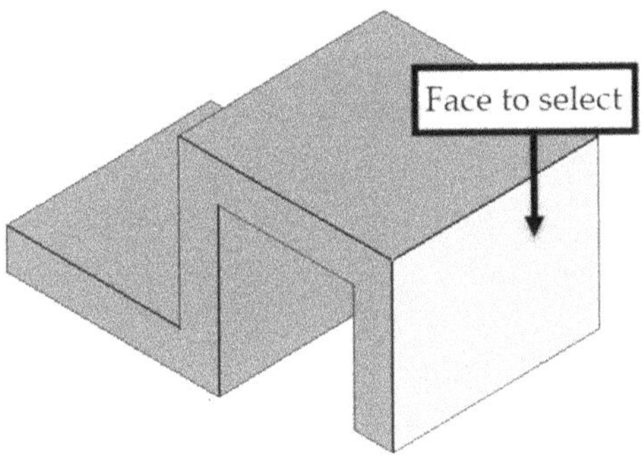

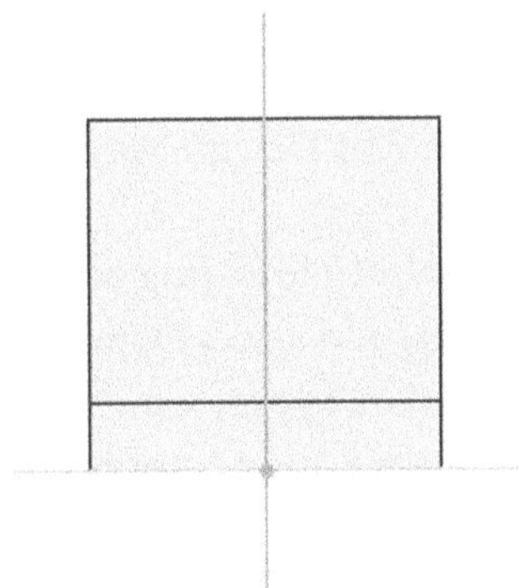

3. Click the **External geometry**  icon on the **Sketcher geometries** toolbar.
4. Select the right vertical edge and the upper horizontal edge of the model, as shown.

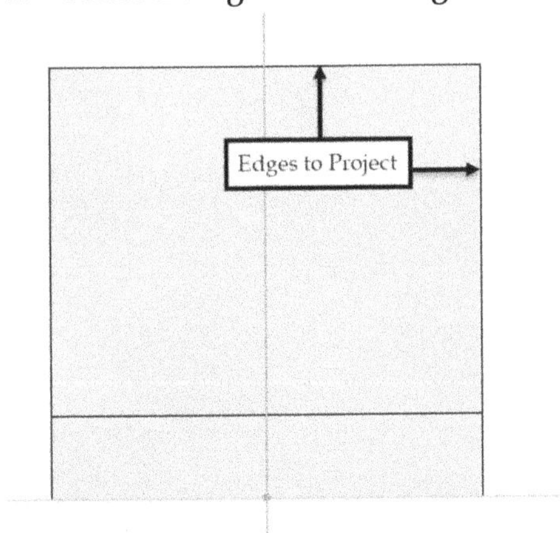

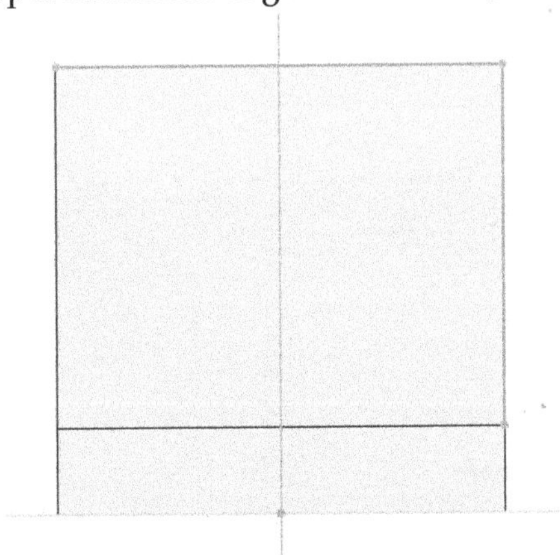

5. Click the **Create circle**  icon on the **Sketcher geometries** toolbar. Next, select the vertical axis of the sketch to define the center point. Move the pointer outward and click to create a circle.

6. Click the **Constrain vertical distance** icon on the **Sketcher constraints** toolbar. Next, select the center point of the circle and the top right corner.
7. Type-in **31** in the **Length** box on the **Insert Length** dialog. Click **OK**.

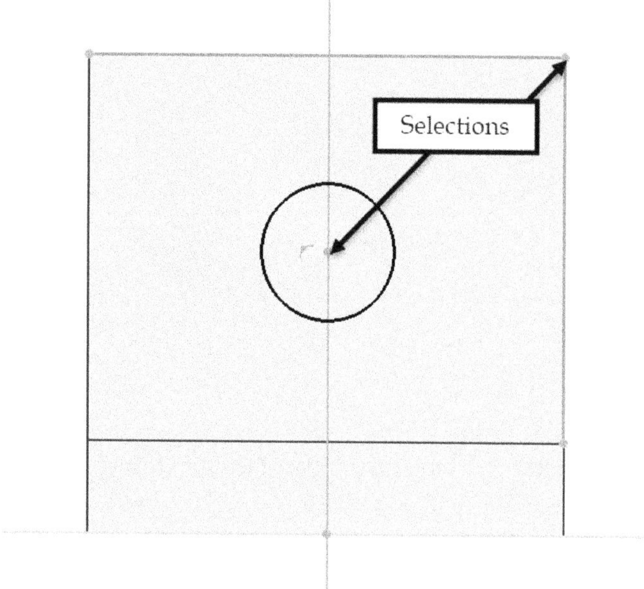

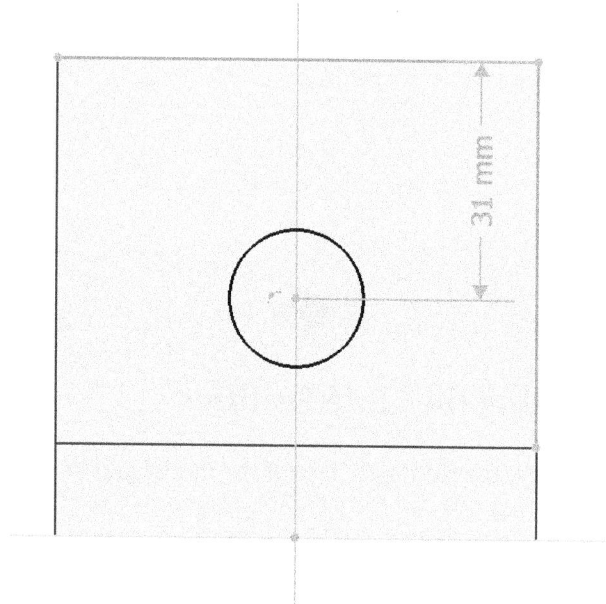

8. Click the **Leave Sketch** icon on the **Part Design Helper** toolbar.
9. Click the **Hole** icon on the **Part Design Modeling** toolbar; the **Hole parameters** section appears on the **Combo View** panel.
10. On the **Combo View** panel, select **Threading and size >Profile > None** under the **Hole Parameters** section.
11. Type-in **21** in the **Diameter** box and select **Depth > Through All**.
12. Under the **Hole cut** section, select **Type > Countersink**.
13. Type-in **24** and **82** in the **Diameter** and **Countersink angle** boxes, respectively.
14. Click **OK** to create the *Hole* feature.

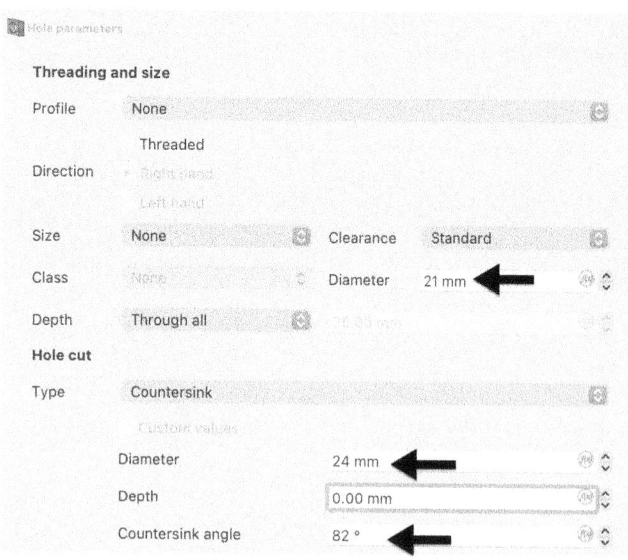

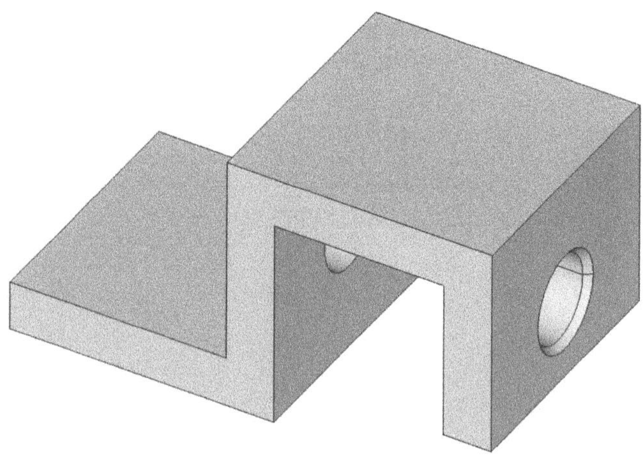

15. Select the top face and click the **Create Sketch** command on the **Part Design Helper** toolbar.
16. On the **Sketcher geometries** toolbar, click the **Create circle** icon and then select the horizontal axis. Move the pointer outward and click to create a circle.
17. Click the **External geometry** icon on the **Sketcher geometries** toolbar.
18. Select the left vertical edge and the upper horizontal edge of the model, as shown.
19. Click the **Constrain horizontal distance** icon on the **Sketcher constraints** toolbar. Next, select the center point of the circle and the bottom left corner.
20. Type-in **33** in the **Length** box on the **Insert Length** dialog. Click **OK.**

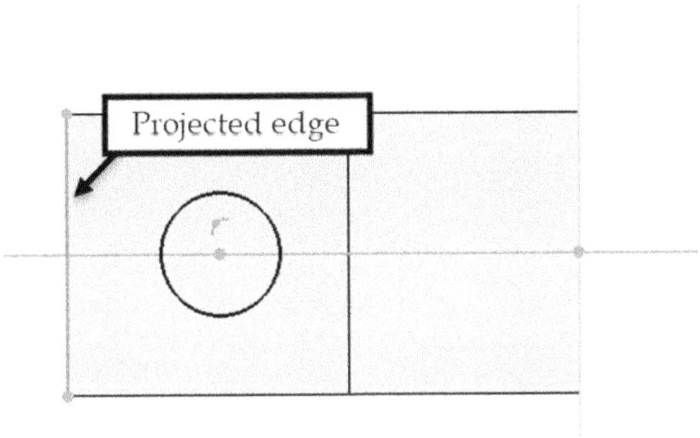

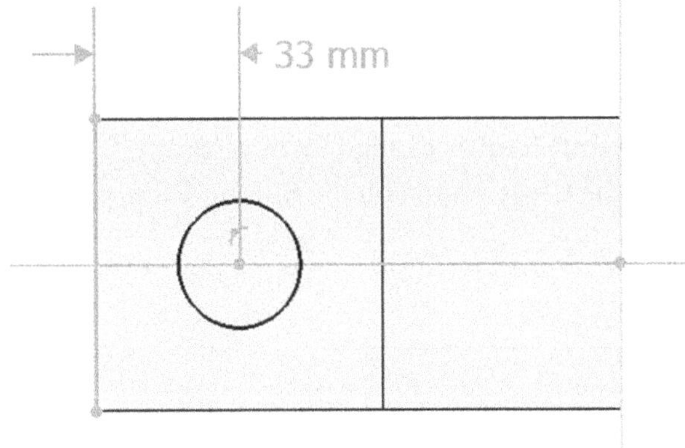

21. Click **Close** on the **Combo View** panel.
22. Click the **Hole** icon on the **Part Design Modeling** toolbar. Next, select **Profile > None** from the **Hole parameters** section.
23. Select **Depth > Through all**. Next, type-in **20** in the **Diameter** box.
24. Click **OK** to create the hole feature.

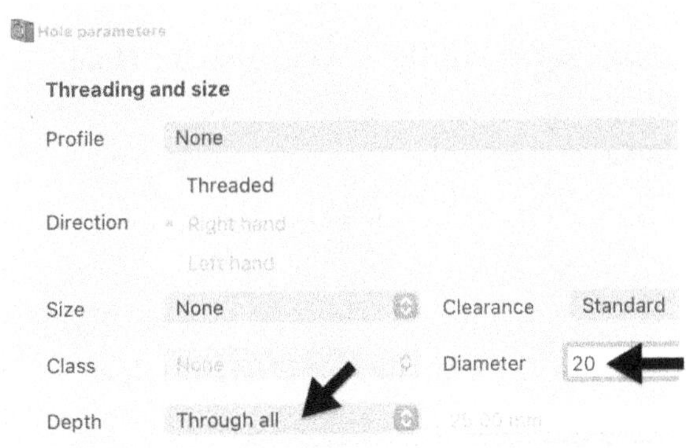

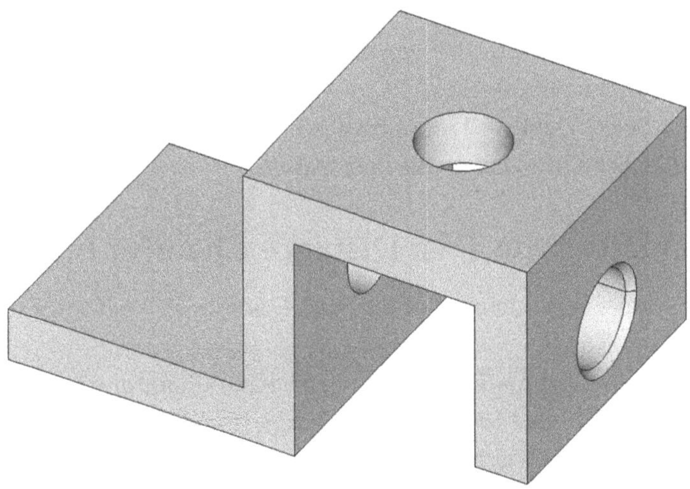

25. Select the lower top face of the model and click the **Create Sketch** icon on the **Part Design Helper** toolbar.
26. Click the **Create circle** icon on the **Sketcher geometries** toolbar and create two circles, as shown.

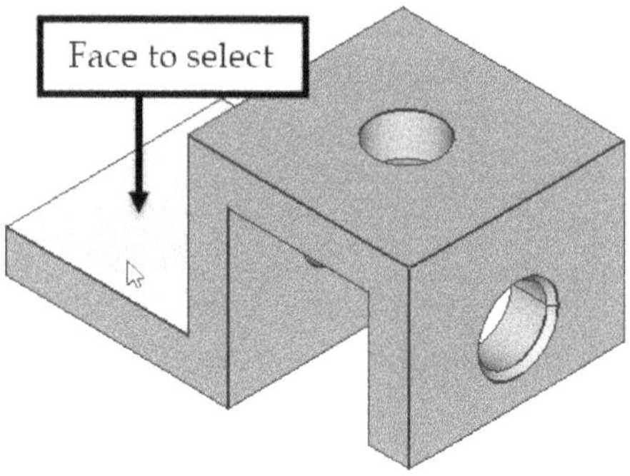

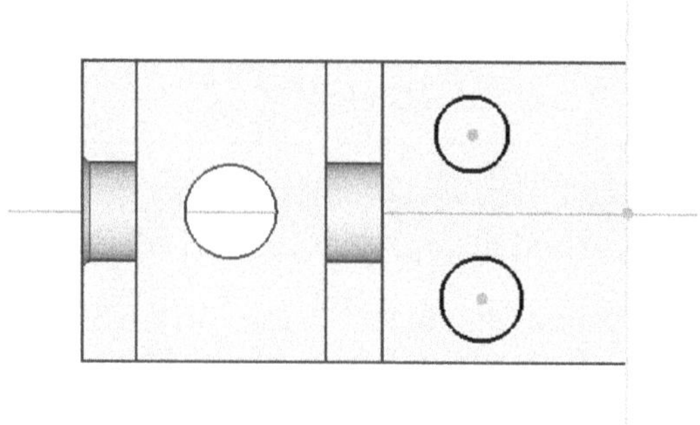

27. Create the dimensional constraints between the circles and the sketch origin.
28. Select the centerpoints of the two circles. Next, click the **Constrain vertical** icon on the **Sketcher constraints** toolbar to align them vertically.
29. Click the **Close** button on the **Combo View** panel.

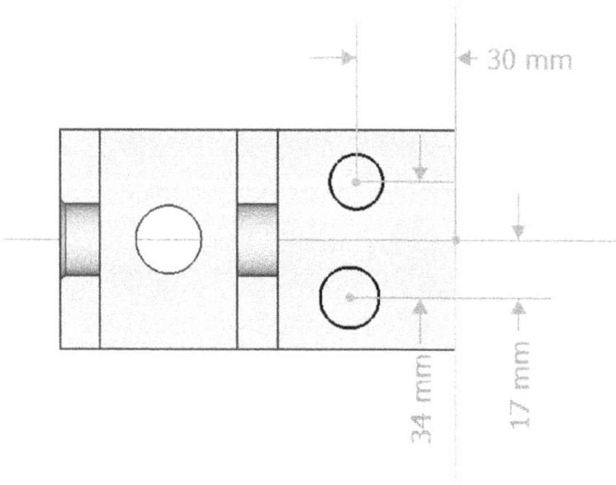

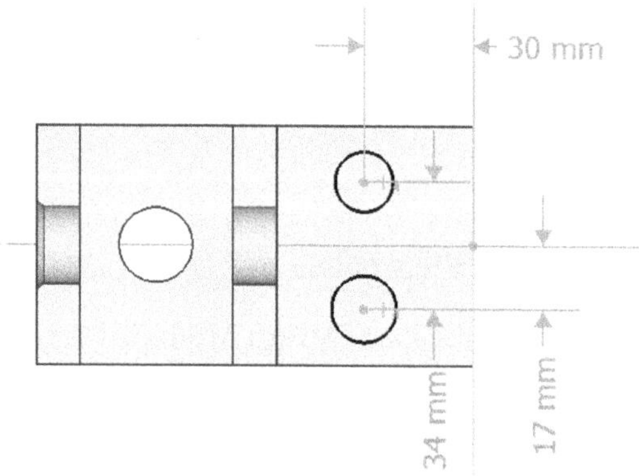

30. Click the **Hole** icon on the **Part Design Modeling** toolbar. Next, select **Profile > None** from the **Hole parameters** section.
31. Select **Depth > Through all**. Next, type-in **10** in the **Diameter** box.
32. Click **OK** to create the hole feature.

Creating the Two Distance Chamfer features

1. Select vertical edge of the model, as shown. Next, click the **Chamfer** icon on the **Part Design Helper** toolbar.
2. Select **Type > Two distances** from the **Chamfer parameters** section of the **Combo View** panel.
3. Type **20** and **10** in the **Size 2** boxes, respectively.
4. Click the **Add** button on the **Chamfer parameters** section. Next, rotate the model and select the vertical edge on the back side, as shown.
5. Click **OK** to create the chamfer feature.

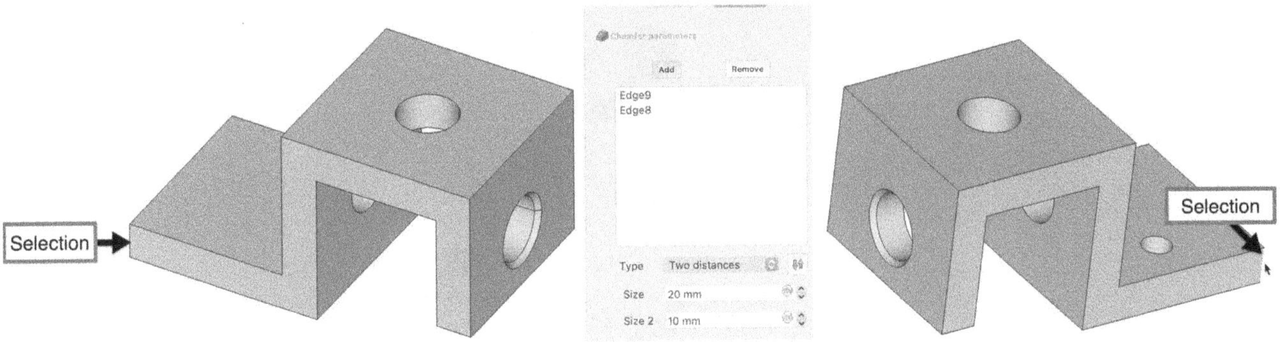

Creating Fillets and Chamfers

1. Press and hold the Ctrl key and select the edges of the model, as shown.
2. Click the **Fillet**  icon on the **Part Design Modeling** toolbar.
3. Type **20** in the **Radius** box available on the **Fillet parameters** section of the **Combo View** panel. Click **OK** to create the fillets.

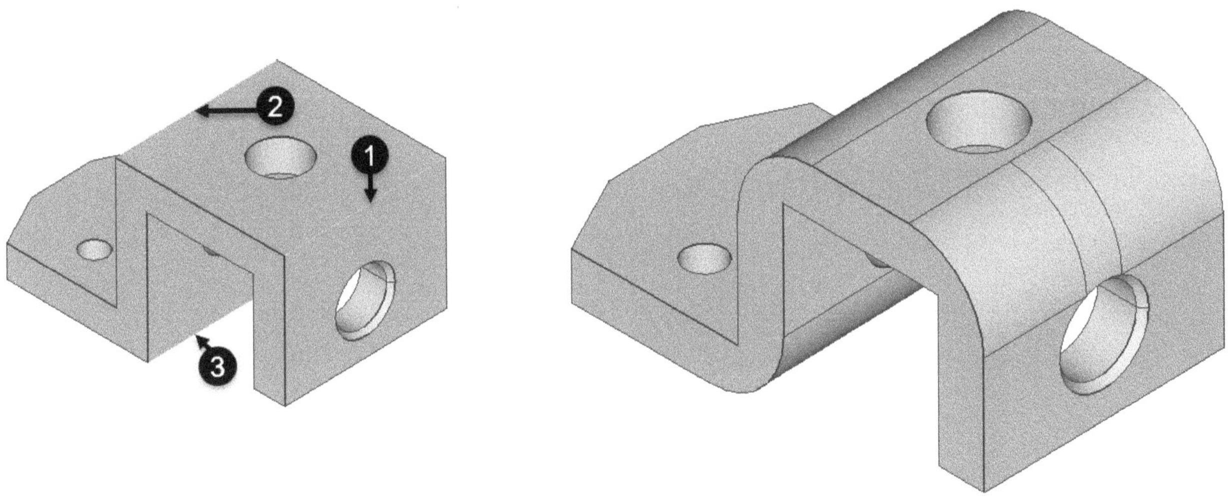

4. On the **View** toolbar, select **Draw style** drop-down > **Wireframe**.
5. Press and hold the Ctrl key and select the horizontal edges of the geometry, as shown.

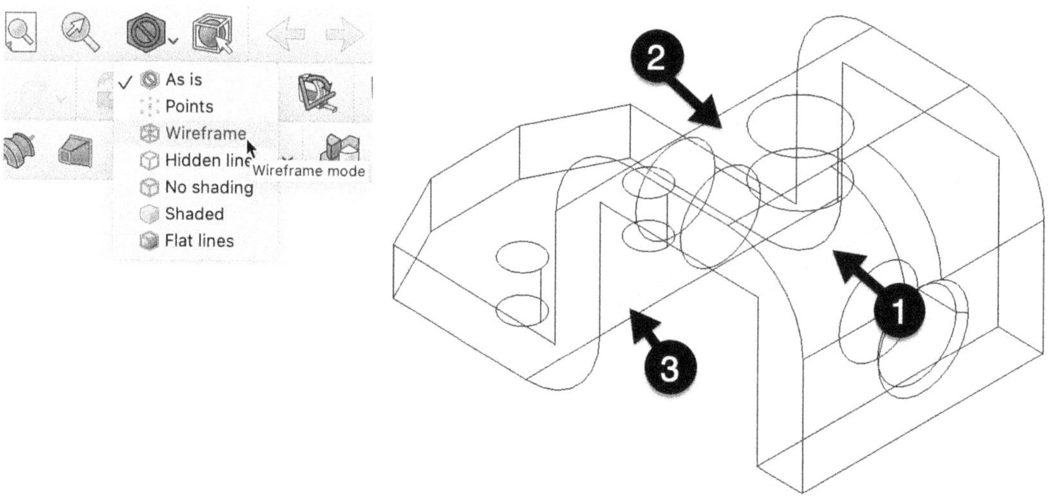

6. Click the **Fillet** icon on the **Part Design Modeling** toolbar. Type **8** in the **Radius** box of the **Fillet parameters** section, and click **OK.**
7. On the **View** toolbar, select **Draw style** drop-down > **Flat lines.**

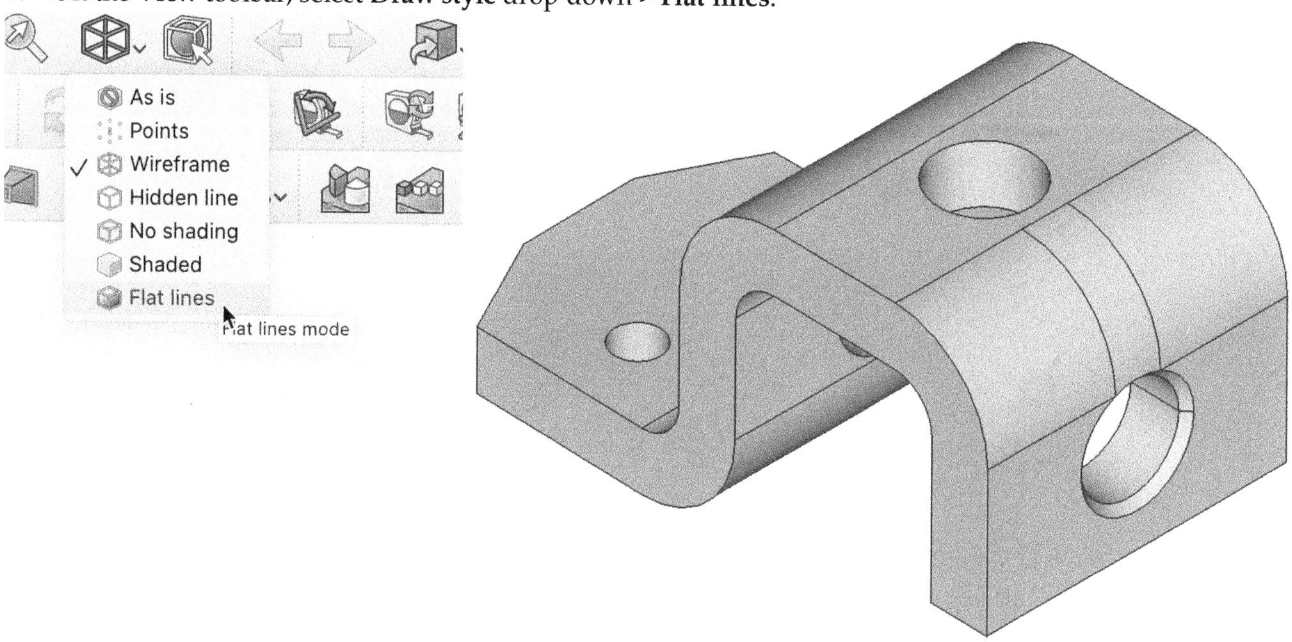

8. Press and hold the middle and right mouse button, and then drag the cursor in the forward direction.
9. Press and hold the Ctrl key and select the edges of the model, as shown.
10. Click the **Chamfer** icon on the **Part Design Modeling** toolbar.
11. Select **Type > Equal** distance from the **Chamfer parameters** section of the **Combo View** panel
12. Type 10 in the **Size** box in the **Chamfer parameters** section. Next, click **OK** to chamfer the edges.
13. Click the **Save** icon on the **File** toolbar located at the top-left corner of the window. Next, type C4_example1 in the **File name** box, and then click **Save.**
14. Click **File > Close** on the menu bar to close the document.

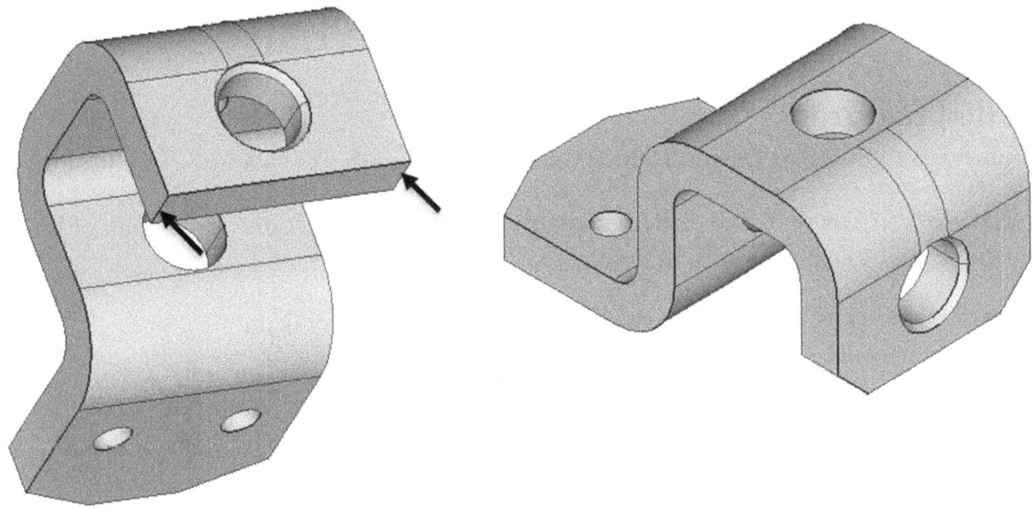

Tutorial 2 (Inches)

In this example, you create the part shown next.

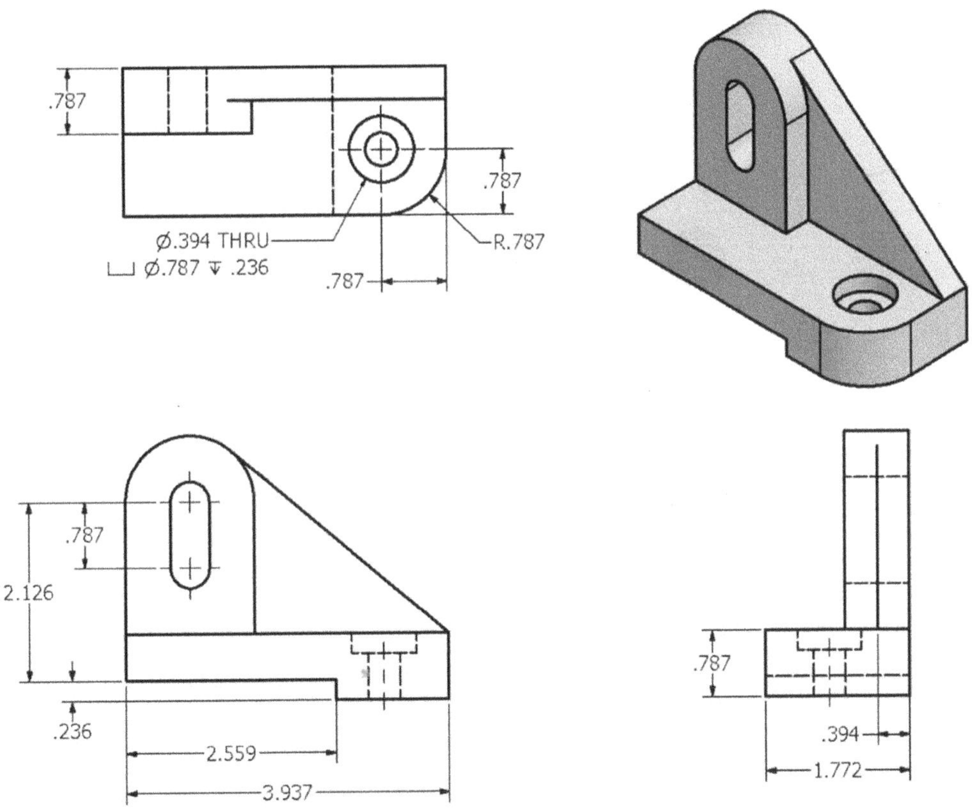

Creating a New document

1. Click **FreeCAD 0.19** on the desktop to start the application.
2. On the menu bar, click **File > New**; it creates a new document.
3. On the **Workbench** toolbar, select **Workbench** drop-down **> Part Design**.

4. Click **Edit > Preferences** on the **Menu** bar; the **Preferences** dialog appears on the screen.
5. Click **Units** tab and select **Unit system > Imperial decimal (in/lb)**.
6. Select **Number of decimals > 3** and click **OK** on the **Preferences** dialog.

Creating the Padded features

1. Click the **Create Sketch** icon on the **Part Design Helper** Toolbar and select the XY_plane. Next, click **OK**.
2. Click the **Create rectangle** icon on the **Sketcher geometries** toolbar and select the origin point of the sketch. Next, move the pointer toward the bottom right corner and click.
3. On the **Sketcher constraints** toolbar, click the **Constrain horizontal distance** icon.
4. Select the lower horizontal line and move the pointer downward.
5. Type-in **3.937** in the **Length** box of the **Insert Length** dialog and click **OK**.
6. On the **Sketcher constraints** toolbar, click the **More** drop-down and click the **Constrain vertical distance** icon.
7. Select the right vertical line and move the pointer towards the right.
8. Type-in **1.772** in the **Length** box on the **Insert Length** dialog and click **OK**.

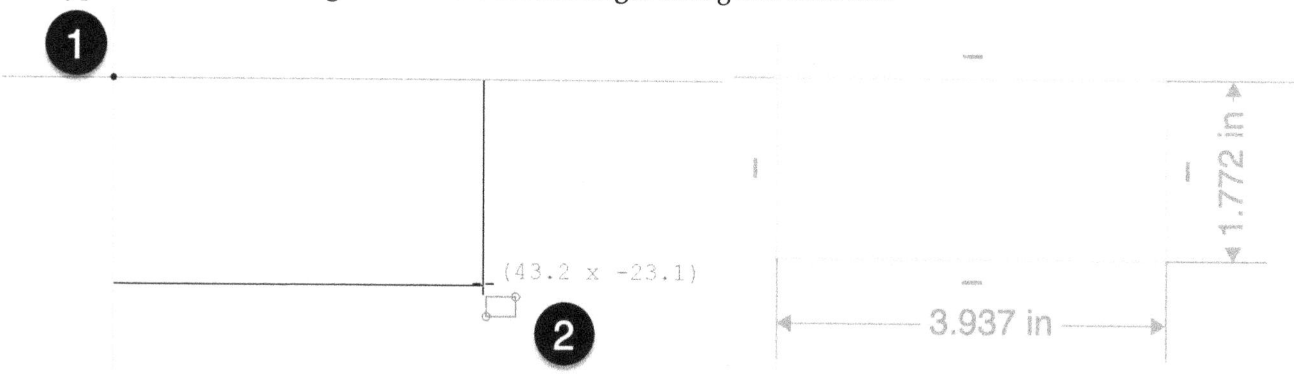

9. Click the **Close** button on the **Combo View** panel.
10. On the **Part Design Modeling** toolbar, click the **Pad** command.
11. On the **Pad parameters** section, select **Type > Dimension** and enter **0.787** in the **Length** box. Click **OK** to create the *Pad* feature.

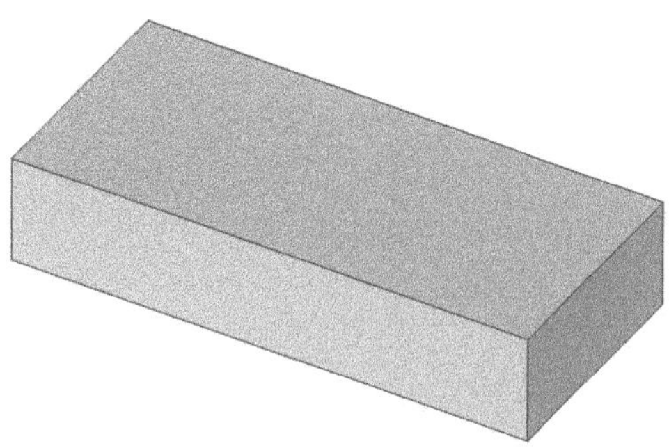

12. Activate the **Create Sketch** command and select the **XZ_plane** on the **Combo View** panel. Next, click **OK**.
13. Click the **Create rectangle** command on the **Sketcher geometries** toolbar and specify the first and second corners of the rectangle, as shown.

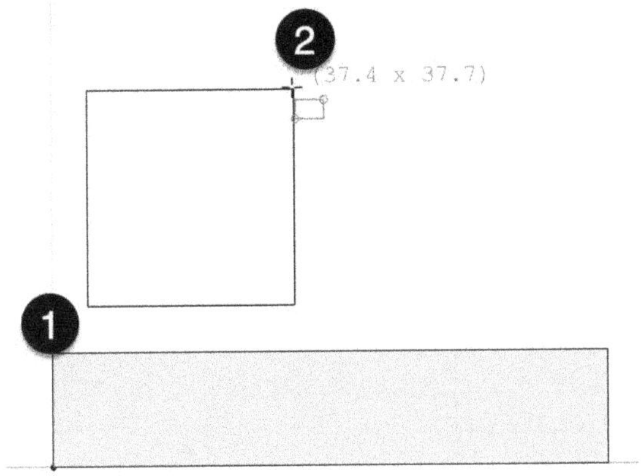

14. On the **Sketcher geometries** toolbar, click **Create an arc** drop-down > **End points and rim point** . Next, create an arc, as shown in the figure.
15. On the **Sketcher constraints** toolbar, click the **Constrain tangent** icon.
16. Select the arc and left vertical line to make the arc tangent to the line.
17. Likewise, select the arc and right vertical line to make the arc tangent to the line.
18. Select the horizontal line below the arc and press **Delete** key on the keyboard.

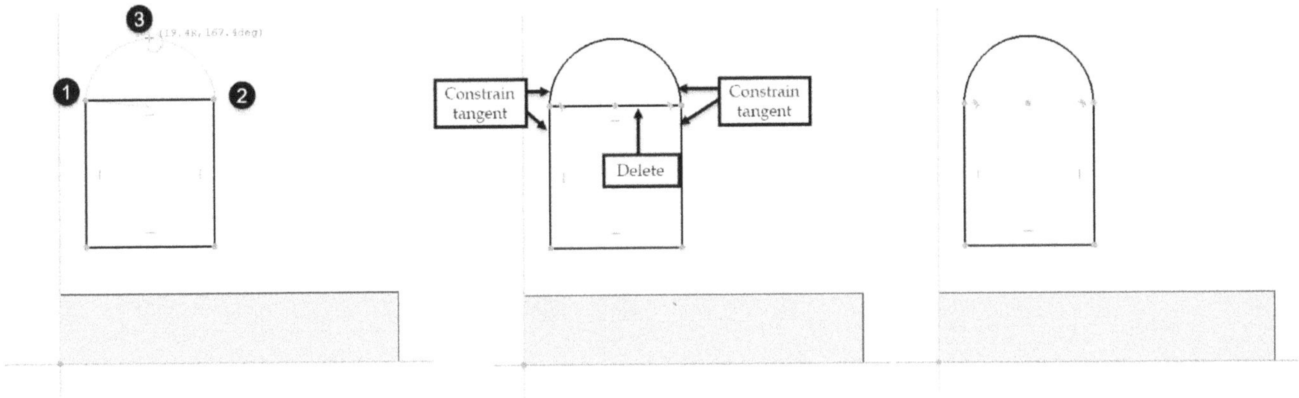

19. Click the **External geometry** command on the **Sketcher geometries** toolbar and select the horizontal line of the part geometry, as shown.
20. On the **Sketcher constraints** toolbar, click the **Constrain coincident** icon and select the endpoint of the sketch. Next, select the endpoint of the projected edge.

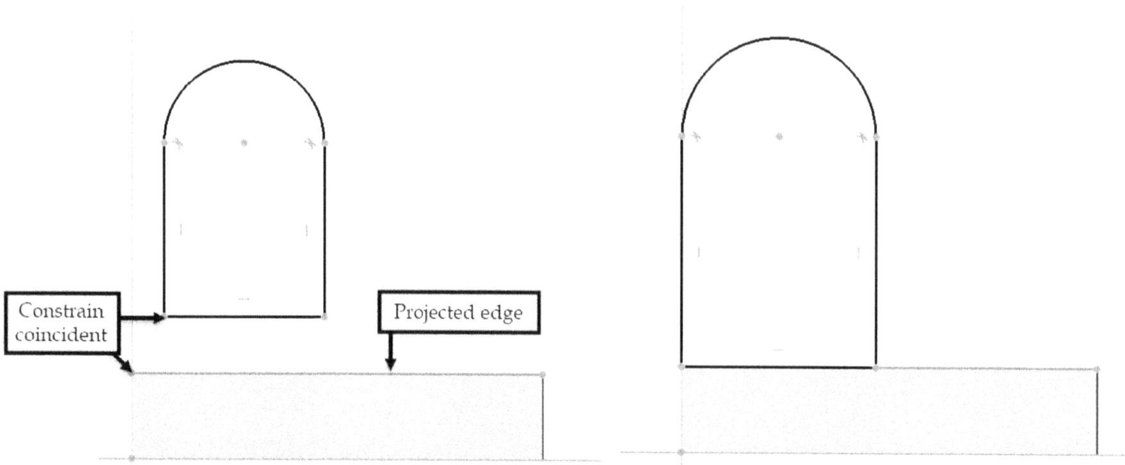

21. Click **Constrain vertical distance** on the **Sketcher constraints** toolbar. Next, select the left vertical line of the sketch.
22. Type-in **1.575** in the **Length** box on the **Insert Length** dialog. Click **OK**.
23. Click the **Constrain radius** icon on the **Sketcher constraints** toolbar. Next, select the arc, type **0.787** in the **Radius** box, and then click **OK**.

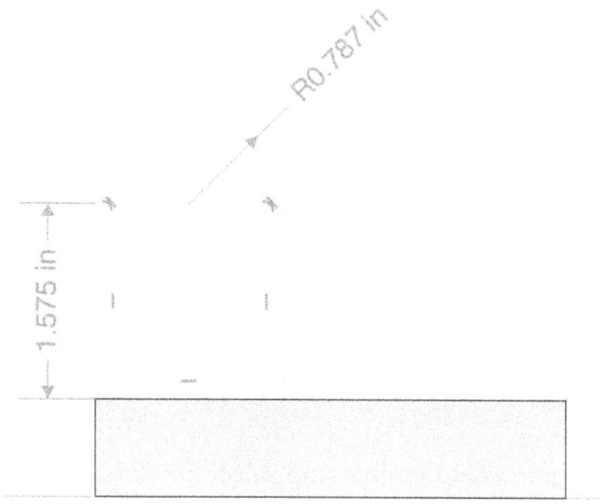

24. Click **Close** on the **Combo View** panel.
25. Activate the **Pad** command and select **Type > Dimension**. Next, enter **0.787** in the **Length** box. Click **OK** to complete the Pad feature.

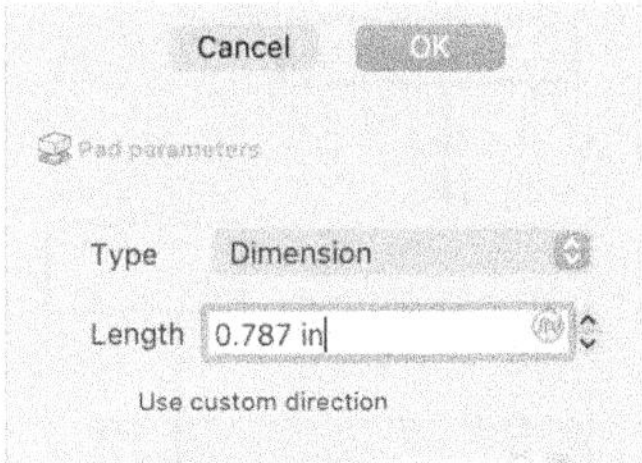

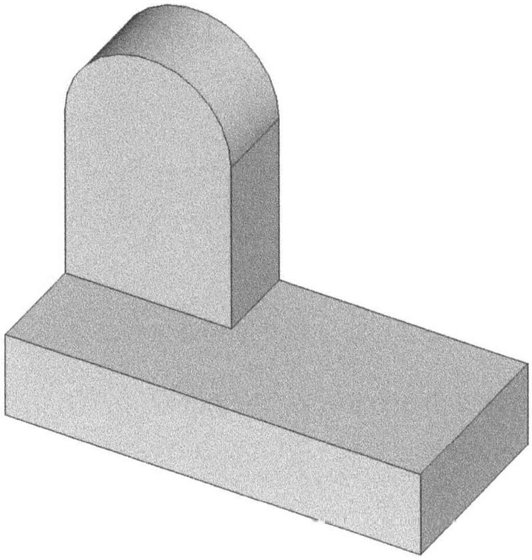

Creating the Rib

1. Activate the **Create Sketch** command and select the XZ_plane. Draw an inclined line, as shown.
2. Click **Sketch > View section** on the menu bar.
3. Click the **External geometry** icon on the **Sketcher geometries** toolbar. Next, select the curved, vertical, and horizontal edges, as shown.

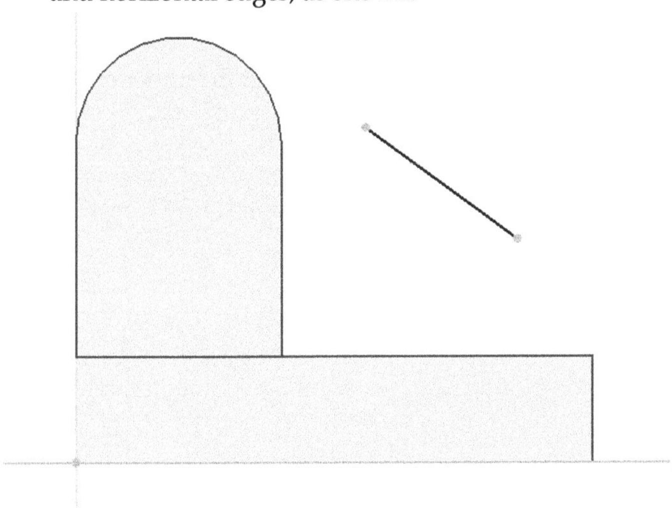

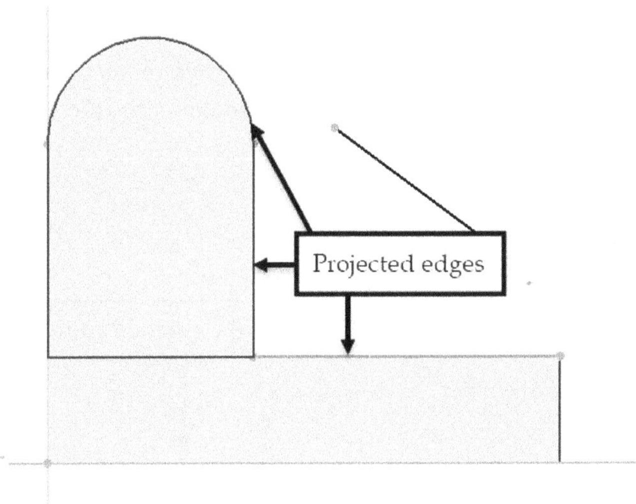

4. Create an arc and two lines, as shown.
5. Click the **Constrain equal** icon on the **Sketcher constraints** toolbar. Next, select the arc and curved edge.
6. Click the **Constrain coincident** icon on the **Sketcher constraints** toolbar. Next, select the endpoint of the arc and the endpoint of the curved edge.
7. Select the endpoint of the horizontal line and the endpoint of the horizontal edge.
8. On the **Sketcher constraints** toolbar, click **Constrain tangent** , and then select the inclined line and the arc; the line is made tangent to the arc.
9. Click the **Constrain coincident** icon on the **Sketcher constraints** toolbar. Next, select the lower endpoint of the inclined line and the endpoint of the horizontal line.

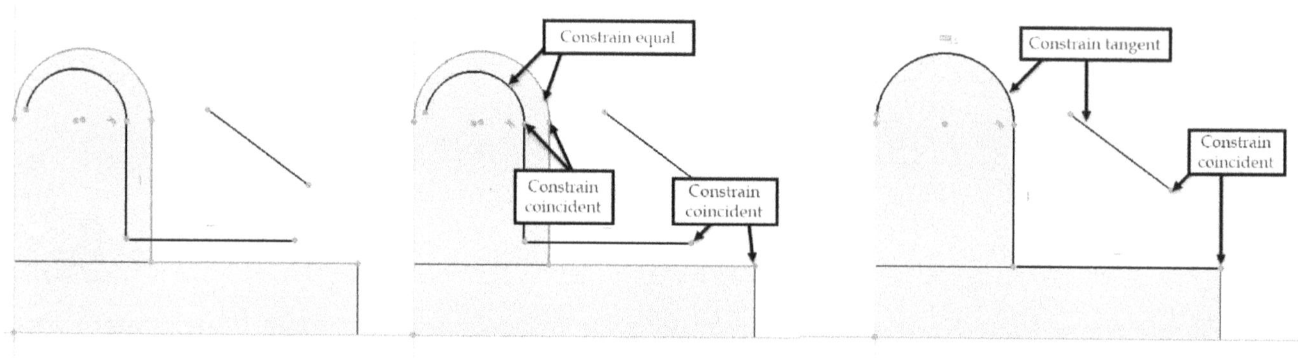

10. Click on the endpoint of the arc and drag it toward the right.

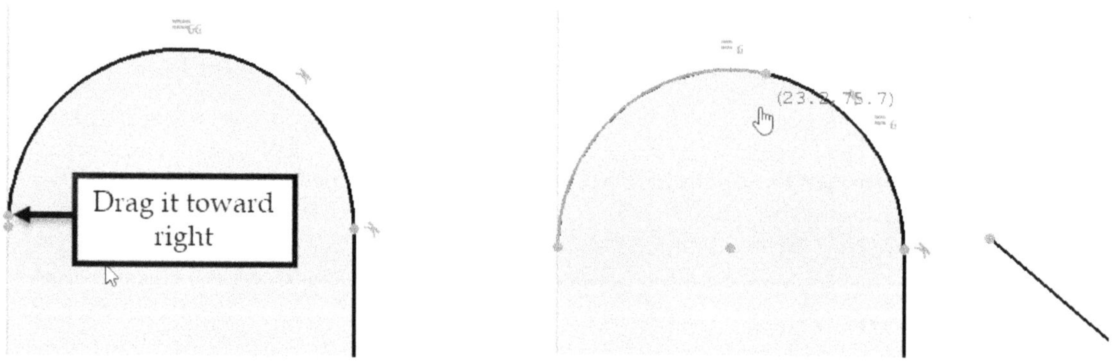

11. Click the **Constrain coincident** icon on the **Sketcher constraints** toolbar. Next, select the upper endpoint of the inclined line and the endpoint of the arc.

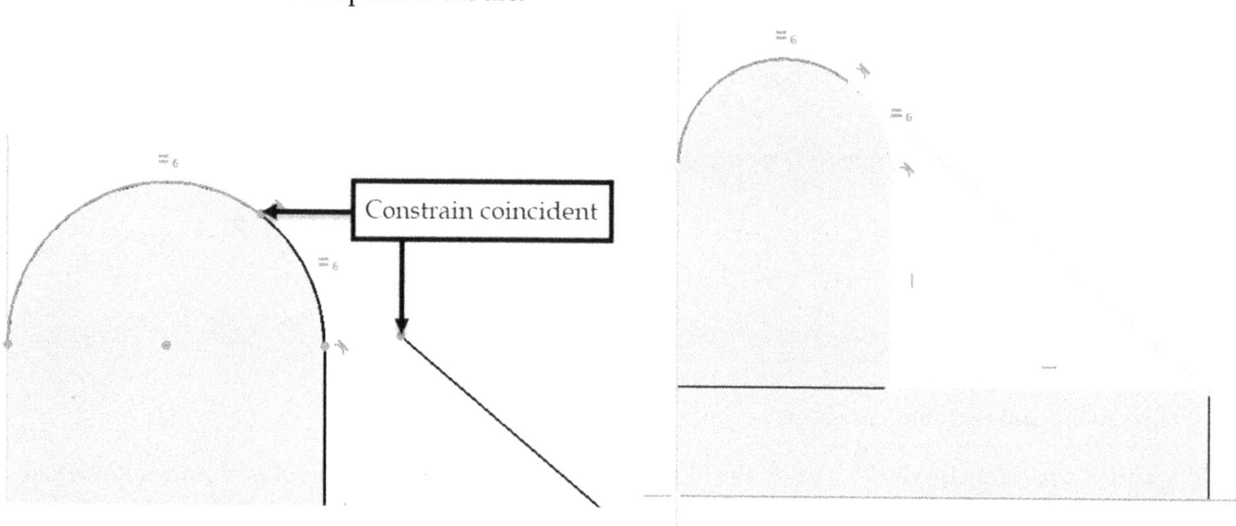

12. Click **Close** on the **Combo View** panel.

13. Click the **Pad** icon on the **Part Design Modeling** toolbar.

14. Type-in **0.394** in the **Length** box. Next, click **OK** to create the *Pad* feature.

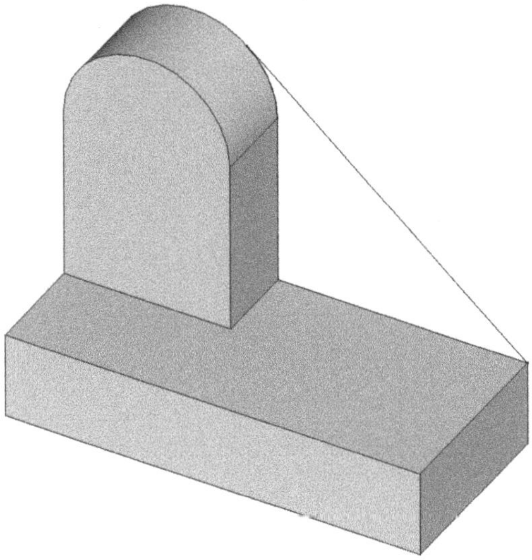

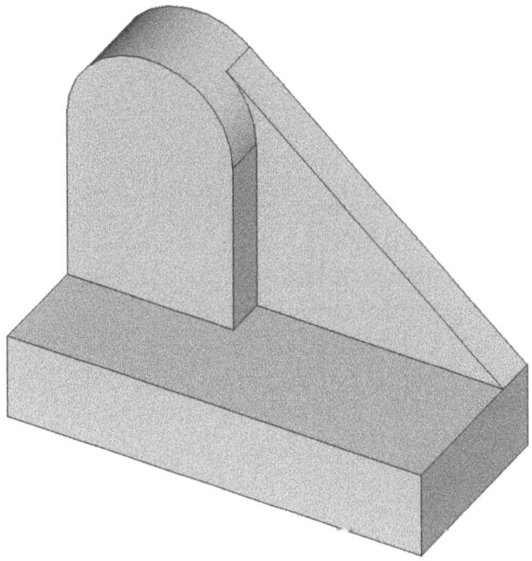

Creating the Pocket features

1. Select the front face of the second feature and click the **Create Sketch** icon on the **Part Design Helper** toolbar.
2. On the **Sketcher geometries** toolbar, click the **Create slot** icon.
3. Click on the sketch plane to specify the first end of the slot. Next, move the pointer downward and click.
4. Click the **External geometry** icon on the **Sketcher geometries** toolbar.

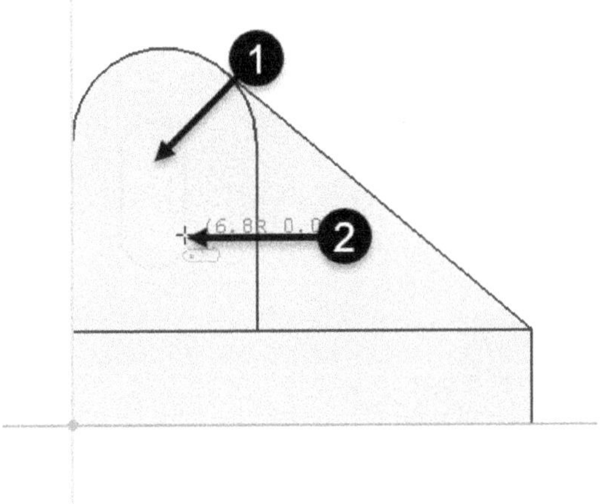

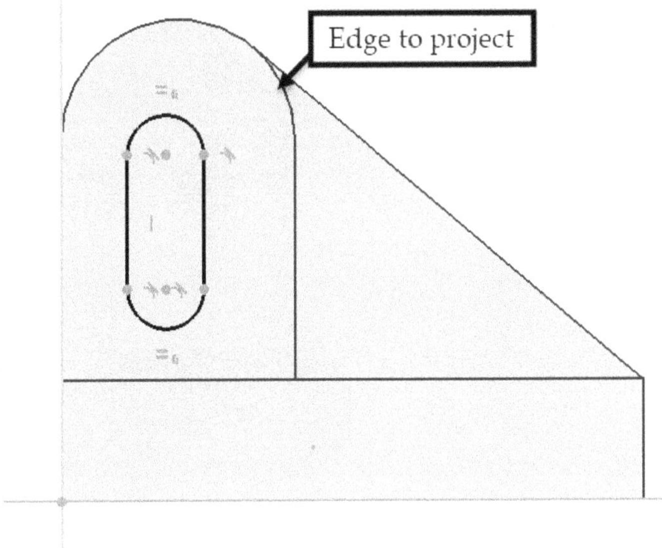

5. Click **Constrain coincident** on the **Sketcher constraints** toolbar. Next, select the center point of the projected curve edge and the centerpoint of the slot endcap.
6. Create the radius and vertical distance constraints, as shown.

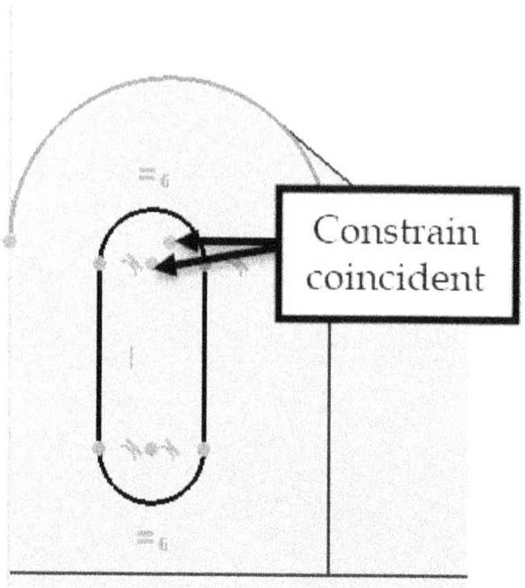

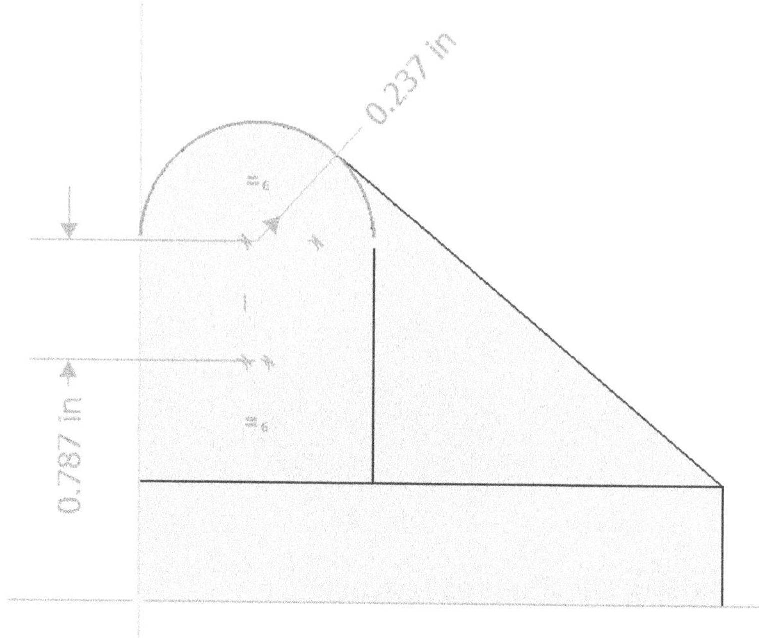

7. Click **Close** on the **Combo View** panel.
8. Click the **Pocket** icon on the **Part Design Modeling** toolbar.
9. On the **Pocket parameters** section, select **Type > Through all**. Next, click **OK**.

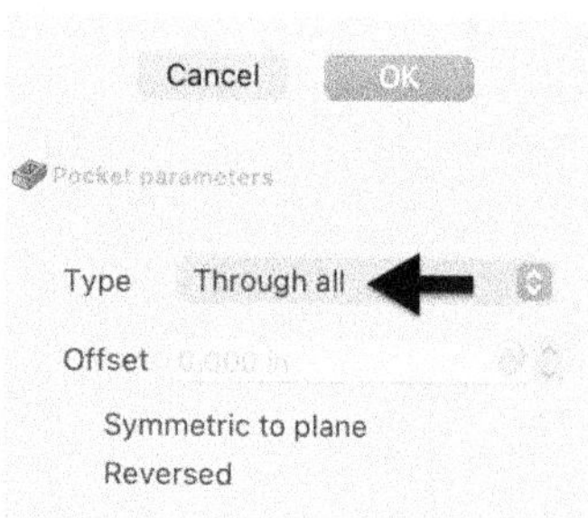

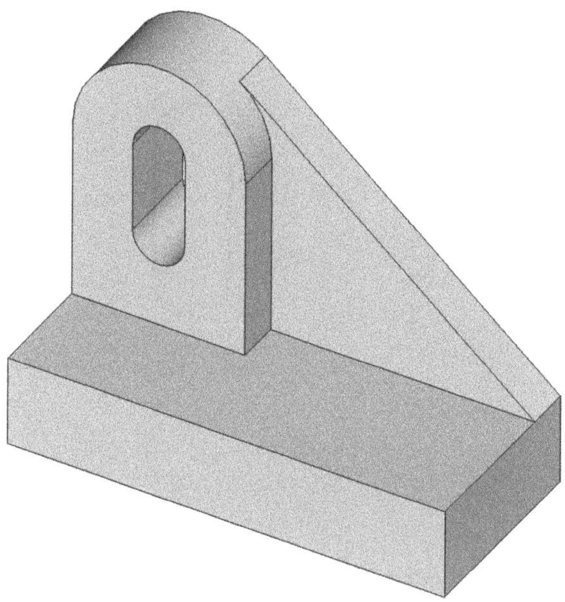

Creating the Fillet and Hole features

1. Select the right vertical edge and click **Create fillet** on the **Part Design Modeling** toolbar.
2. Type **0.787** in the **Radius** box and click **OK**.
3. Select the top face of the first feature and click the **Create Sketch** icon on the **Part Design Helper** toolbar.
4. Click **External geometry** on the **Sketcher geometries** toolbar. Next, select the curved edge, as shown.

5. Click the **Create circle** 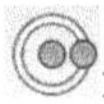 icon on the **Sketcher geometries** toolbar. Next, select the center point of the projected edge. Move the pointer outward and click to create a circle.
6. Click **Close** on the **Combo View** panel.

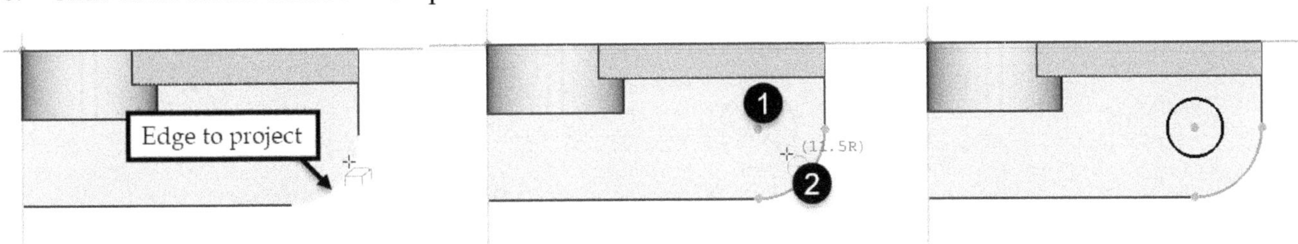

7. Click the **Hole** icon on the **Part Design Modeling** toolbar.
8. On the **Hole parameters** section, specify the settings, as shown. Next, click **OK** to create the hole.

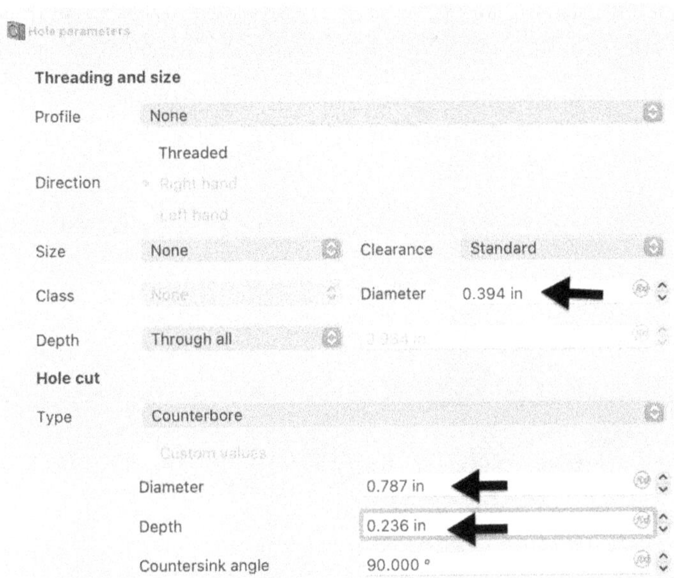

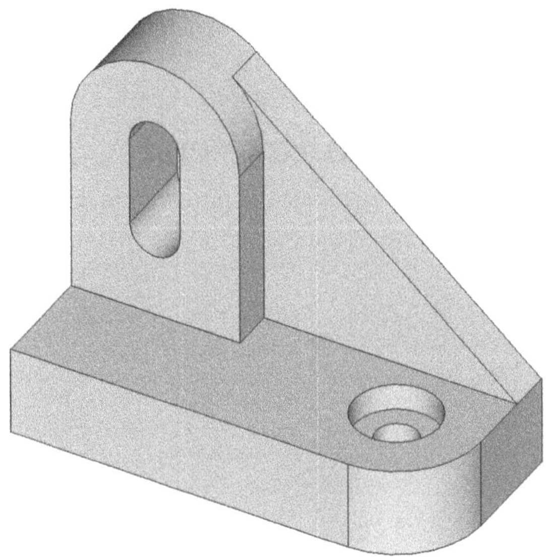

9. Select the front face of the rectangular base and click the **Create Sketch**  icon on the **Part Design Helper** toolbar.
10. Draw a sketch and add dimensions to it. Click **Close** on the **Combo View** panel.
11. Create a *Pocket* feature using the sketch.

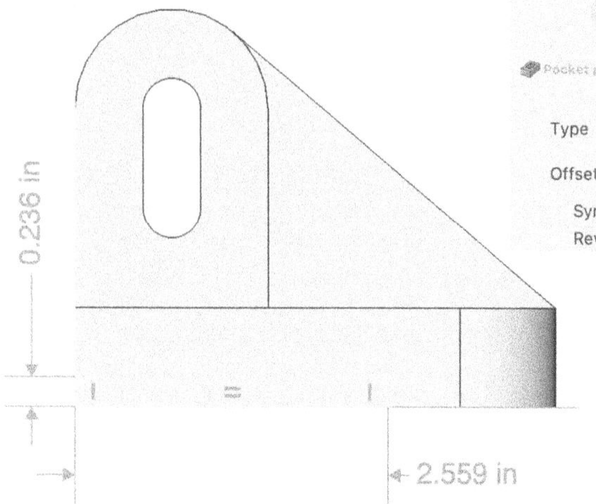

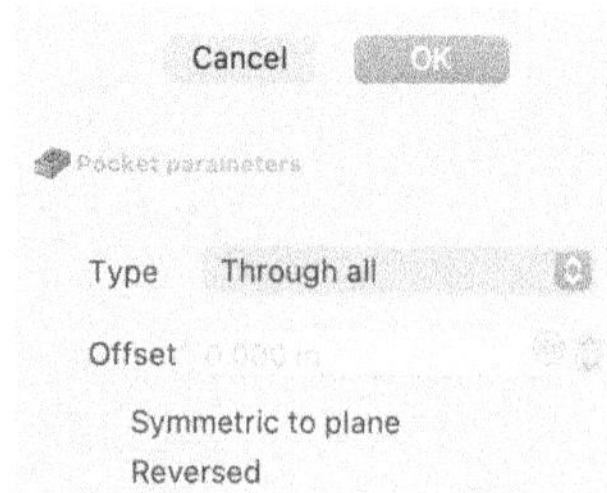

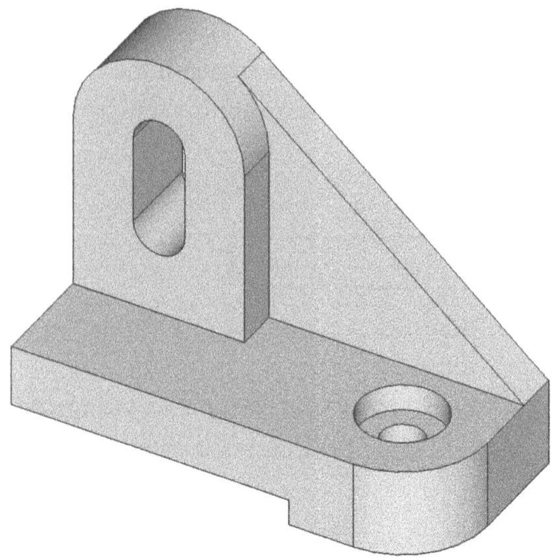

12. Save and close the part file.

Tutorial 3 (Millimeters)

In this example, you create the part shown next.

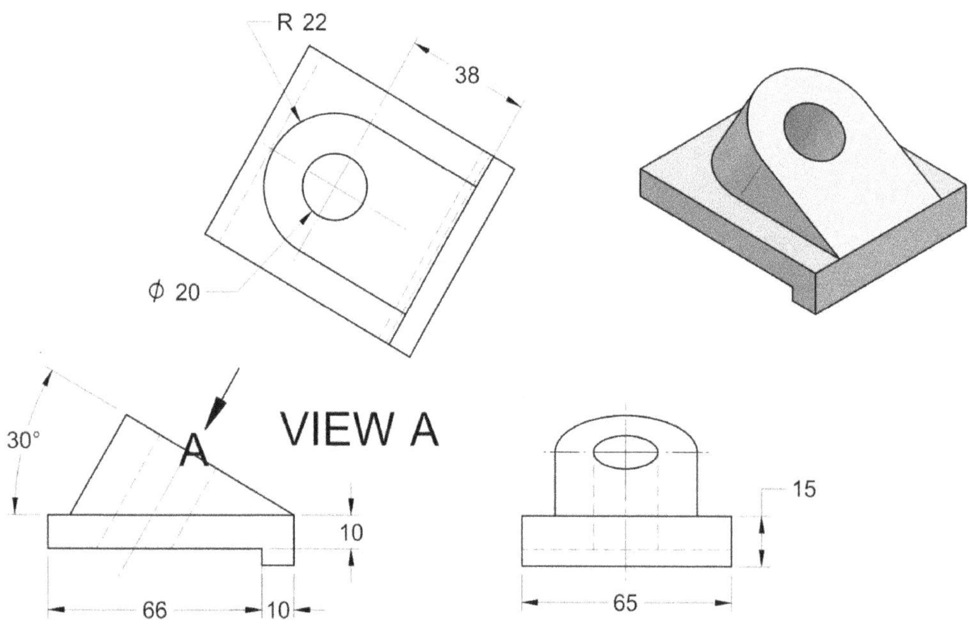

Creating a New document

1. Click **FreeCAD 0.19** on the desktop to start the application.
2. On the menu bar, click **File > New**; it creates a new document.
3. On the **Workbench** toolbar, select **Workbench** drop-down **> Part Design**.
4. Click **Edit > Preferences** on the **Menu** bar; the **Preferences** dialog appears on the screen.
7. Click **Units** tab and select **Unit system > Standard (mm/kg/s/degree)**.
8. Select **Number of decimals > 2** and click **OK** on the **Preferences** dialog.

Creating the Padded features

1. Click the **Create Sketch** icon on the **Part Design Helper** Toolbar and select the XZ_plane. Next, click **OK**.
2. Create the closed sketch and add dimensional and geometric constraints, as shown.

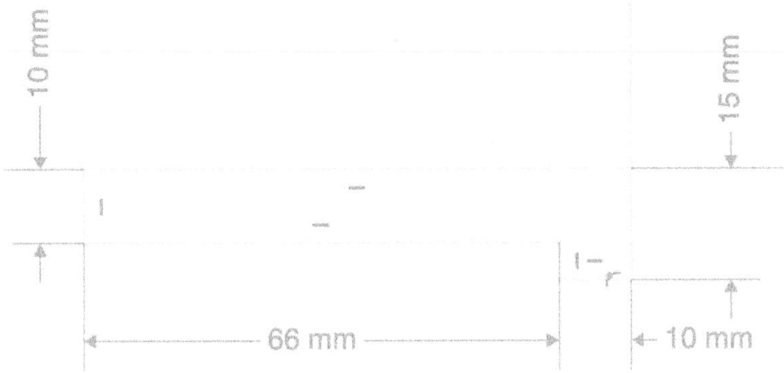

3. Click the **Close** button on the **Combo View** panel.
4. On the **Part Design Modeling** toolbar, click the **Pad** icon.
5. On the **Pad parameters** section, select **Type > Dimension** and enter **65** in the **Length** box.
6. Check the **Symmetric to plane** option and click **OK** to create the *Pad* feature.

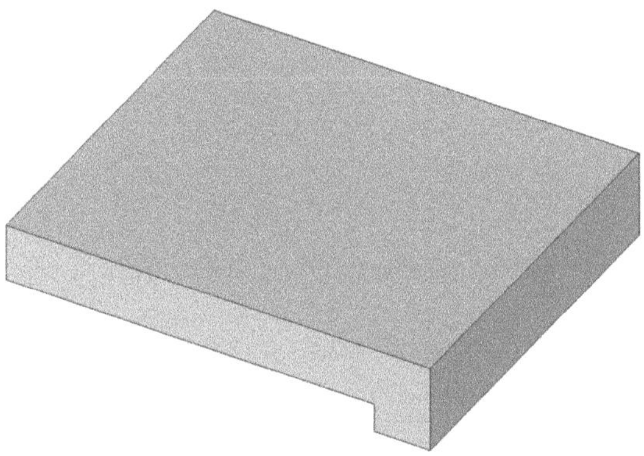

Creating the Angled Plane

1. Click the **Create a datum plane** icon on the **Part Design Helper** Toolbar and select the top face of the model.
2. Type -30 in the **Around y-axis** box. Next, click **OK**.

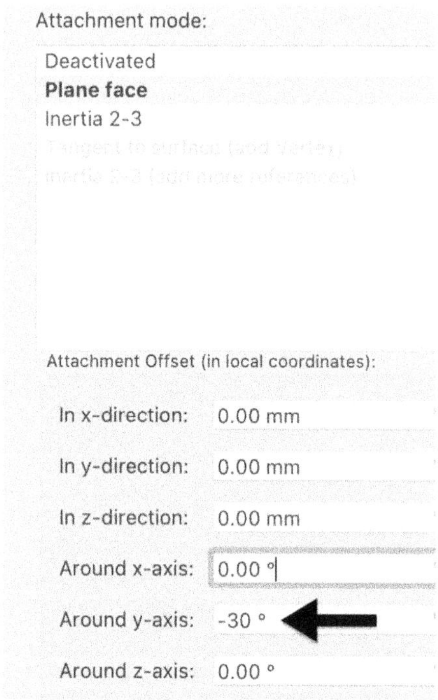

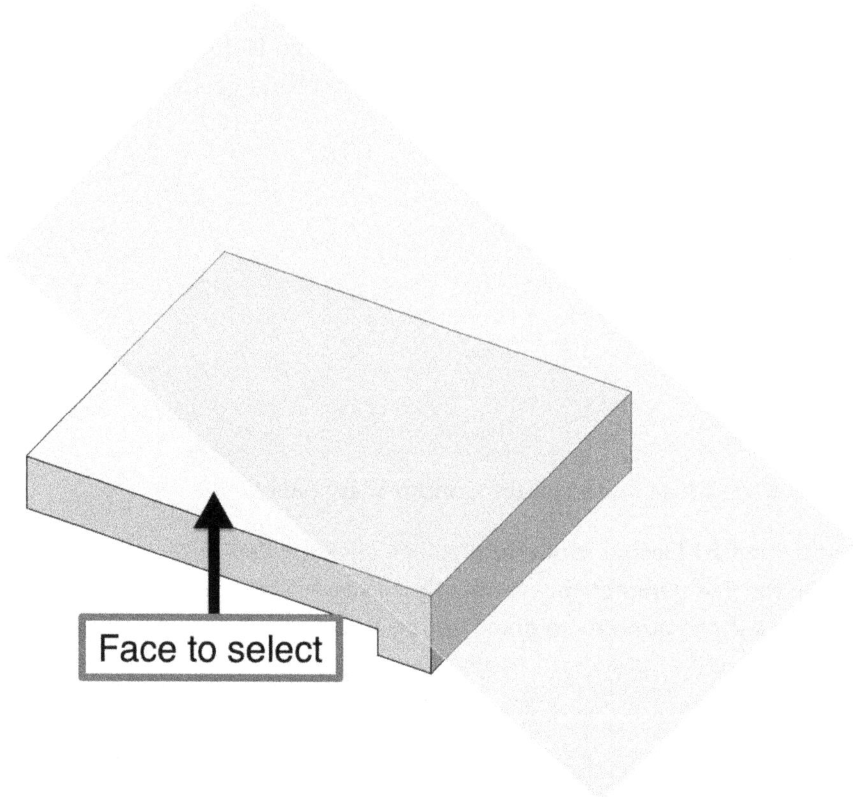

Creating the Padded feature

1. Click the **Create Sketch** icon on the **Part Design Helper** Toolbar and select the newly created datum plane. Next, click **OK**.
2. Create the closed sketch and add dimensional and geometric constraints, as shown.

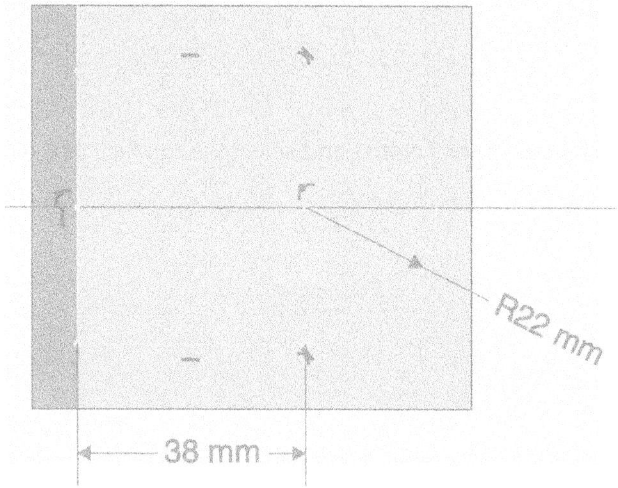

3. Click the **Close** button on the **Combo View** panel.
4. On the **Part Design Modeling** toolbar, click the **Pad** command.
5. Check the **Reversed** option on the **Pad Parameters** section of the **Combo View** panel.
6. Select **Type > To Last** from the **Pad Parameters** section and click **OK**.

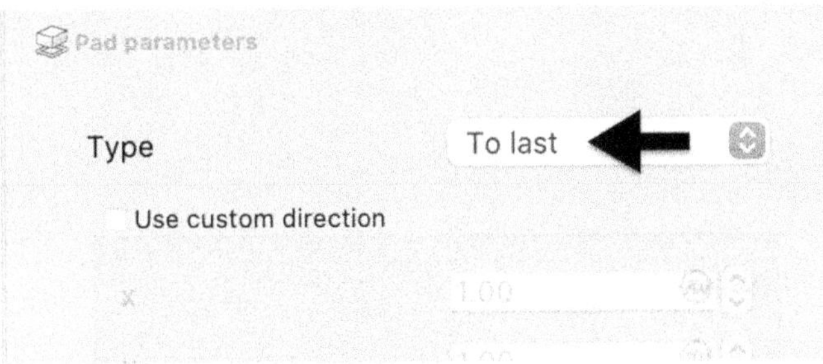

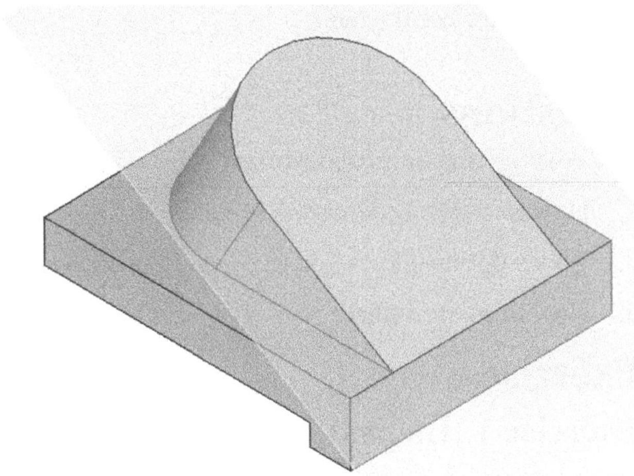

Creating the Hole feature

1. Select inclined face and click the **Create Sketch** icon on the **Part Design Helper** toolbar.
2. Click **External geometry** on the **Sketcher geometries** toolbar. Next, select the curved edge, as shown.
3. Click the **Create circle** icon on the **Sketcher geometries** toolbar. Next, select the center point of the projected edge. Move the pointer outward and click to create a circle.

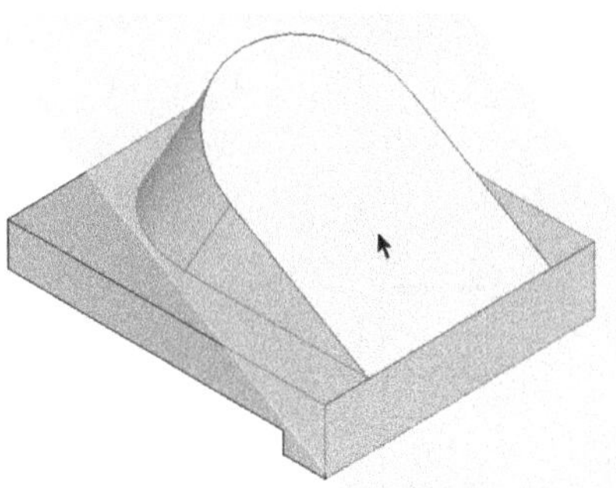

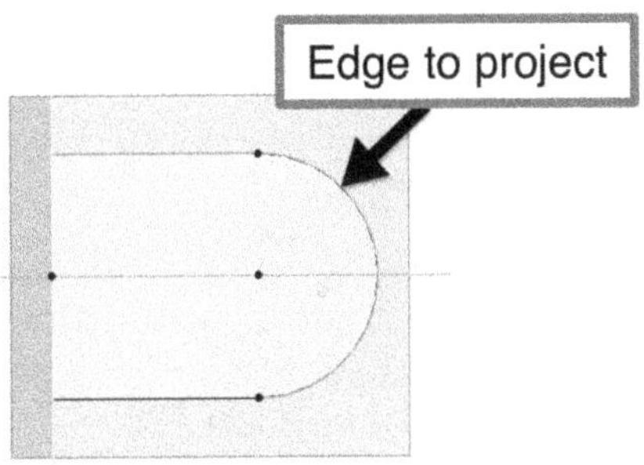

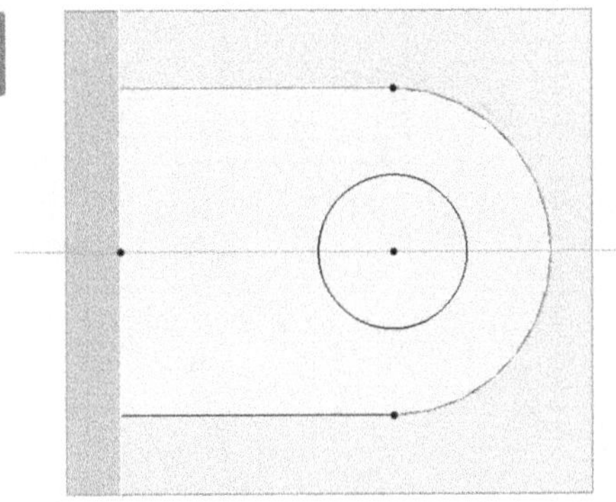

4. Click **Close** on the **Combo View** panel.
5. Click the **Hole** icon on the **Part Design Modeling** toolbar.
6. On the **Hole parameters** section, specify the settings, as shown. Next, click **OK** to create the hole.

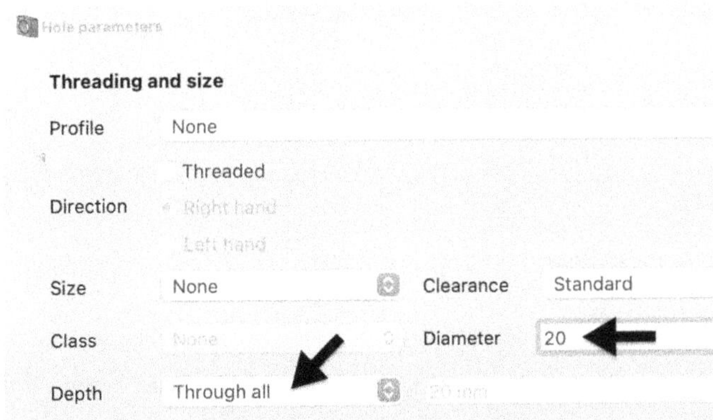

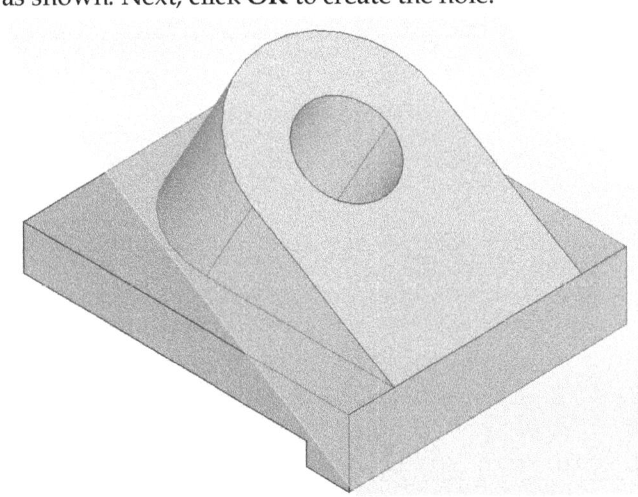

7. Save and close the file.

Questions

1. What are the dress up features?
2. How to create a counterbored hole?
3. Which option allows you to create chamfer with unequal setbacks?
4. How to create a fillet?

Exercises

Exercise 1 (Inches)

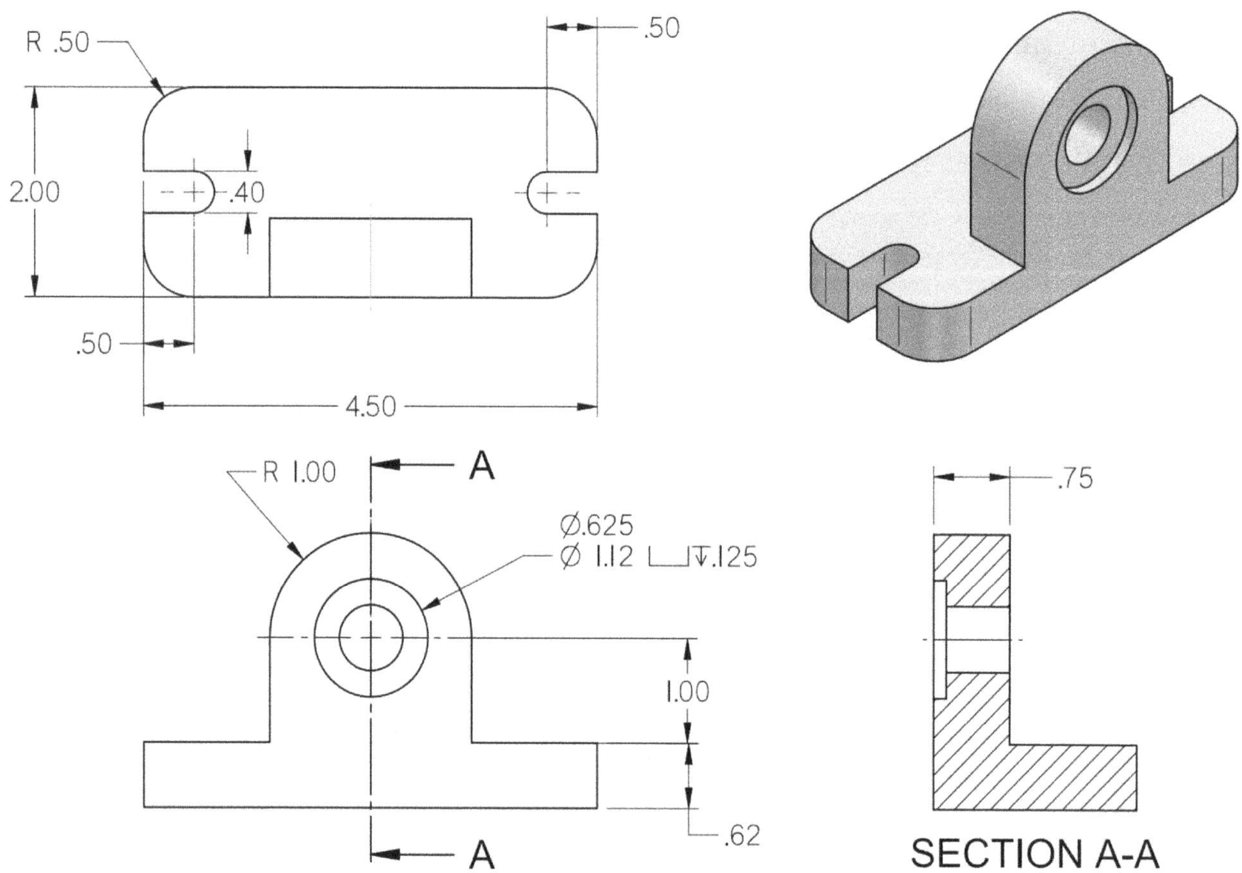

Exercise 2 (Millimeters)

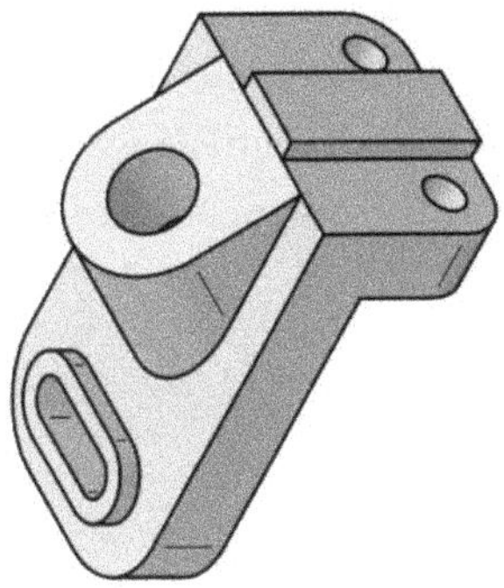

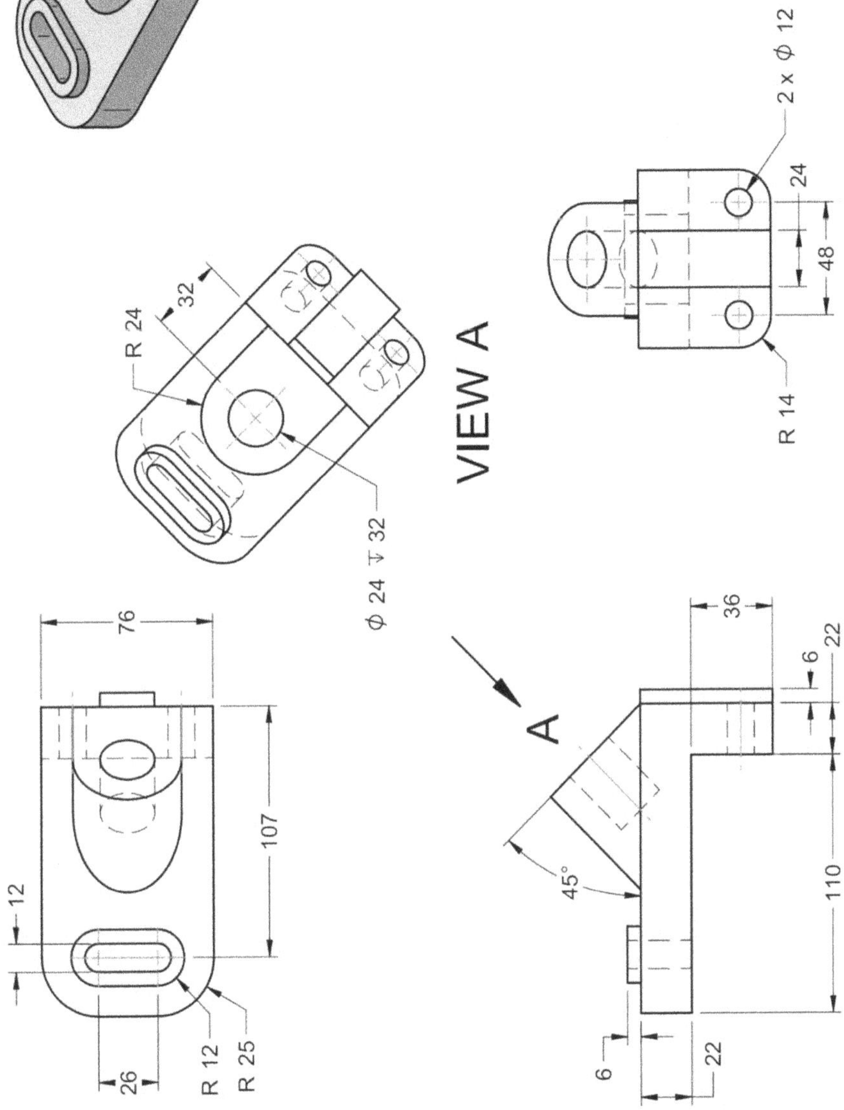

Chapter 5: Patterned Geometry

When designing a part geometry, most of the time, there are elements of symmetry in each part, or there are at least a few features that are repeated multiple times. In these situations, FreeCAD 0.19 offers some commands that save you time. For example, you can use mirror features to design symmetric parts, which makes designing the part quicker. This is because you only have to design a portion of the part and use the mirror feature to create the remaining geometry.

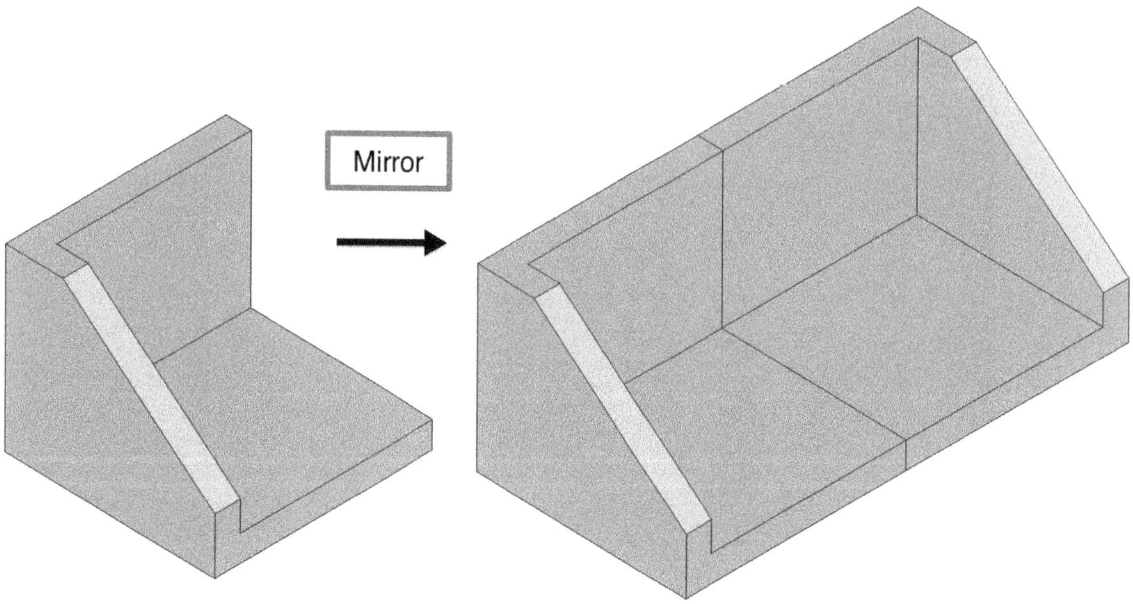

In addition, there are some pattern commands to replicate a feature throughout a part quickly. They save you time from creating additional features individually and help you modify the design easily. If the design changes, you only need to change the first feature; the rest of the pattern features will update automatically. In this chapter, you will learn to create mirrored and pattern geometries using the commands available in FreeCAD 0.19.

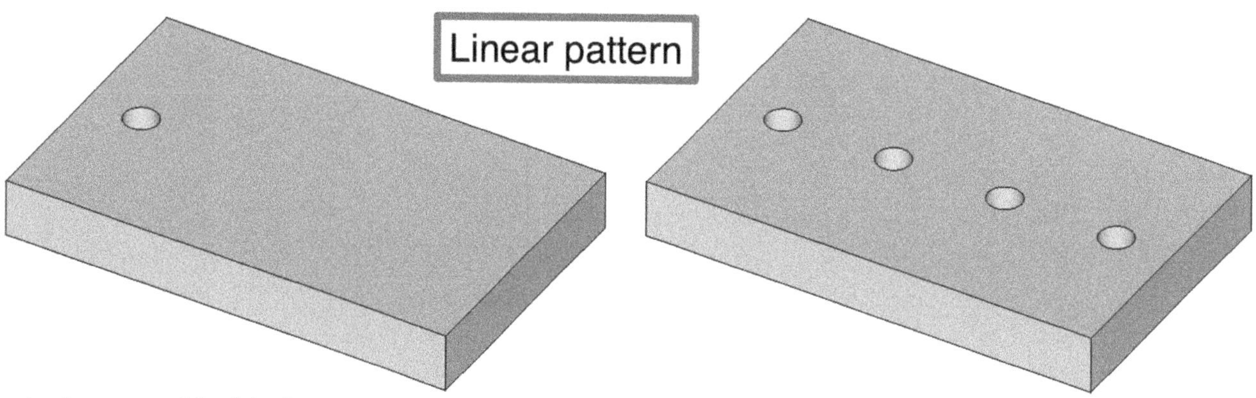

The topics covered in this chapter are:

- *Mirrored* feature
- *Linear Pattern*
- *Polar Pattern*
- *Multi Transform*

Mirrored

If you are designing a part that is symmetric, you can save time by using the **Mirrored** command. Using this command, you can replicate the individual features of the entire body. To mirror features (3D geometry), you need to have a face or plane to use as a reference. You can use a model face, default plane, or create a new plane if it does not exist where it is needed.

On the Menu bar, click **Part Design > Create a pattern > Mirrored** (or) click the **Mirrored** icon on the **Part Design Modeling** toolbar. Press and hold the Shift key and select the features to mirror from the **Combo View** panel. Next, click **OK** and select a mirroring plane from the **Plane** drop-down. You can select a sketch axis or a base plane from the **Plane** drop-down. Alternately, you can select a user-defined Datum plane from the graphics window to define the mirroring plane. To do this, select the **Select reference** option from the **Plane** drop-down and select the datum plane about which the features are to be mirrored. Click **OK** to mirror the selected features.

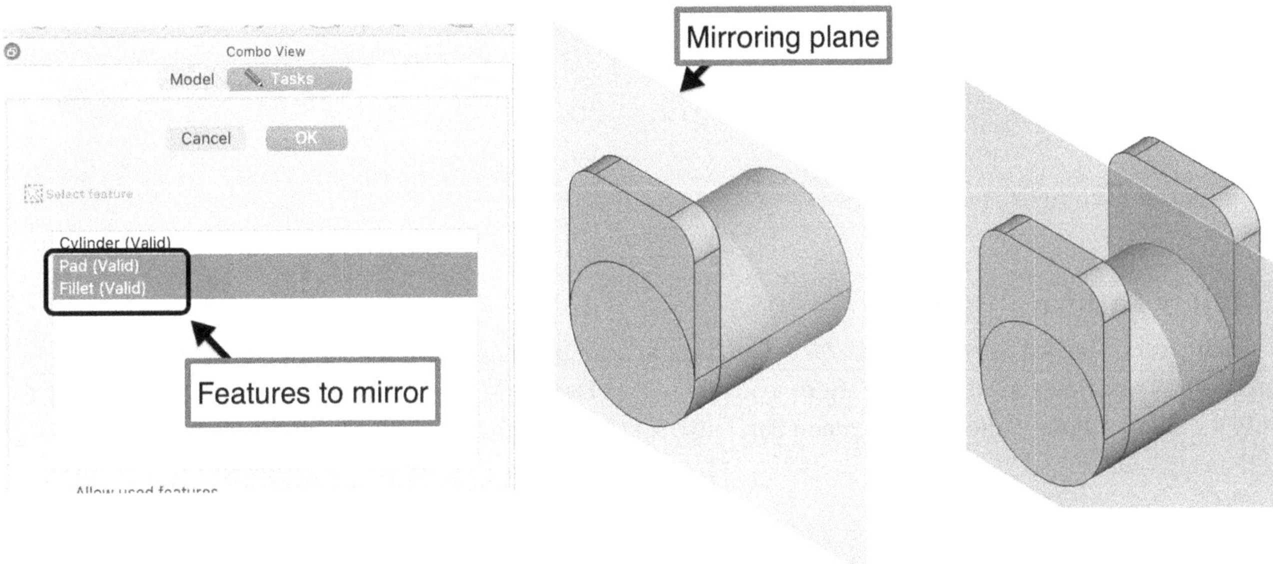

Create Patterns

FreeCAD 0.19 allows you to replicate a feature using the pattern commands: **Linear Pattern**, **Polar Pattern**, and **MultiTransform**. The following sections explain the different patterns that can be created using the three pattern commands.

Linear Pattern

To create a pattern in a linear fashion, you must first click the **Linear Pattern** icon on the **Part Design Modeling** toolbar (or) click **Part Design > Apply a pattern > Linear Pattern** on the menu bar. Select the feature to pattern from the **Combo View** panel and click **OK**. Next, select an axis from the **Direction** drop-down. You can select a sketch axis or a base X, Y, or Z axis from the **Direction** drop-down. Alternately, you can select an edge or face from the graphics window to define the pattern direction. To do this, select the **Select reference** option from the **Direction** drop-down and select an edge along which the features are to be patterned.

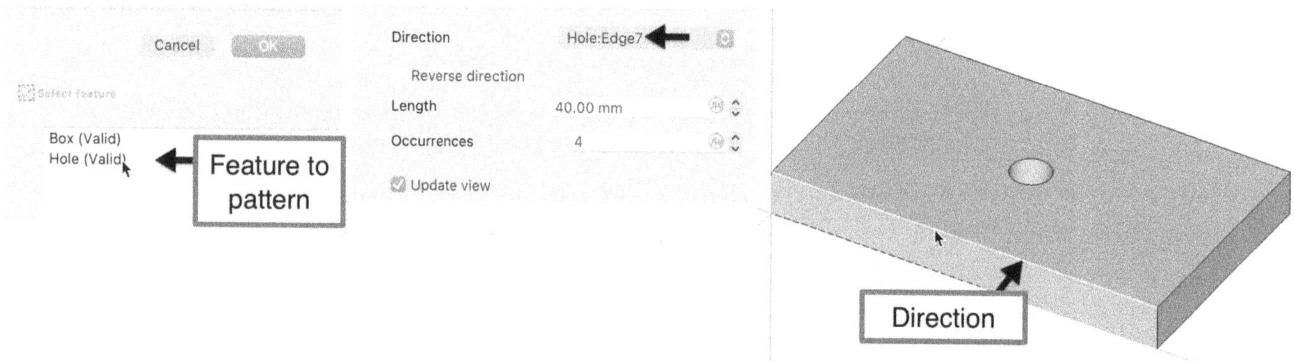

Next, type-in values in the **Length** and **Occurrences** boxes; the value entered in the **Length** box defines the total length of the pattern. Check the **Reverse direction** option, if you want to reverse the pattern direction.

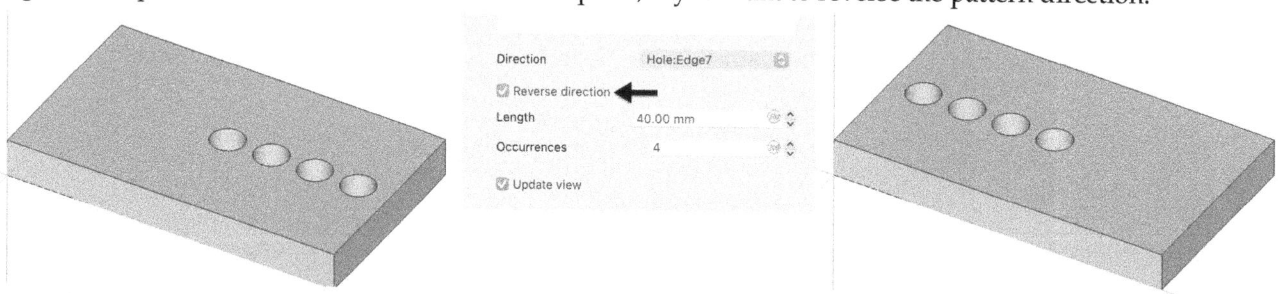

Next, click **OK** to complete the pattern.

Polar Pattern

The polar pattern is used to pattern the selected features in a circular fashion. To create a pattern in a polar fashion, you must first click the **Polar Pattern** icon on the **Part Design Modeling** toolbar (or) click **Part Design > Apply a pattern > Polar Pattern** on the menu bar. Next, select the feature to pattern from the **Combo View** panel, and then click **OK**.

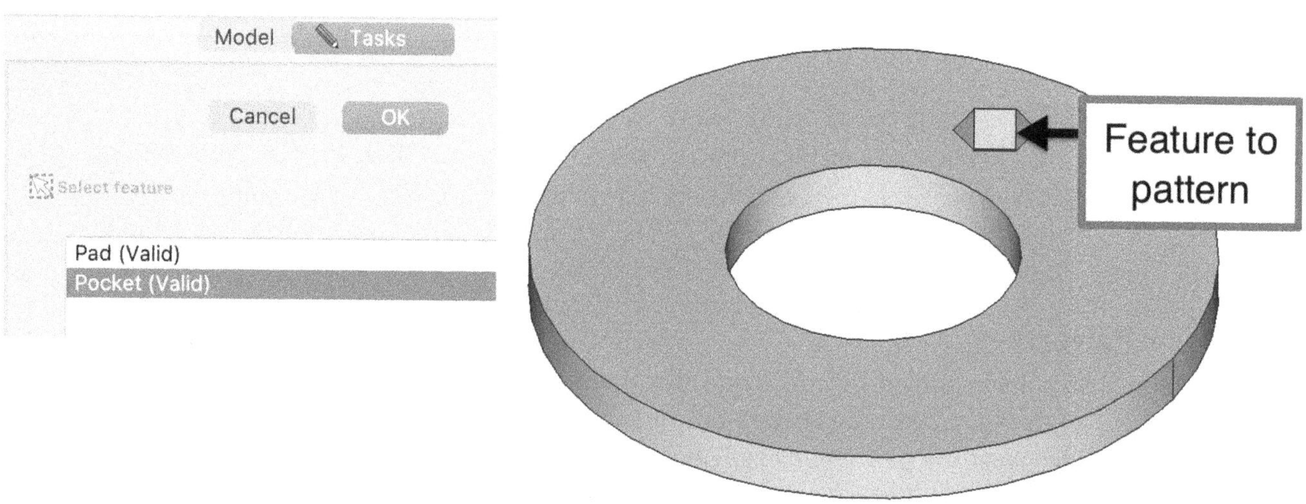

Next, select an axis from the **Axis** drop-down. You can select a sketch axis or a base X, Y, or Z axis from the **Direction** drop-down. Alternately, you can select the axis from the graphics window. To do this, select the **Select reference** option from the **Axis** drop-down and select axis of the rotation. Next, specify the angle and number of occurrences values in the **Angle** and **Occurrences** boxes, respectively. Click **OK** to create a polar pattern.

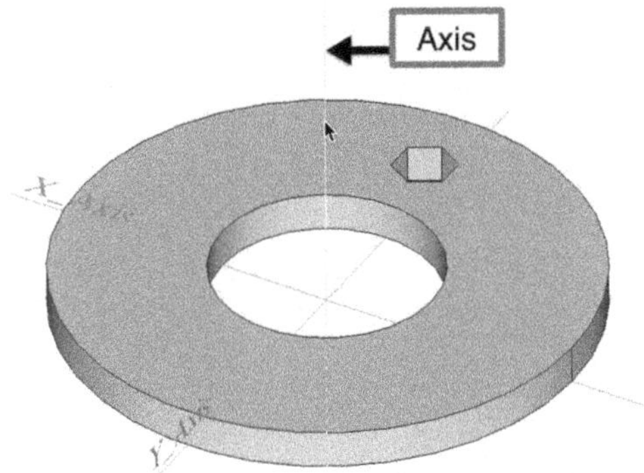

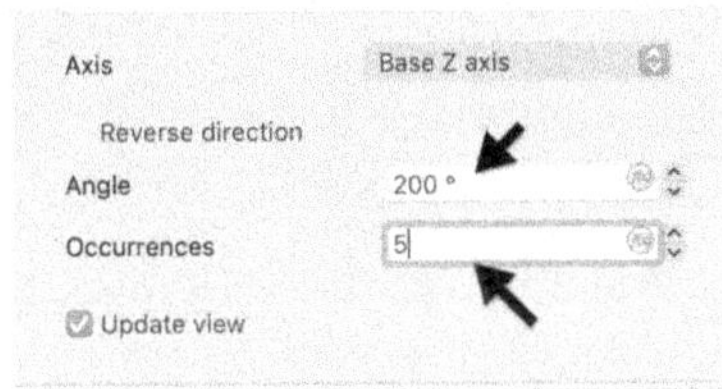

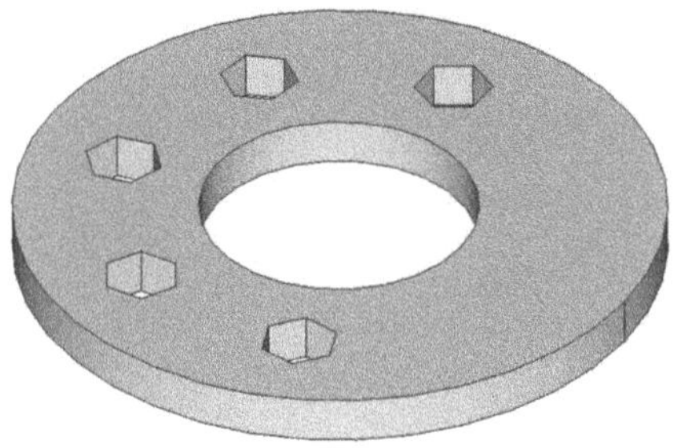

Multi Transform

The **MultiTransform** command is used to create patterns of multiple patterns. For example, you can pattern the pocket feature shown in the following image using three types of patterns (Linear Pattern, Scaled Pattern, and Polar Pattern) at a time.

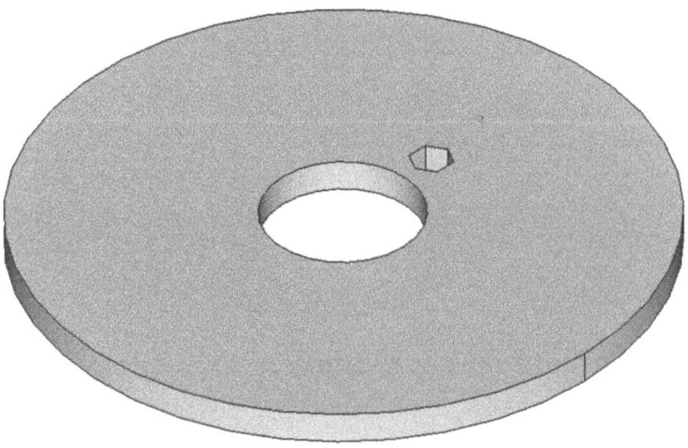

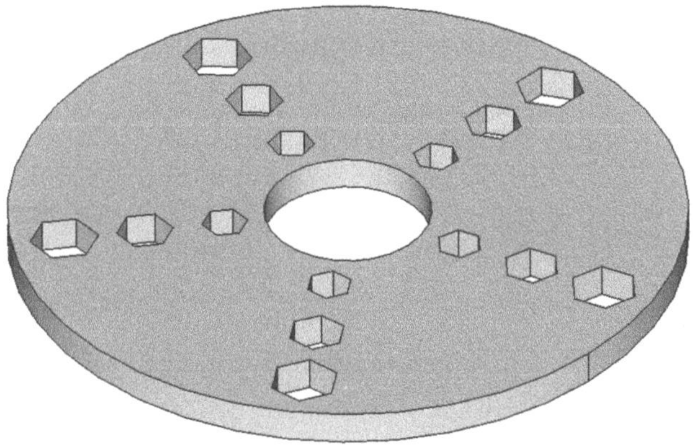

Click the **MultiTransform** icon on the **Part Design Modeling** toolbar (or) click **Part Design > Apply a pattern > Create MultiTransform** on the menu bar. Next, select the feature to pattern from the **Combo View** panel, and then click **OK**. Next, right-click in the **Transformations** section and select **Add linear pattern**. Specify the **Direction, Length**, and **Occurences**.

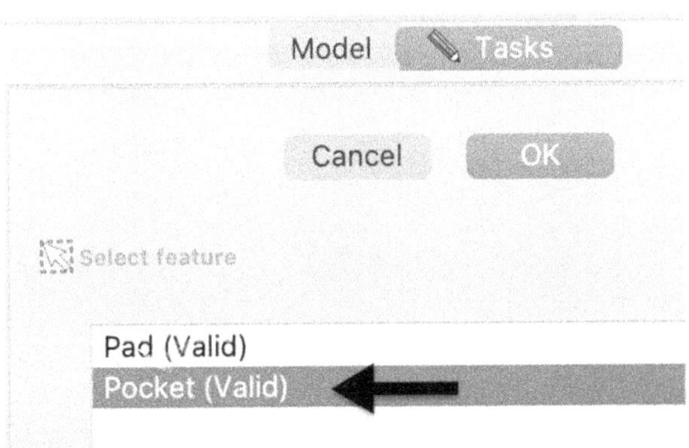

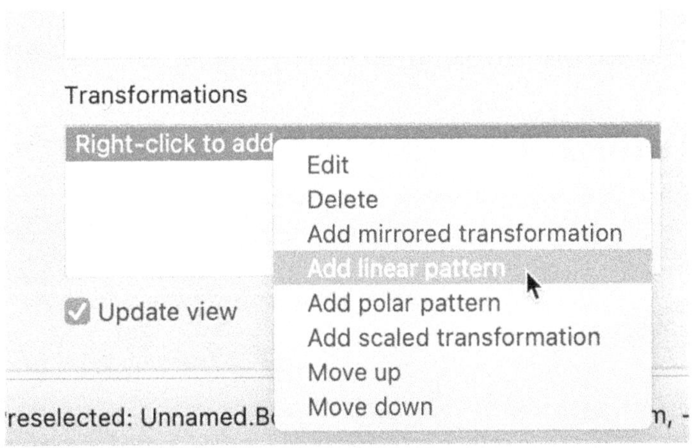

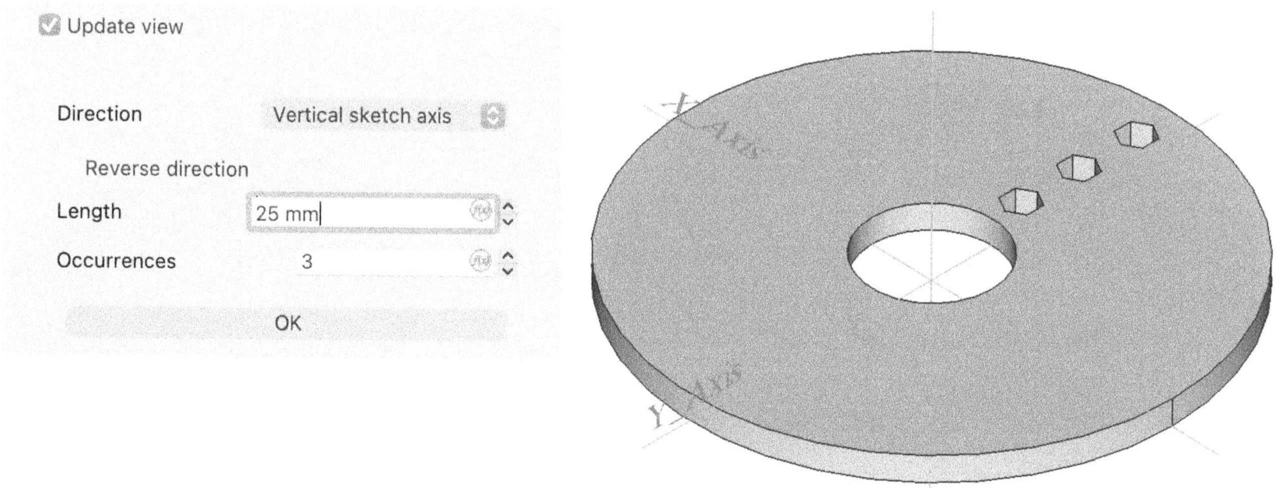

Right-click in the **Transformations** section and select **Add scaled transformation**. Specify the scale factor and occurrences to be scaled in the **Factor** and **Occurrences** boxes, respectively.

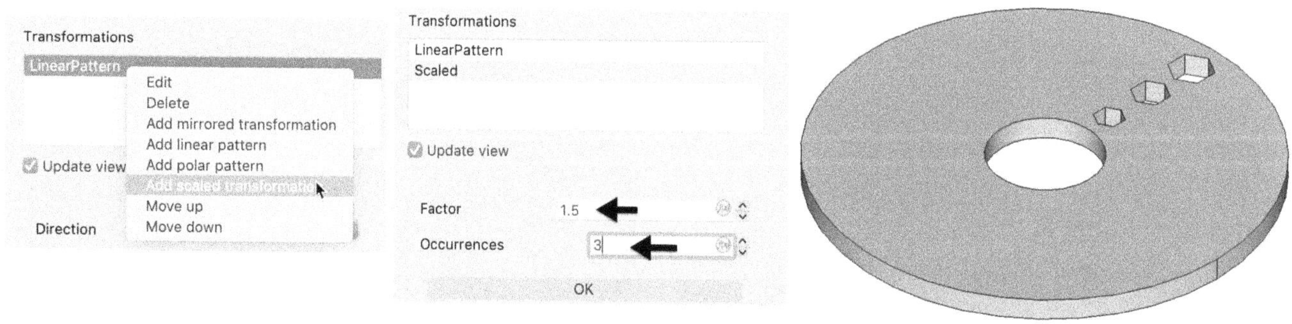

Right-click in the **Transformations** section and select **Add polar pattern**. Specify the rotation **Axis**, **Angle**, and **Occurrences**. Next, click **OK** twice.

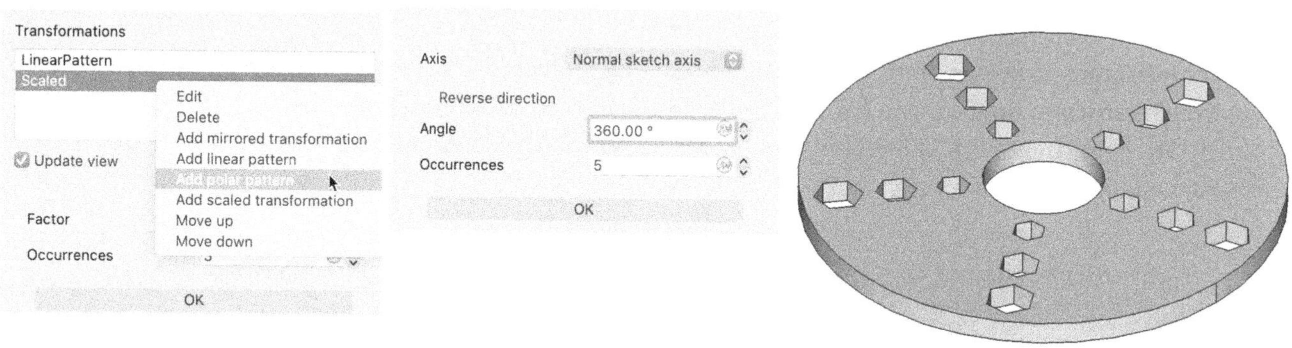

Tutorial 1

In this example, you create the part shown next.

.

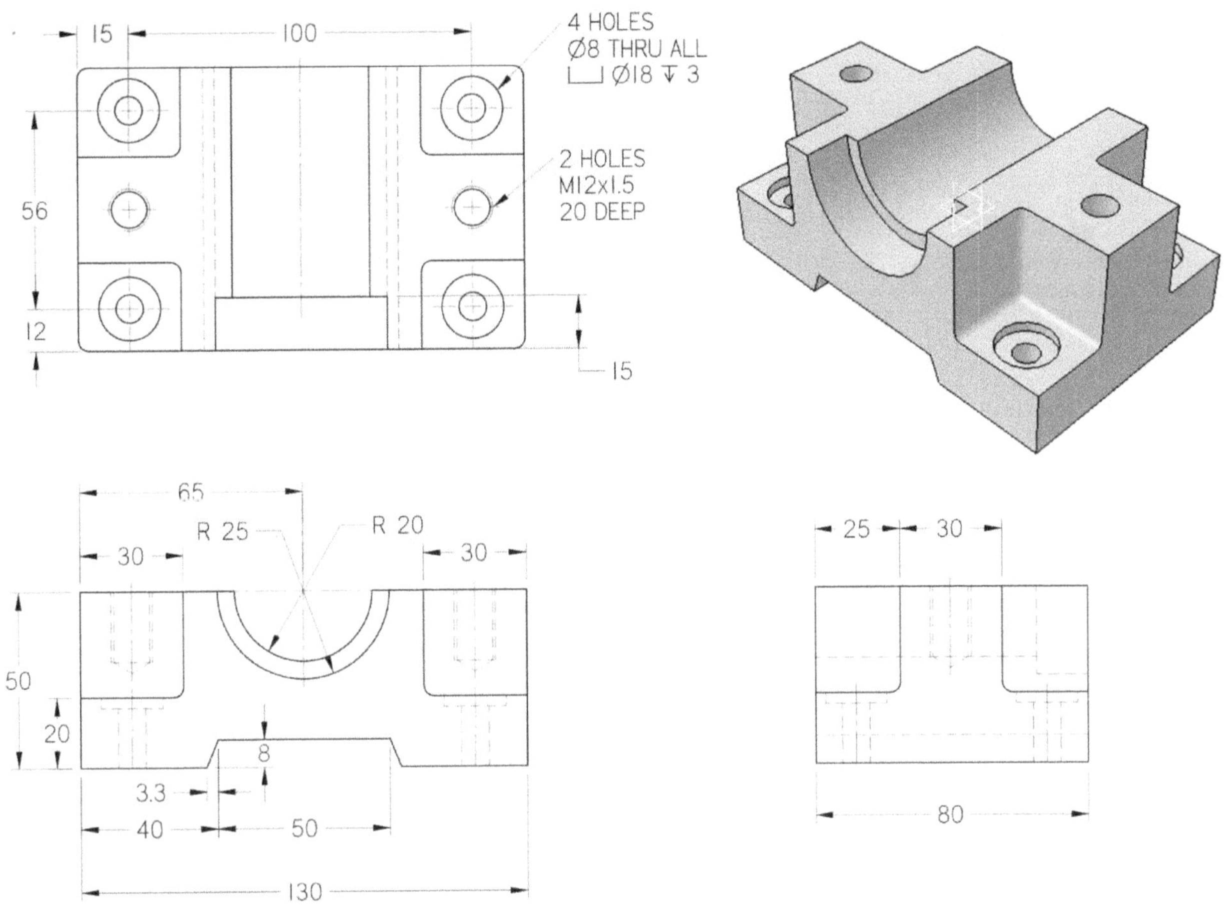

Creating a New document

1. Click **FreeCAD 0.19** on the desktop.
2. On the menu bar, click **File> New**; it creates a new document.
3. On the **Workbench** toolbar, select **Workbench** drop-down **> Part Design**.
4. Click **Edit > Preferences** on the **Menu** bar; the **Preferences** dialog appears on the screen.
5. Click **Units** tab and select **Unit System > Standard (mm/kg/s/degree)**.
6. Select **Number of decimals > 2** and click **OK** on the **Preferences** dialog.

Creating the Padded features

1. To start a sketch, click **Create Sketch** on the **Part Design Helper** toolbar.
2. Click on the XZ_Plane and click **OK** on the **Combo View** panel; the selected plane orients normal to the screen.
3. On the **Sketcher geometries** toolbar, click **Rectangle** . Next, specify the two corners of the rectangle.
4. Click the **Constrain symmetrical** icon on the **Sketcher constraints** toolbar. Next, select the lower right corner point of the rectangle.
5. Select the lower-left corner of the rectangle and the sketch origin point; the two corners of the rectangle are made symmetric about the sketch origin.

6. Add vertical and horizontal distance constraints to the rectangle.

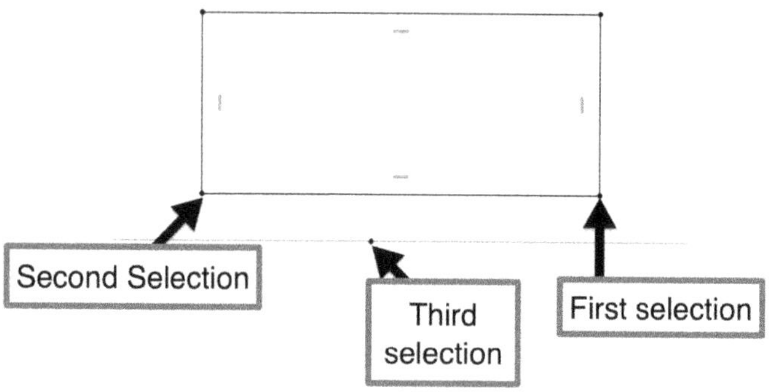

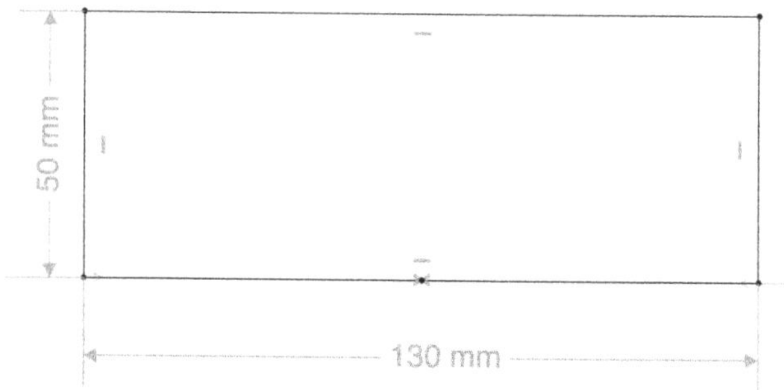

7. Click the **Close** button on the **Combo View** panel.
8. Click the **Pad** tool on the **Part Design Modeling** toolbar.
9. On the **Pad parameters** section, check the **Symmetry to plane** option. Type **80** in the **Length** box and click **OK** to create the *Pad* feature.

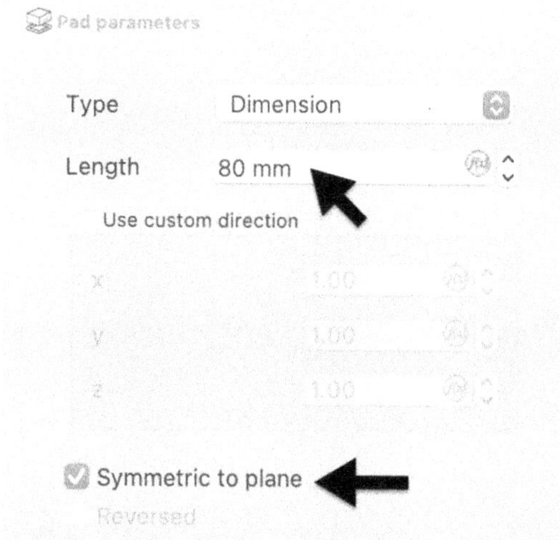

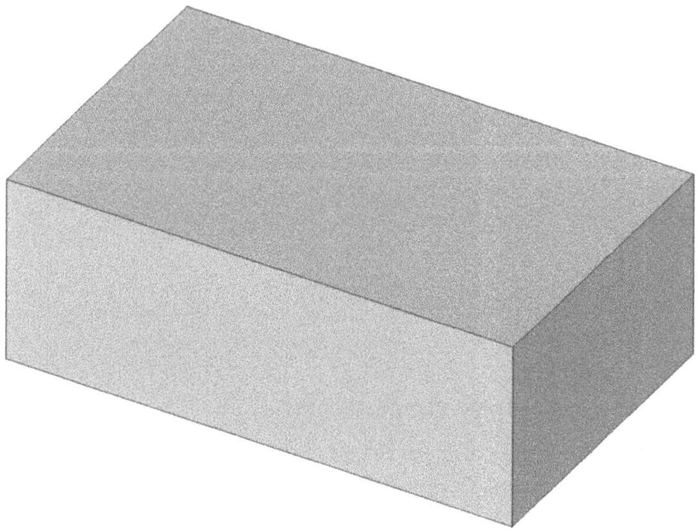

10. Select the top face of the model and click the **Create Sketch** icon on the **Part Design Helper** toolbar.
11. Click the **External geometry** icon on the **Sketcher geometries** toolbar.
12. Select the right vertical edge of the top face of the *Pad* feature, as shown. Next, create a rectangle.
13. Click the **Constrain coincident** icon on the **Sketcher constraints** toolbar.
14. Select the bottom-right corner point of the rectangle. Next, select the bottom endpoint of the projected edge; the two selected points are constrained coincidentally.

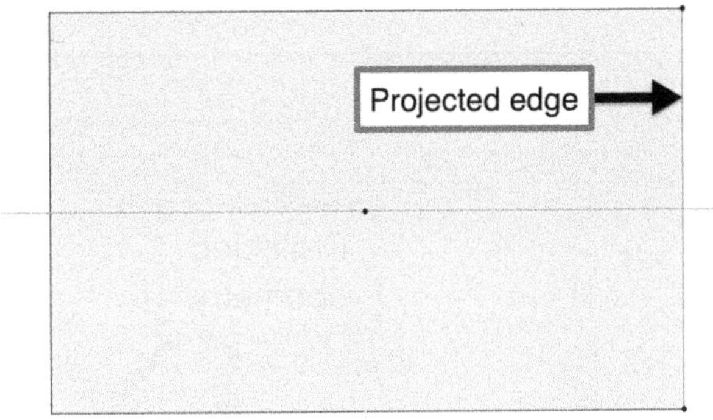

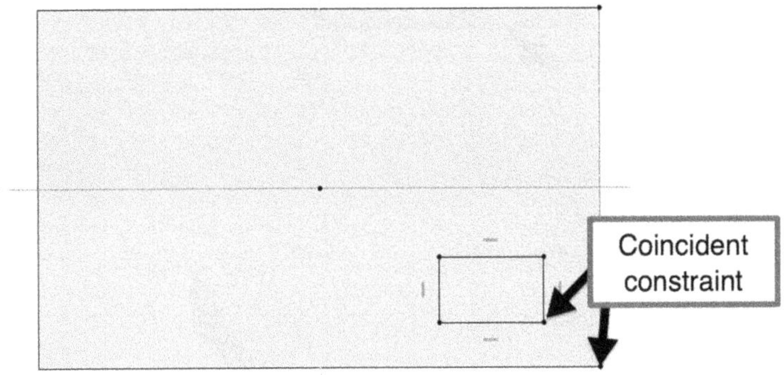

15. On the **Sketcher constraints** toolbar, click the **Constrain horizontal distance** icon.
16. Select the horizontal line and type-in **30** in the **Length** box on the **Insert Length** dialog, and then click **OK**.
17. Likewise, click the **Constrain vertical distance** icon on the **Sketcher constraints** toolbar.
18. Select the vertical line and type-in **25** in the **Length** box on the **Insert Length** dialog, and click **OK**.
19. Click the **Close** button on the **Combo View** panel.
20. Click the **Pocket** icon on the **Part Design Modeling** toolbar; the **Pocket parameters** section appears on the **Combo View** panel.
21. Type-in **30** in the **Length** box on the **Combo View** panel and click **OK** to create the *Pocket* feature.

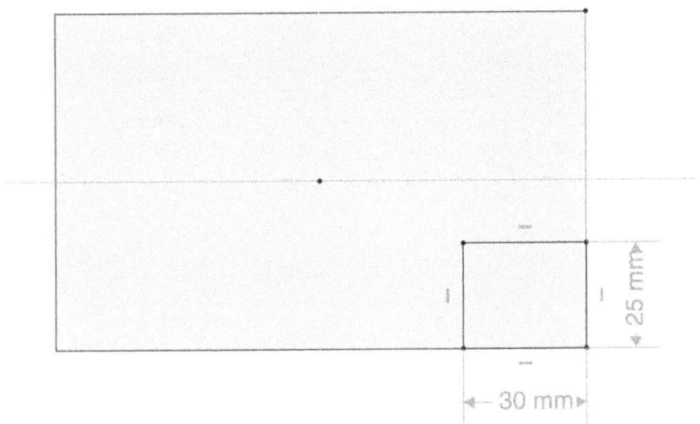

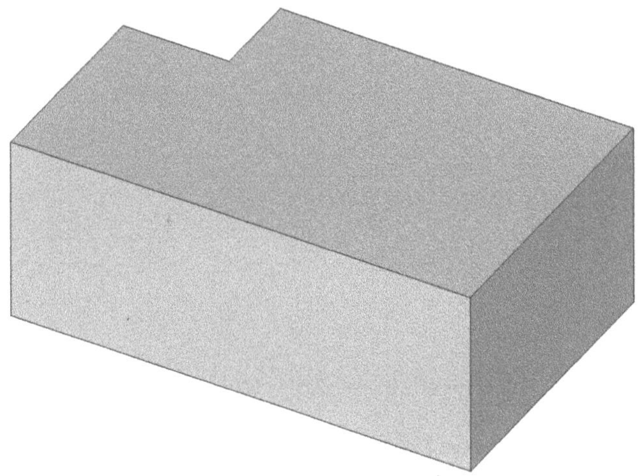

Creating the Hole features

1. Select the horizontal face of the *Pocket* feature and click the **Create Sketch** on the **Part Design Helper** toolbar.
2. On the **Sketcher geometries** toolbar, click the **External geometry** icon to project the edges.
3. Select the horizontal and vertical edges of the *Pocket* feature, as shown.

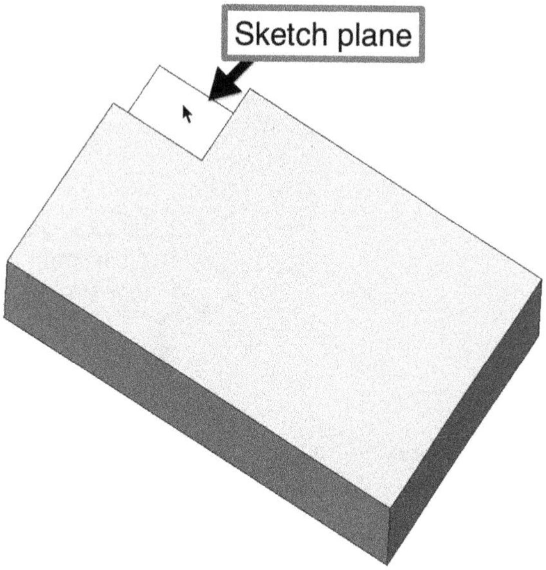

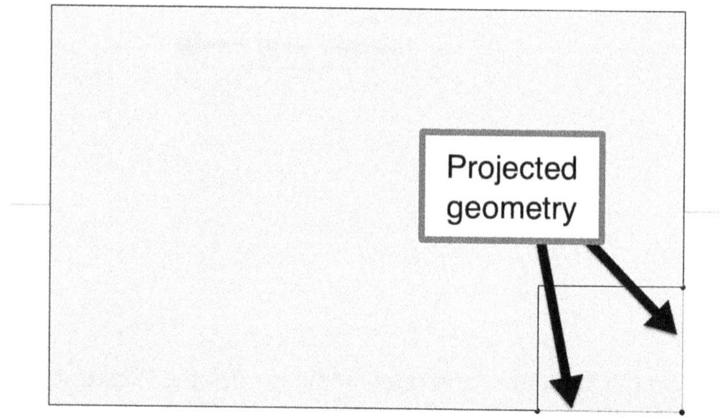

4. Click the **Create circle** icon on the **Sketcher geometries** toolbar.
5. Create a circle and horizontal and vertical distance constraints, as shown.

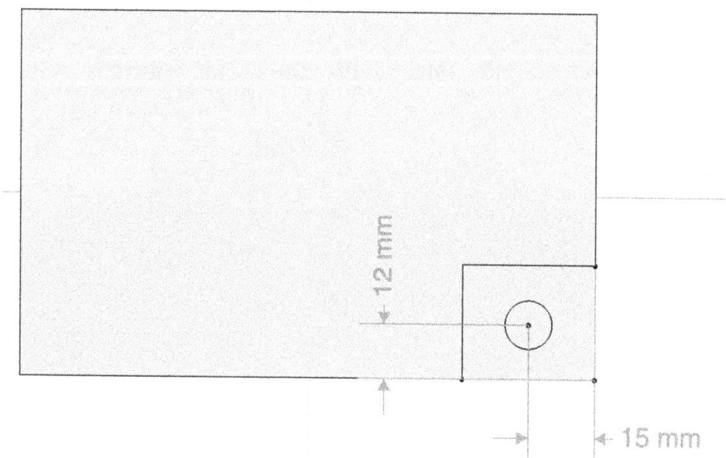

6. Click the **Close** button on the **Combo View** panel.
7. Click the **Hole**  tool on the **Part Design Modeling** toolbar; the sketch is selected automatically.
8. Type-in **8** in the **Diameter** box and select **Depth > Through All.**
9. On the **Hole Parameters** section, under the **Hole cut**, select **Type > Counterbore.**
10. Type-in **18** and **3** in the **Diameter** and **Depth** boxes, respectively.
11. Under the **Drill point**, select **Type > Flat.**
12. Click **OK** on the **Combo View** panel; the counterbore hole is created.

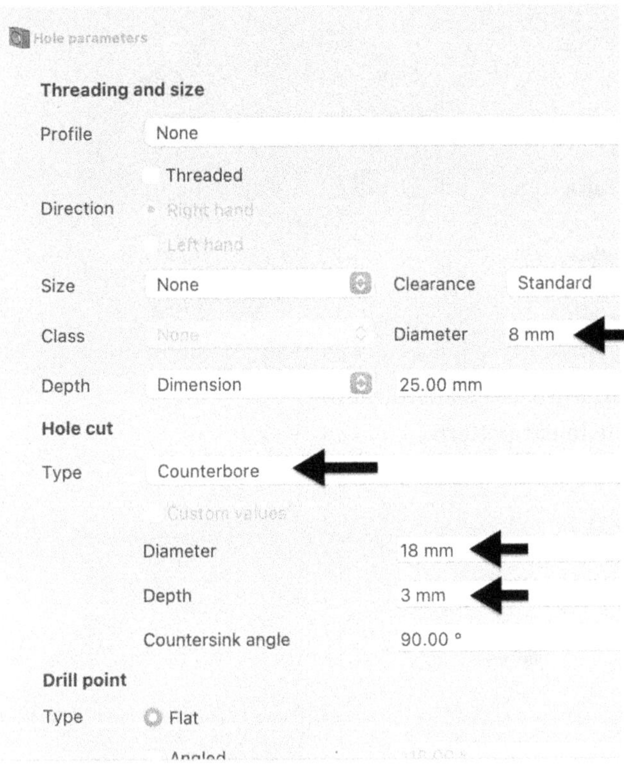

Creating the Multi transform feature

1. Click the **Create MultiTransform** icon on the **Part Design Modeling** toolbar.
2. Select the **Pocket** feature from the **Combo View** panel and click **OK**.
3. Click the **Add feature** button on the **MultiTransform parameters** section and select the cylindrical face of the counterbore hole.

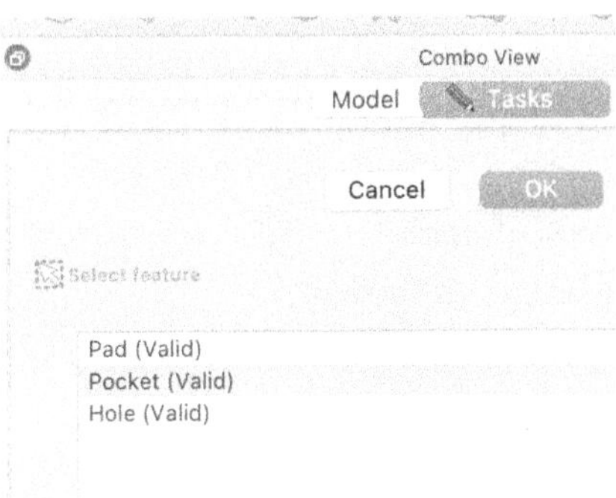

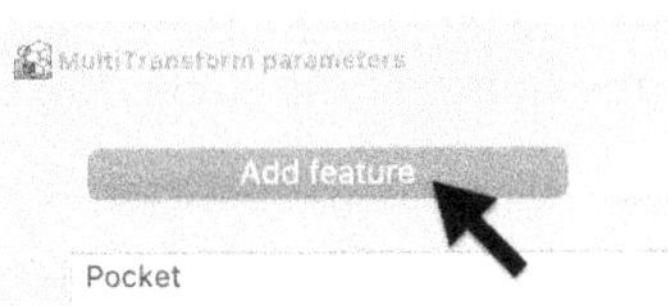

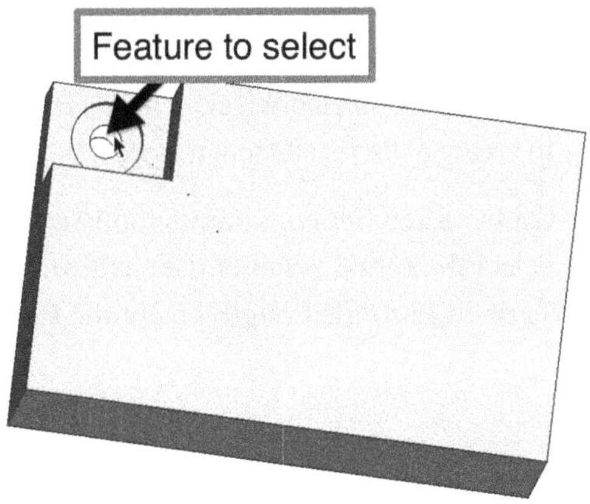

4. Right-click in the **Transformations** section and select **Add linear pattern**.
5. On the **MultiTransform parameters** section, select **Direction > Base X axis**.
6. Type **100** and **2** in the **Length** and **Occurrences** boxes, respectively.

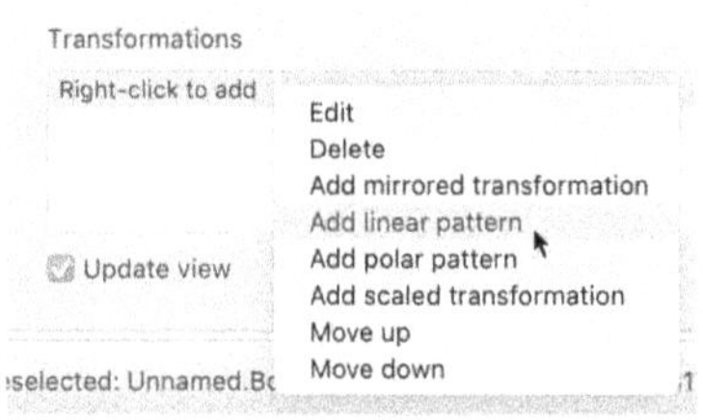

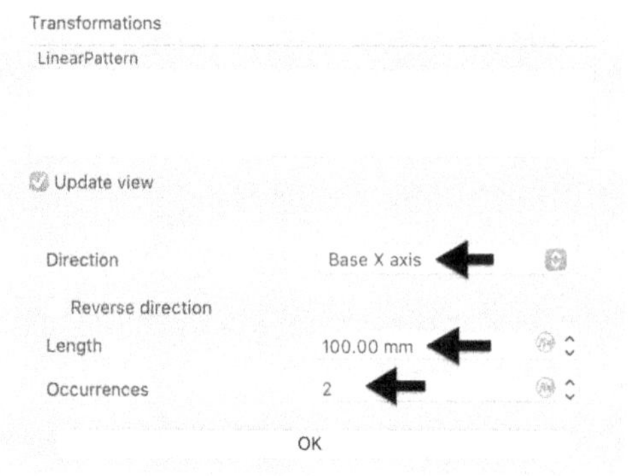

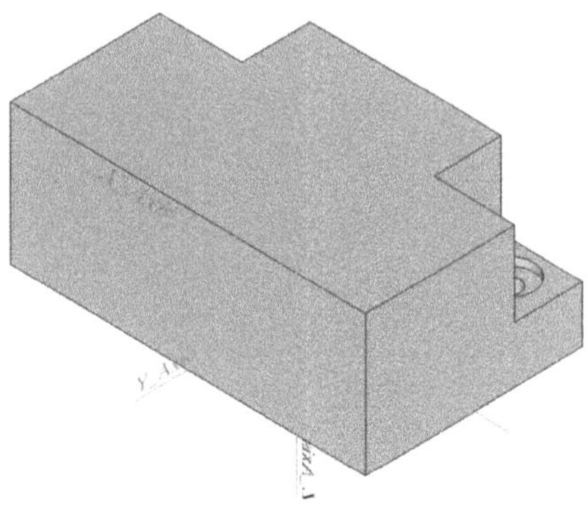

7. Right-click in the **Transformations** section and select **Add mirrored transformation**.
8. Select **Plane > Base XZ plane**. Next, click **OK** to create the linear pattern.

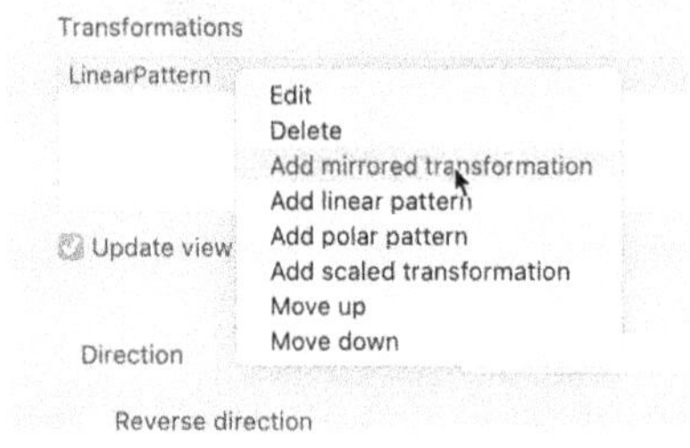

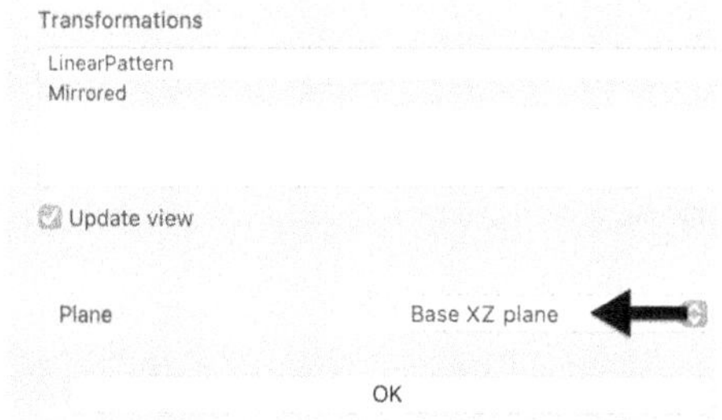

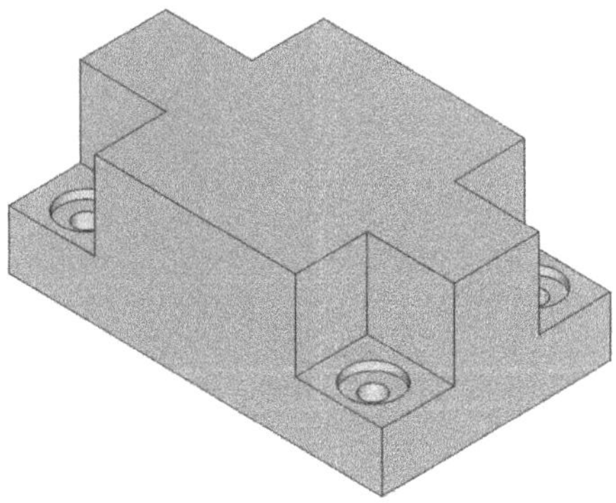

Creating the threaded Hole feature

1. Select the top face of the model and click the **Create Sketch** on the **Part Design Helper** toolbar.
2. Click the **External geometry** icon on the **Sketcher geometries** toolbar.
3. Select the left vertical edge of the part, as shown. Press **Esc** to deactivate the tool.
4. Click the **Create circle** icon on the **Sketcher geometries** toolbar.
5. Click on the horizontal sketch axis to define the center of the circle. Next, move the pointer outward and click to create a circle, as shown.
6. On the **Sketcher constraints** toolbar, click the **Constrain horizontal distance** tool.
7. Select the center point of the circle and the endpoint of the projected vertical edge, as shown.
8. Type-in **15** in the **Length** box of the **Insert Length** dialog and click **OK.**

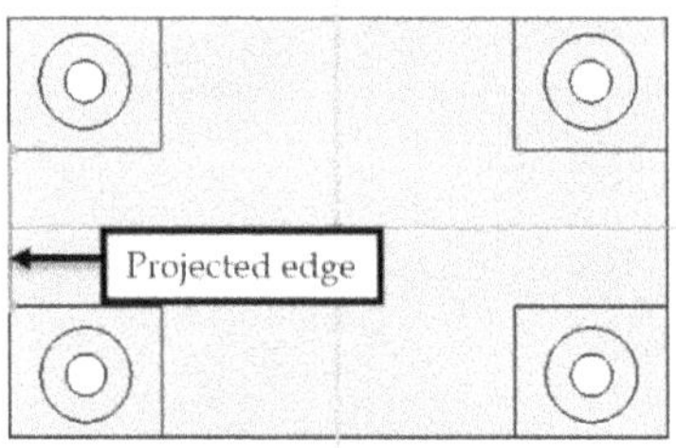

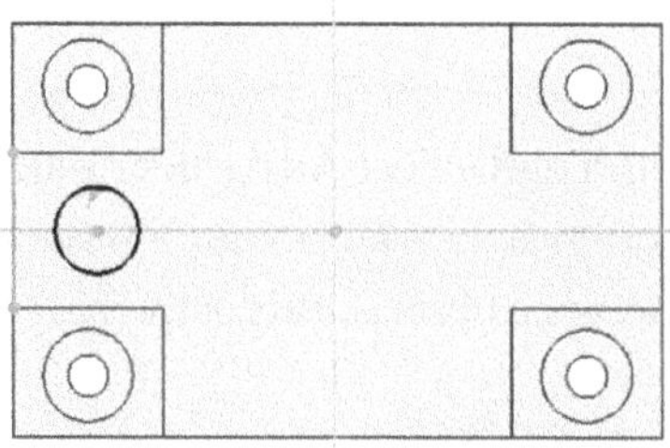

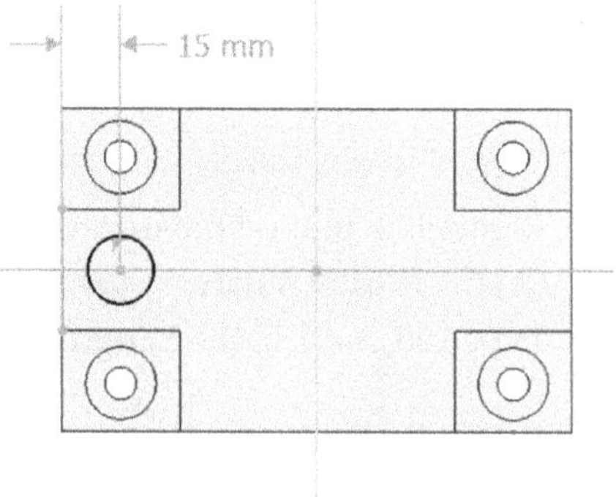

9. Click **Close** on the **Combo View** panel.

10. Click the **Hole** icon on the **Part Design Modeling** toolbar; the **Hole parameters** section appears on the screen.
11. Under the **Threading and size**, select **Profile > ISO metric regular profile.**
12. Check the **Threaded** option, and then select **Direction > Right hand**.
13. Select **Size > M12** and **Class > 4G**.
14. Select **Depth > Dimension** and type-in **20** in the dimension box.
15. Under the **Drill point** section, select the **Type > Angled** option. Next, type **118** in the angle box.
16. Click **OK** on the **Combo View** panel.

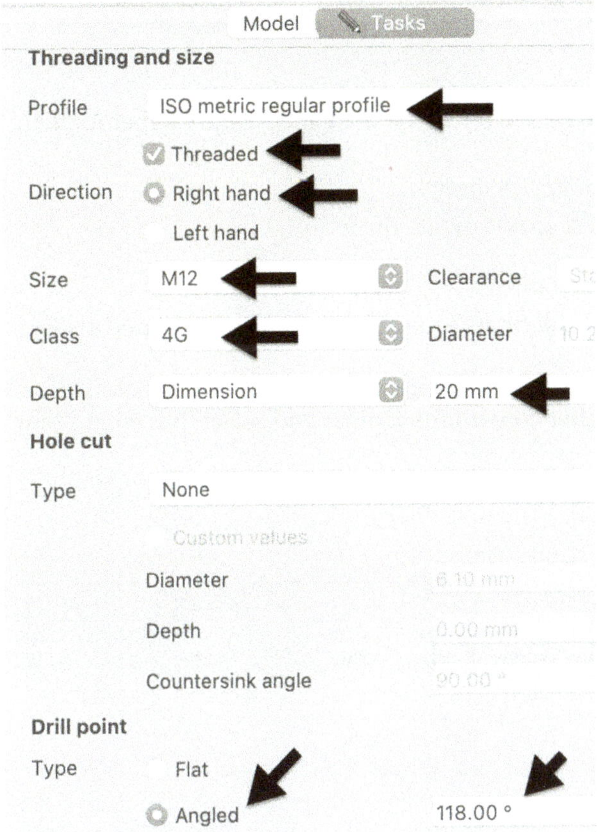

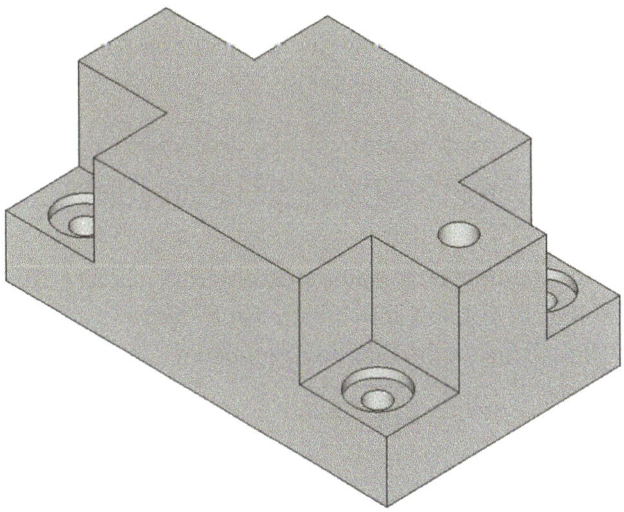

Mirroring the Features

1. On the **Part Design Modeling** toolbar, click the **Mirrored** icon.
2. Under the **Select feature** on the **Combo View**, select **Hole001** and click **OK**.
3. On the **Combo View** panel, select **Plane> Vertical sketch axis** option under the **Mirrored parameters** section.
4. Click **OK** to mirror the *Hole* feature.

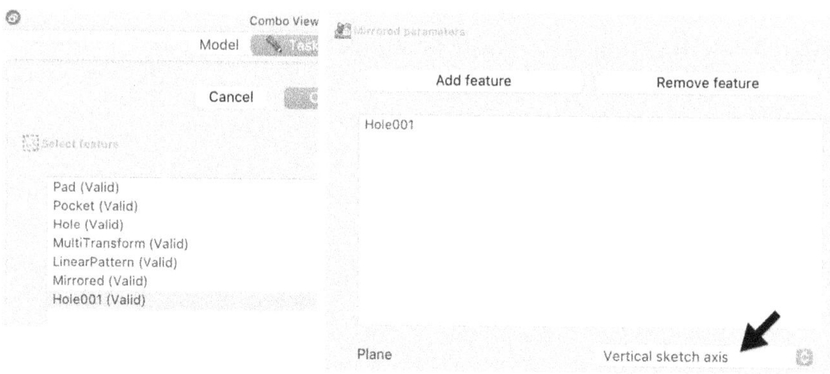

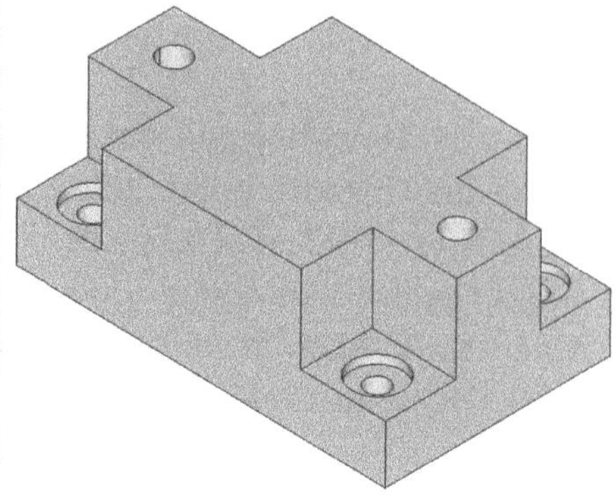

Creating the Hole and Pocket features

1. Select the front face of the part geometry and click the **Create Sketch** icon on the **Part Design Helper** toolbar.
2. Click the **External geometry** icon on the **Sketcher geometries** toolbar.
3. Select the top horizontal edge of the part, as shown. Press **Esc** to deactivate the tool.
4. Click the **Create circle** icon on the **Sketcher geometries** toolbar.
5. Click on the vertical sketch axis to define the center point of the circle. Move the pointer outward and click to create a circle.
6. Click the **Constrain point onto object** icon on the **Sketcher constraints** toolbar and select the center point of the circle. Next, select the projected edge.
7. Click **Close** on the **Combo View** panel.

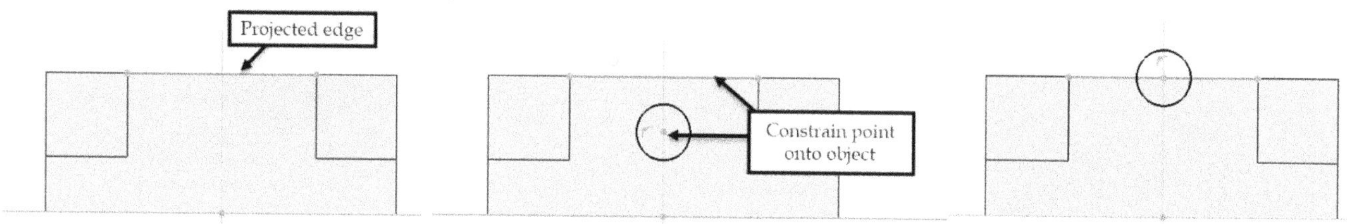

8. Click the **Hole** icon on the **Part Design Modeling** toolbar.
9. On the **Hole parameters** section, select **Profile > None.**
10. Type 40 in the **Diameter** box. Next, select the **Depth > Through all.**
11. In the **Hole cut** section, select **Type > Counterbore.**
12. Type-in **50** and **15** in the **Diameter** and **Depth** boxes, respectively.
13. Click **OK** to create the counterbore hole.

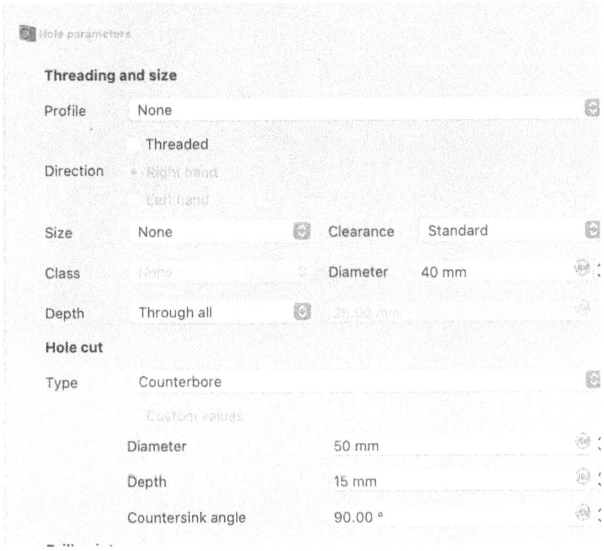

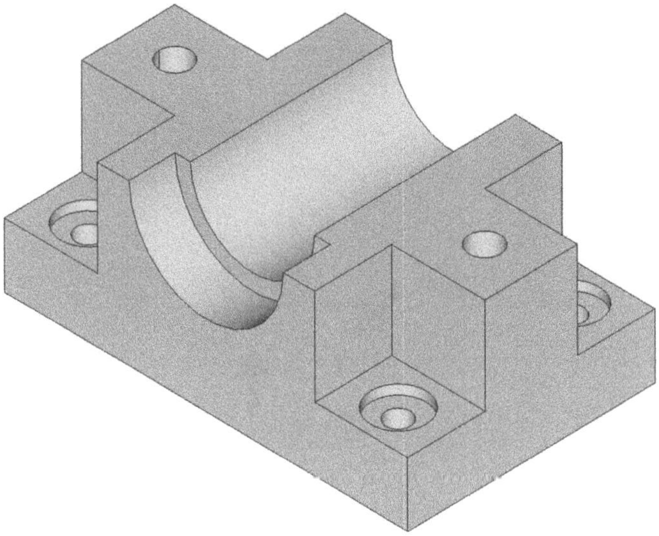

14. Select the front face of the part geometry and click the **Create Sketch** icon on the **Part Design Helper** toolbar.
15. Create a closed sketch using the **Polyline** command. Next, add vertical and horizontal distance constraints to it.
16. Click the **Constrain symmetrical** icon on the **Sketcher constraints** toolbar and select an endpoint of the lower horizontal line. Next, select the other endpoint of the lower horizontal line.
17. Select the sketch origin; the two endpoints of the lower horizontal line are made symmetric about the sketch origin.
18. Click the **Constrain equal** icon on the **Sketcher constraints** toolbar. Next, select the two inclined lines, as shown.

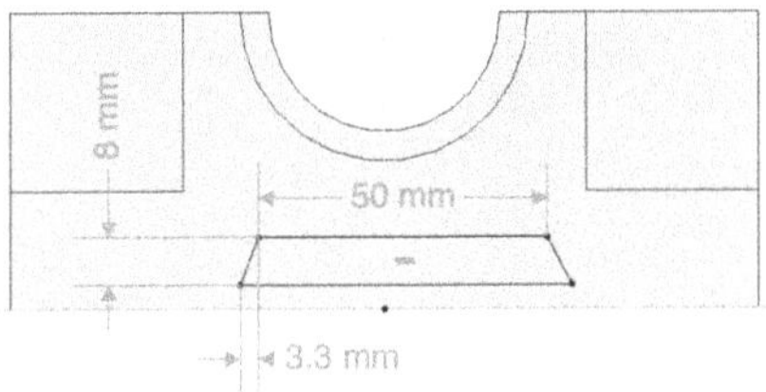

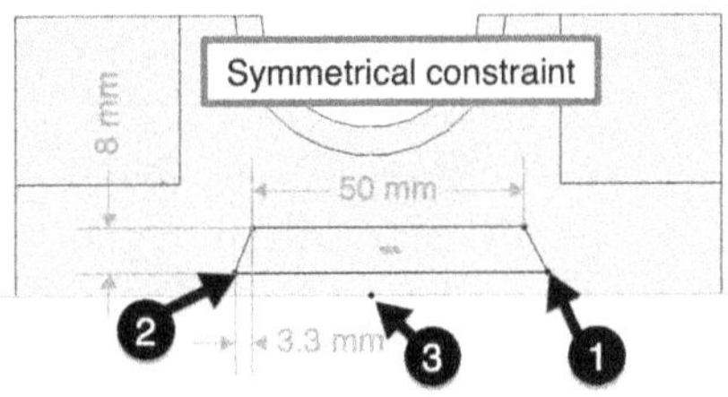

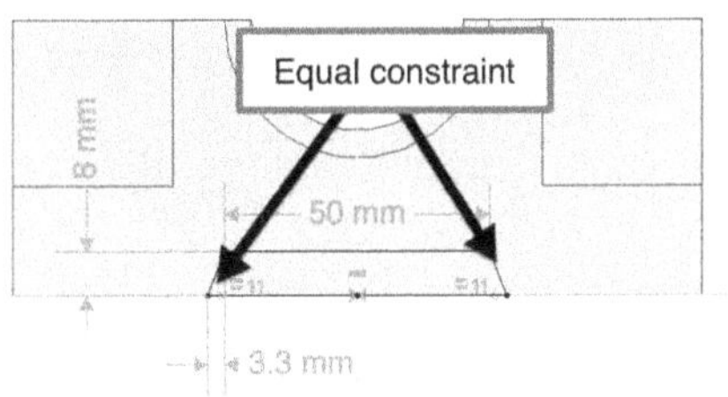

19. Click **Close** on the **Combo View** panel.
20. Click the **Pocket** icon on the **Part Design Modeling** toolbar.
21. Select **Type > Through all** from the **Pocket parameters** section. Next, click **OK**.
22. Press and hold the Ctrl key and select the internal edges of the pocket feature, as shown.
23. Press and hold the middle and right mouse buttons, and then drag the cursor.
24. Press and hold the Ctrl key and select the internal edges of the remaining pocket features.

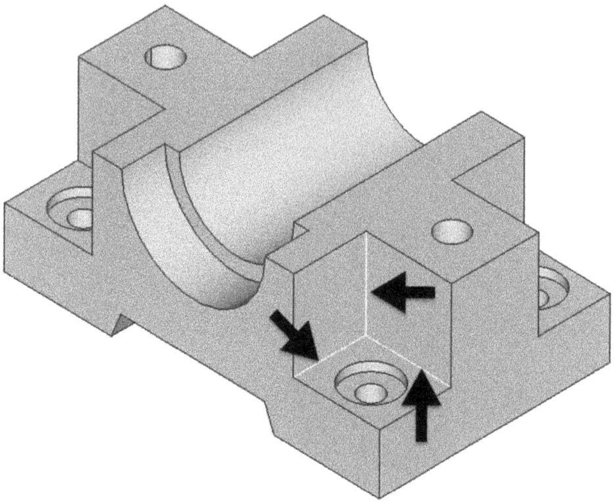

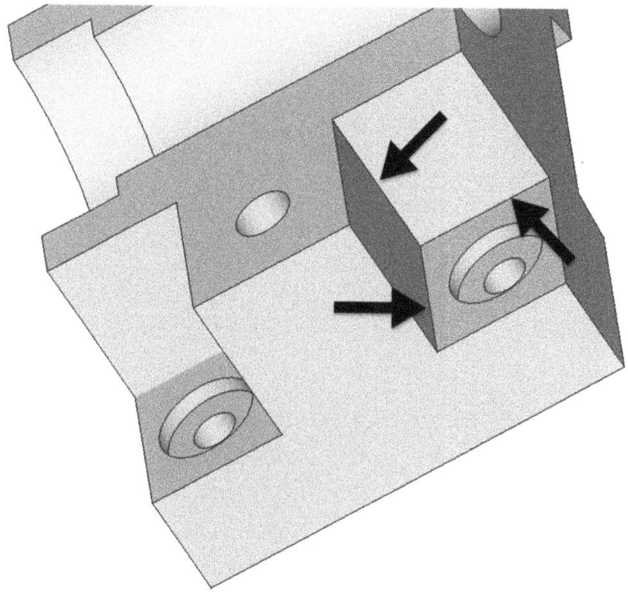

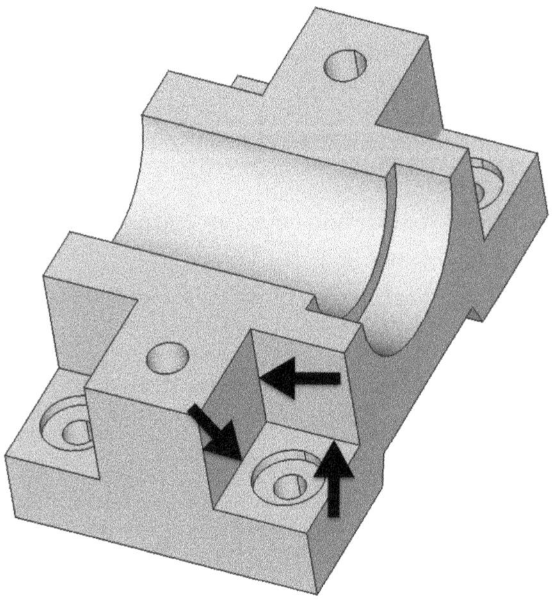

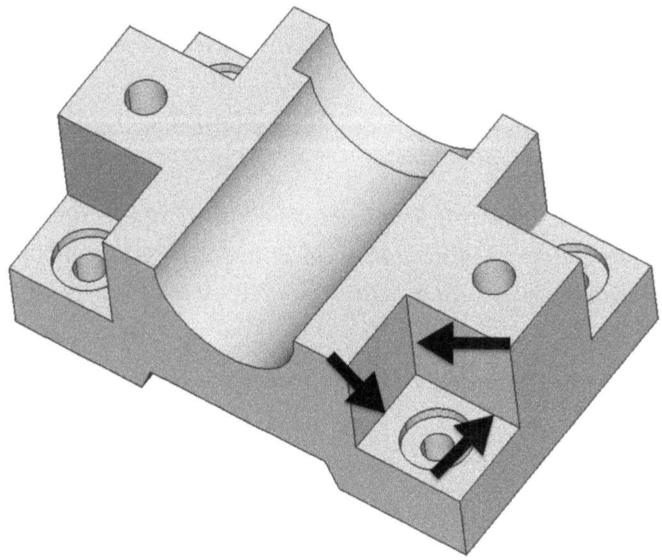

25. Click the **Fillet** icon on the **Part Design Modeling** toolbar. Next, type 2 in the **Radius** box and click **OK** to create the fillets.

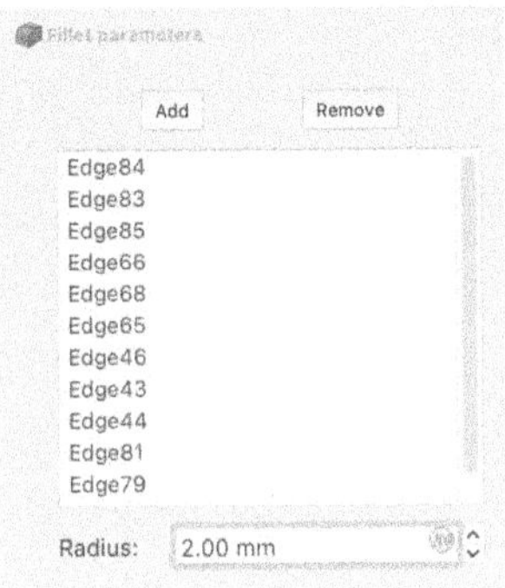

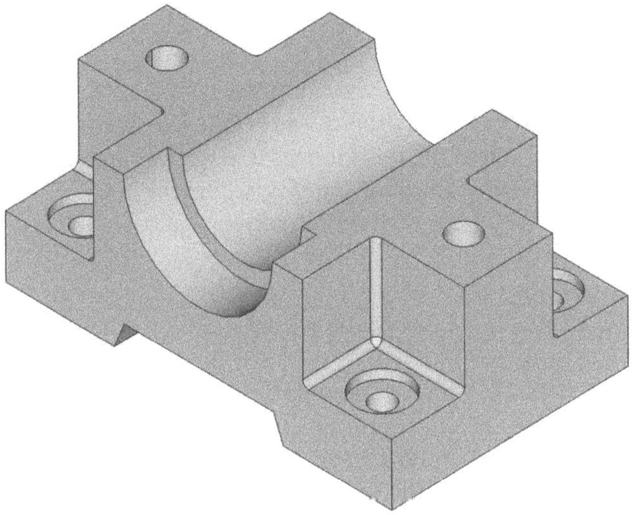

26. Click the **Save** icon on the **File** toolbar located at the top-left corner of the window. Next, type C5_exampl1 in the **File name** box, and then click **Save**.
27. Click **File > Close** on the menu bar to close the document.

Tutorial 2 (Millimetres)

In this example, you create the part shown next.

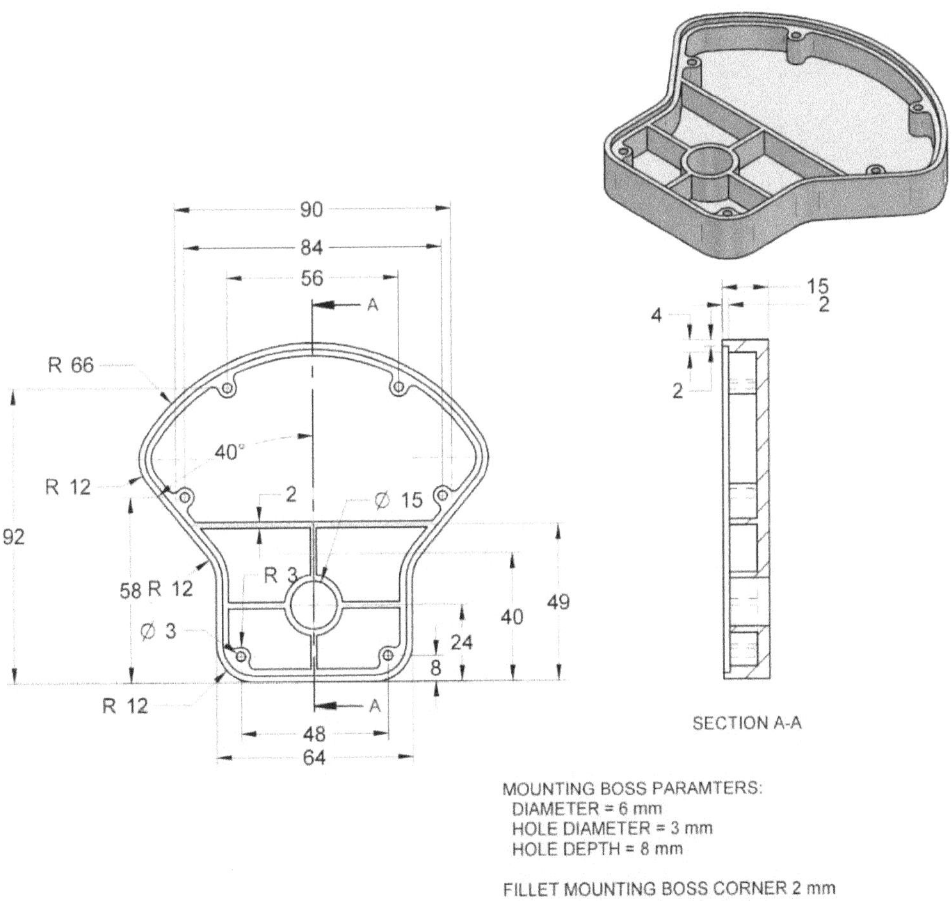

Creating a New document

1. Click **FreeCAD 0.19** on the desktop.
2. On the Menu bar, click **File > New**; it creates a new document.
3. On the **Workbench** toolbar, select **Workbench** drop-down **> Part Design**.
4. Click **Edit > Preferences** on the **Menu** bar; the **Preferences** dialog appears on the screen.
5. Click **Units** tab and select **Unit System > Standard (mm/kg/s/degree)**.
6. Select **Number of decimals > 2** and click **OK** on the **Preferences** dialog.

Creating the Padded Feature

1. Click the **Create Sketch** command on the **Part Design Helper** toolbar and start a new sketch on the **XY** plane.
2. On the **Sketcher geometries** toolbar, click the **Create polyline** command and draw the sketch, as shown in the figure below.
3. On the **Sketcher geometries** toolbar, click the **Arc** drop-down **> Create arc by three points** command, and then create an arc by specifying the points in the sequence, as shown.

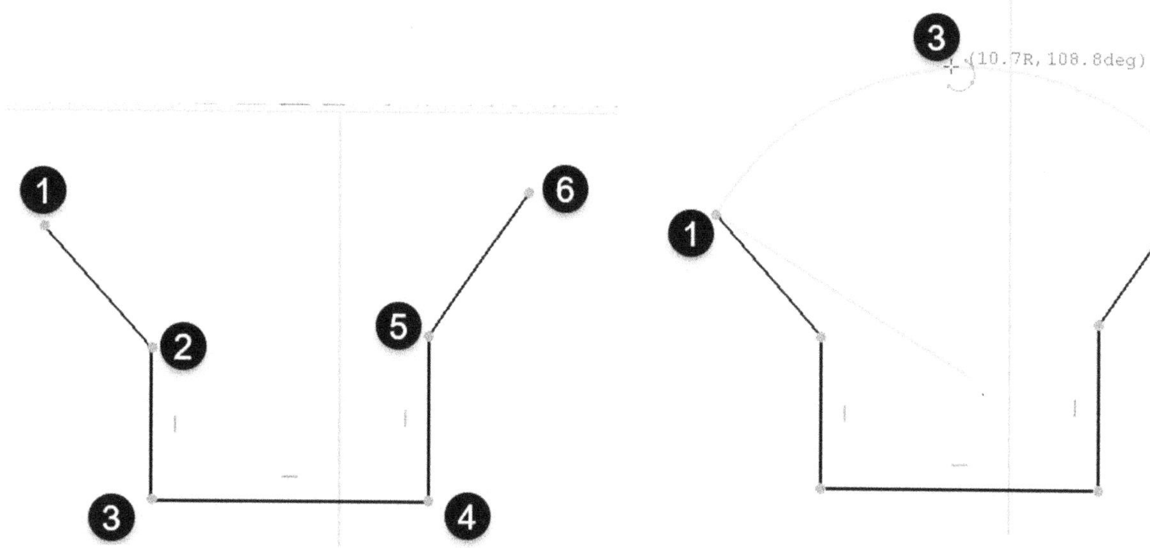

4. Click the **Sketch Fillet** command on the **Sketcher geometries** toolbar, and then select the sharp corners to fillet.
5. Select the arc and the inclined line connected to it; the fillet is created at the corner.
6. Likewise, select the arc and the inclined line connected to it at the other end.

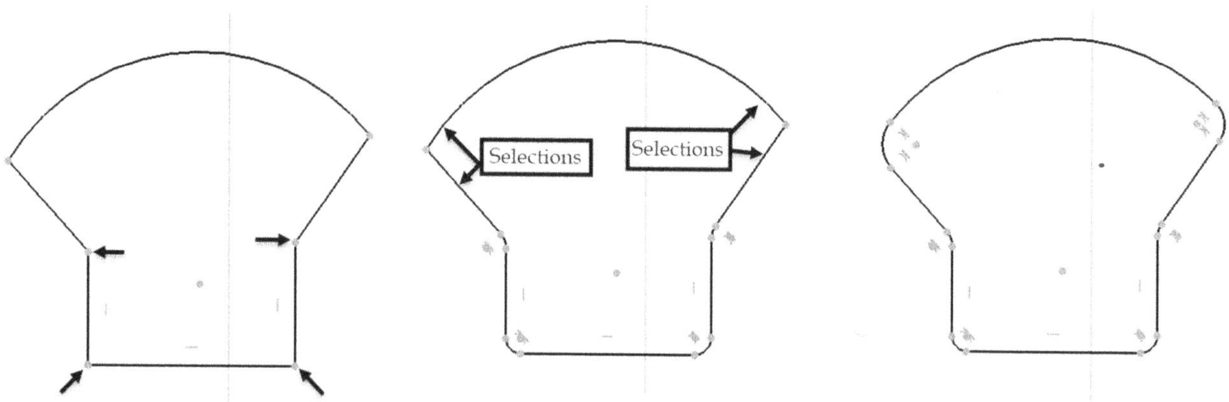

7. Click the **Constrain equal** icon on the **Sketcher constraints** toolbar. Next, apply the **Equal** constraint between the fillets, as shown.
8. Apply the **Equal** constraint between the fillets, as shown.

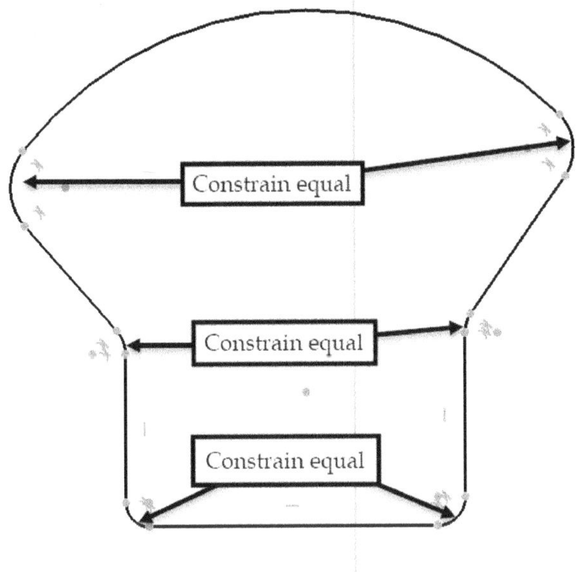

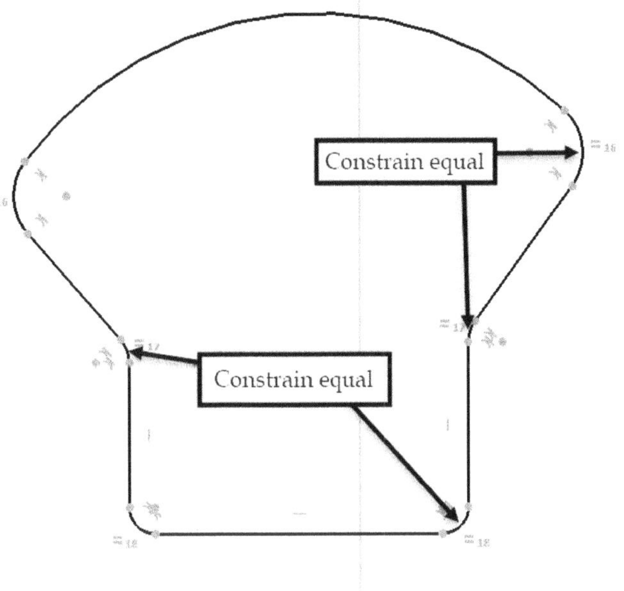

9. Apply the **Equal** constraint between the two vertical lines.
10. Apply the **Equal** constraint between the two inclined lines. Next, right click in the graphics window.
11. Select the centerpoints of the two fillets, as shown. Next, click the **Constrain horizontal** icon on the **Sketcher constraints** toolbar.

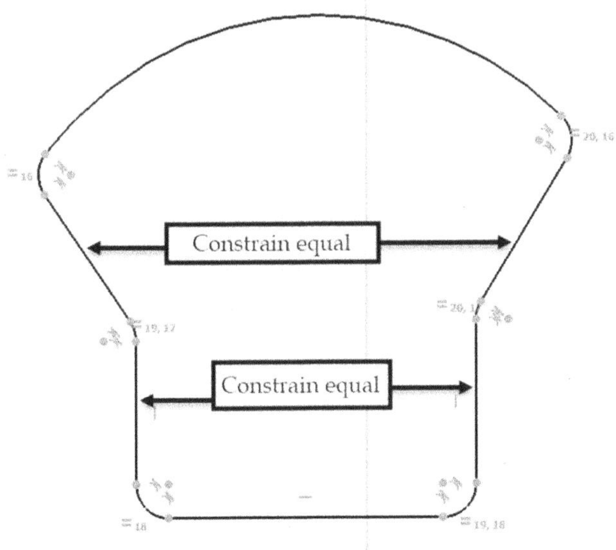

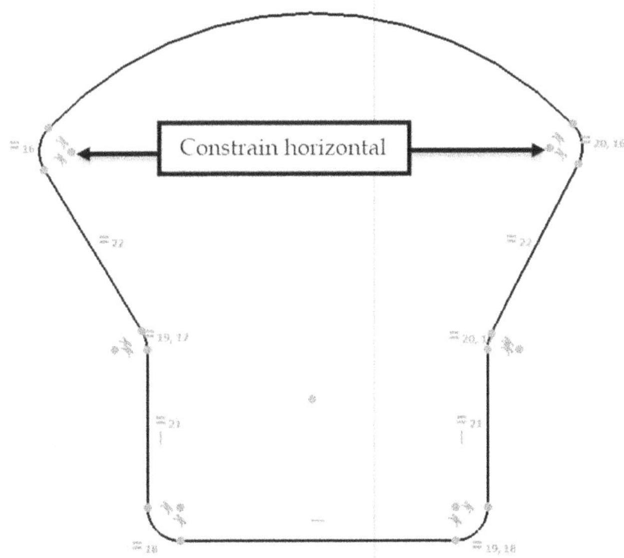

12. Click the **Constrain symmetrical** icon on the **Sketcher constraints** toolbar and select the points in the sequence, as shown.

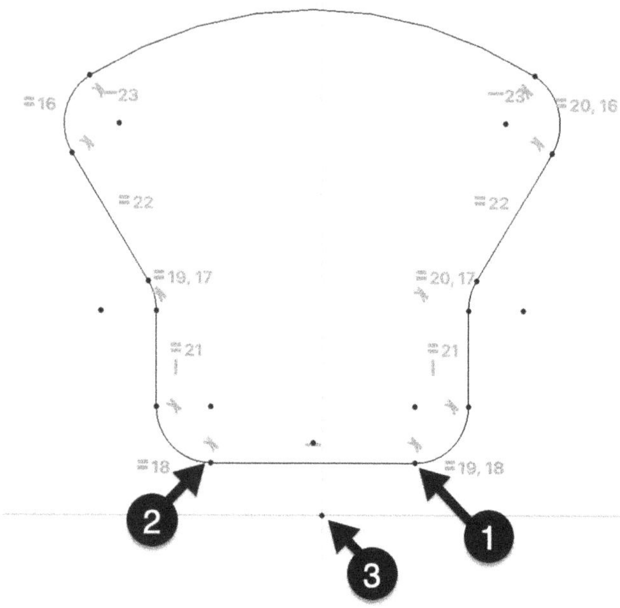

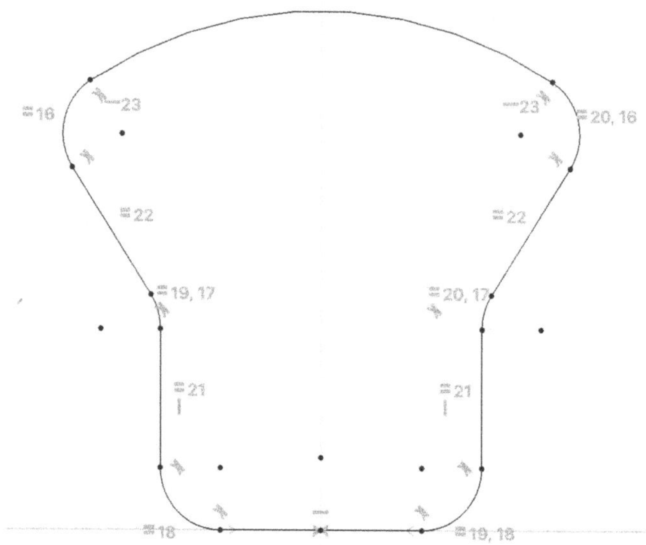

13. Click the **Constrain angle** icon on the **Sketcher constraints** toolbar. Next, select the left inclined and vertical lines, as shown. Type **140** in the **Angle** box and click **OK**.
14. Apply the dimensional constraints to the sketch in the sequence, as shown.

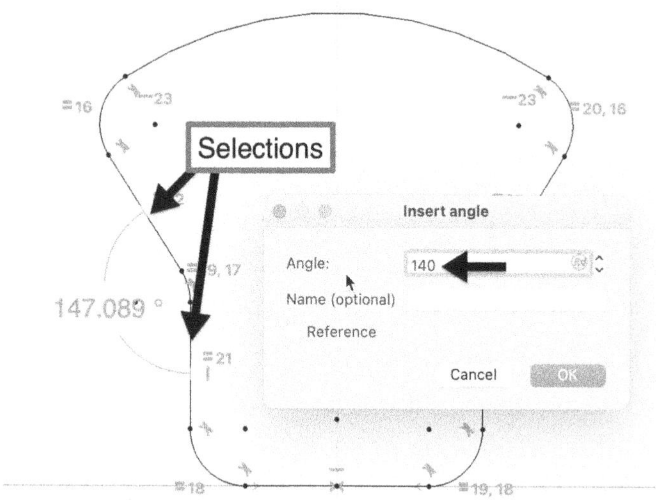

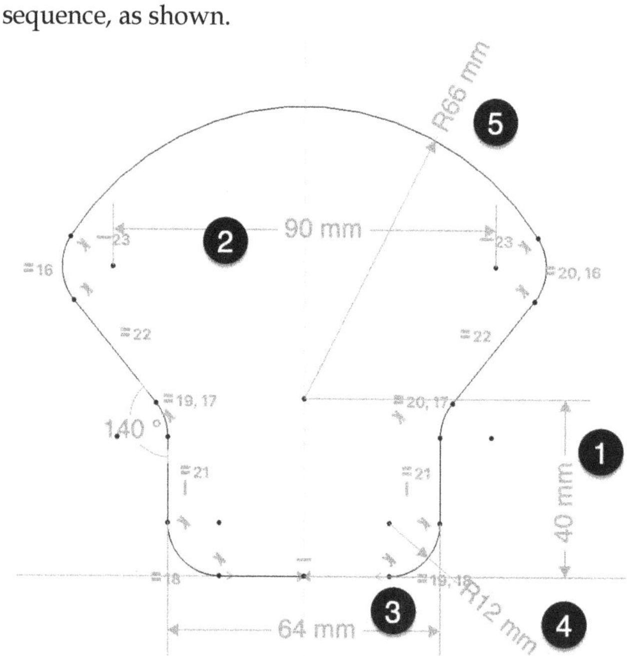

15. Click the **Close** button on the **Combo View** panel.
16. Activate the **Pad** command. Next, type-in **14** in the **Length** box under the **Pad parameters** section of the **Combo View** panel.
17. Click **OK** to create the *Pad* feature.
18. Select the top face of the pad feature and click the **Thickness** icon on the **Part Design Modeling** toolbar.
19. Type-in **4** in the **Thickness** box on the **Thickness parameters** section.

20. Check the **Make thickness inwards** option and click **OK**.

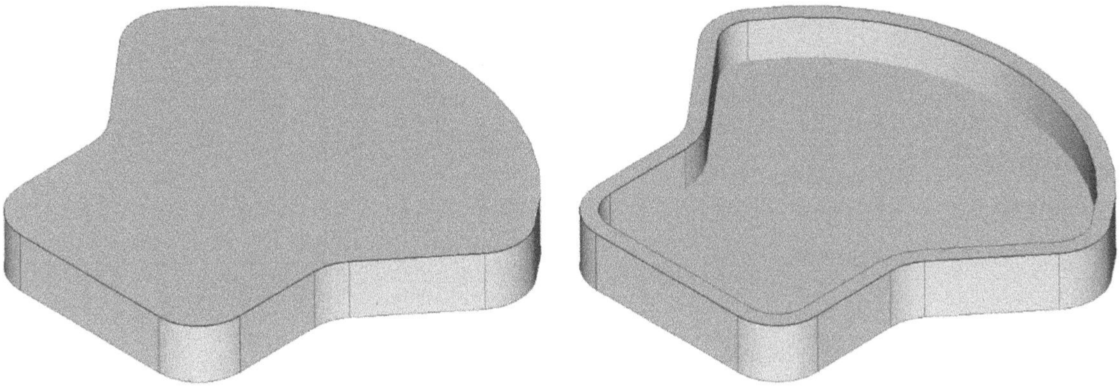

Adding a Lip to the model

1. In the **Combo View** panel, expand the **Origin** node and select the YZ Plane.
2. Click the **Create Sketch**  icon on the **Part Design Helper** toolbar.
3. Click the **View Section**  icon on the **Part Design Helper** toolbar.
4. Click **External geometry**  on the **Sketcher geometries** toolbar, and then select the internal vertical edge.
5. Click the **Create rectangle** icon on the **Sketcher geometries** toolbar.
6. Select the top endpoint of the projected edge. Next, move the pointer toward the bottom left corner and click.
7. Add Vertical distance and Horizontal distance constraints to the rectangle, as shown.

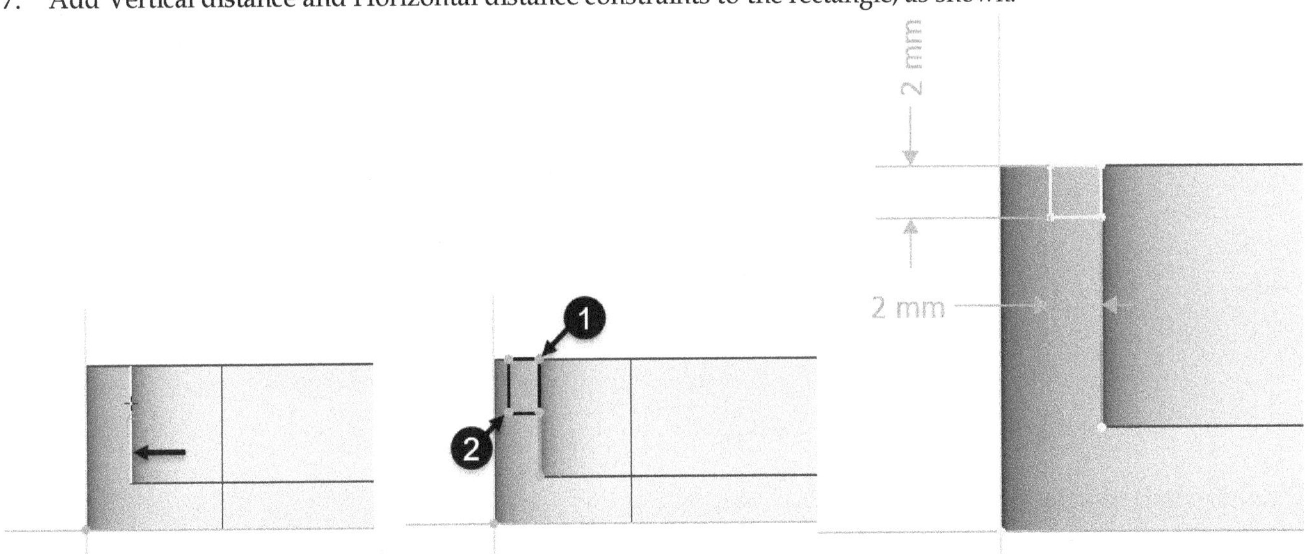

8. Click **Close** on the **Combo View** panel.
9. In the **Combo View** panel, expand the **Pad** feature and select the **Sketch**.
10. In the **Properties** panel, scroll down and click the drop-down next to the **Visibility** property. Next, select **true** from the drop-down; the sketch used for the Pad feature is visible in the graphics window.

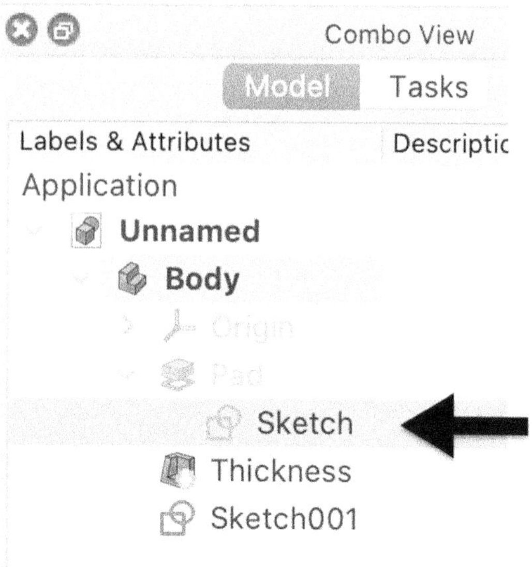

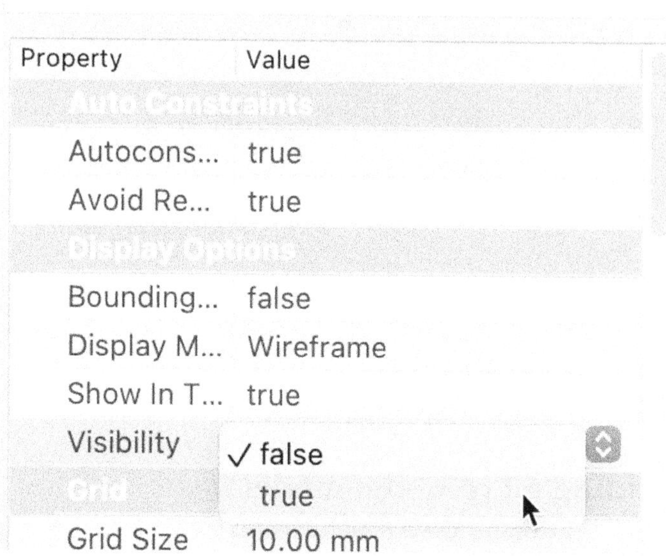

11. Click in the graphics area to deselect the sketch of the Pad feature.
12. Click the **Subtractive Pipe** icon on the **Part Design Modeling** toolbar; the sketch created on the YZ plane is selected as the profile.
13. Press and hold the middle and right mouse buttons. Next, drag the cursor such that the bottom face of the model is displayed.
14. On the **Pipe parameters** section, click the **Object** button under **Pipe to sweep along** section.
15. Select the sketch used for the *Pad* feature. Next, click **OK** to create the subtractive pipe feature.
16. Click the **Set to isometric view** icon on the **View** toolbar to change the view to isometric.

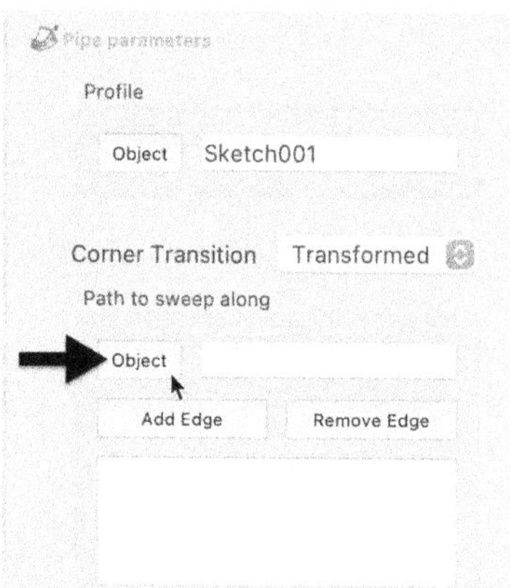

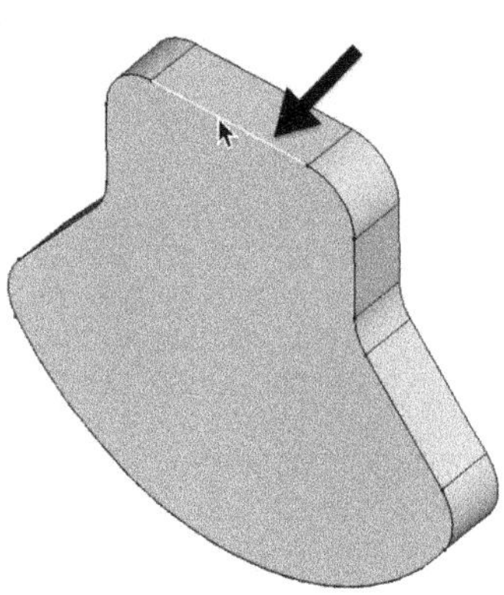

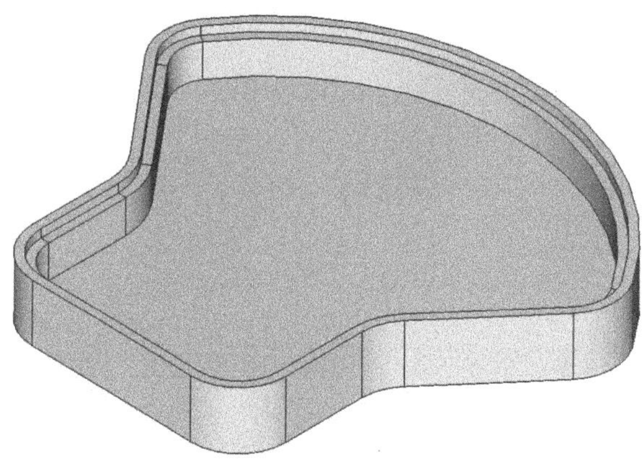

Creating Bosses

1. Click on the horizontal face created by the subtractive pipe feature. Next, click the **Create sketch** icon on the **Part Design Helper** toolbar.
2. Draw the three circles on the sketch plane. Next, click **Constrain equal** on the **Sketcher constraints** toolbar.
3. Select the first two circles to make them equal in diameter. Next, select the second and third circles to make them equal.

4. Add dimensional constraints to them, as shown.

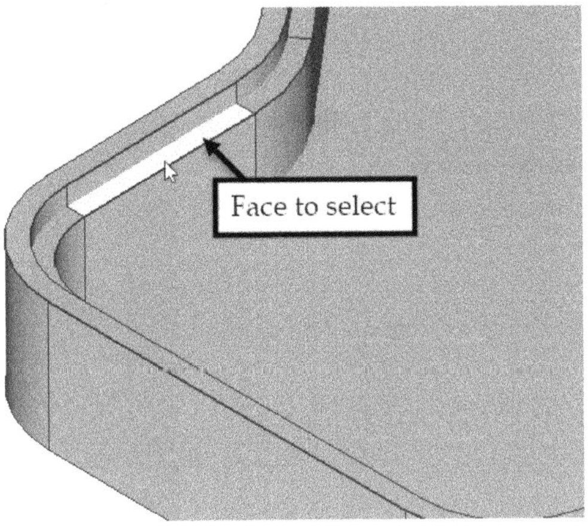

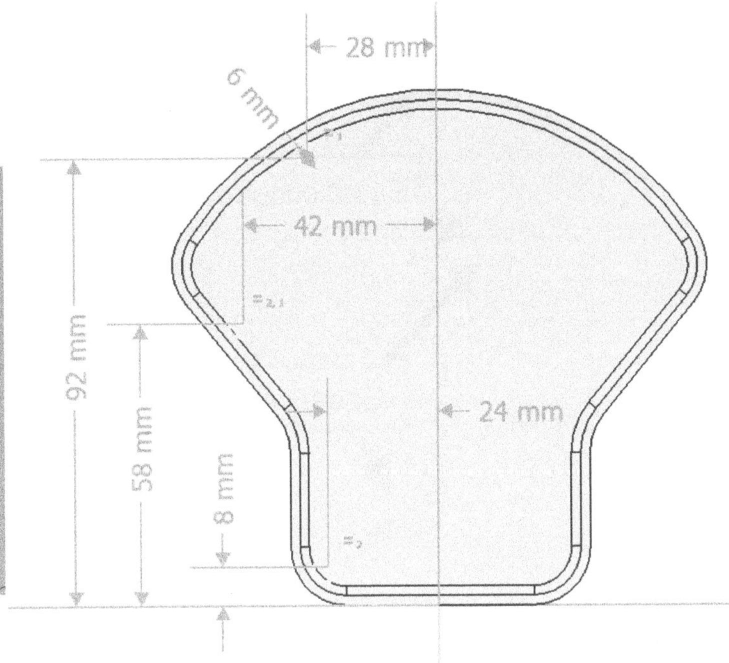

5. Click the **Create circle** icon on the **Sketcher geometries** toolbar. Next, select any one of the centerpoints of the circles. Move the pointer outward and click to create a circle.
6. Likewise, create two more circles concentric to the other two circles.

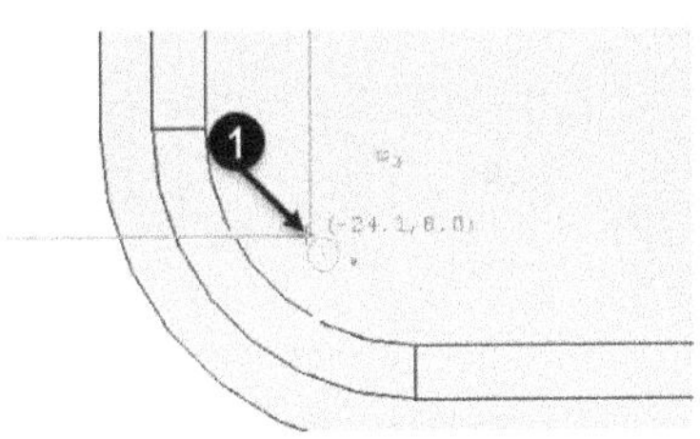

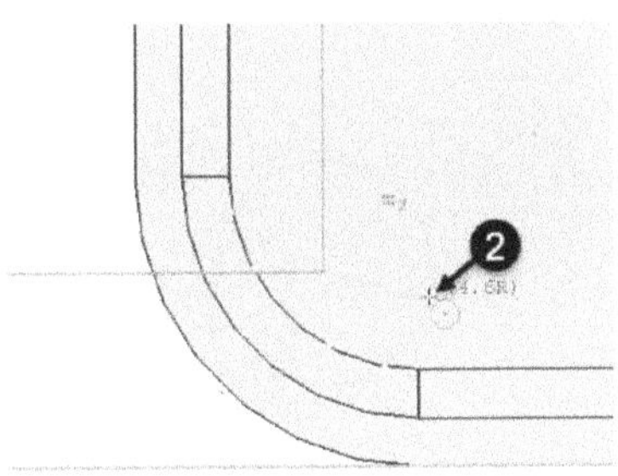

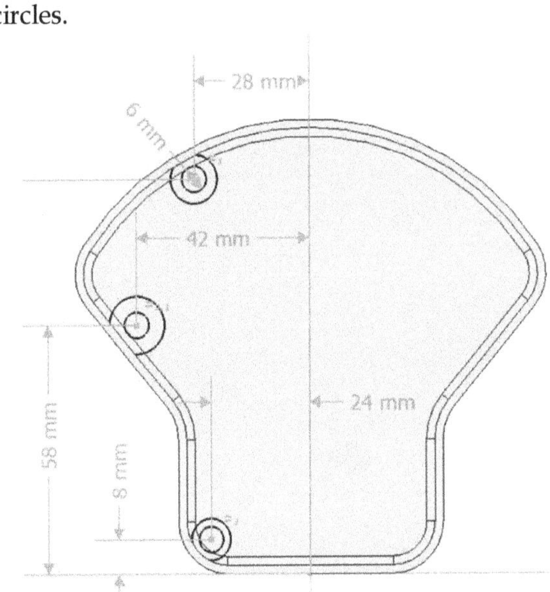

7. Click the **Constrain equal** icon on the **Sketcher constraints** toolbar and select two new circles; the two circles are made equal in diameter.
8. Likewise, make the third new circle equal to the second new circle.
9. Click the **Constrain diameter** icon on the **Sketcher constraints** toolbar and select any one of the new circles.
10. Type 3 in the **Diameter** box and click **OK**.

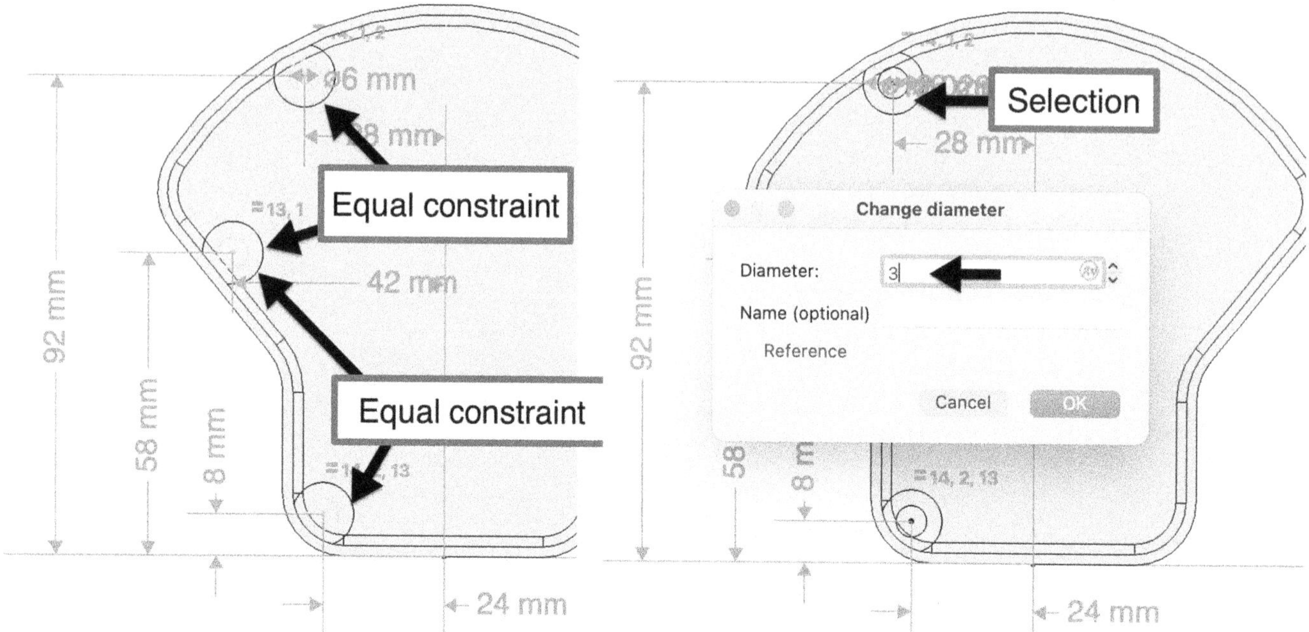

11. Click the **Close** button on the **Combo View** panel.
12. On the **Part Design Modeling** toolbar, click the **Pad** icon.
13. On the **Pad parameters** section, enter **8** in the **Length** box. Next, check the **Reversed** option and click **OK**.

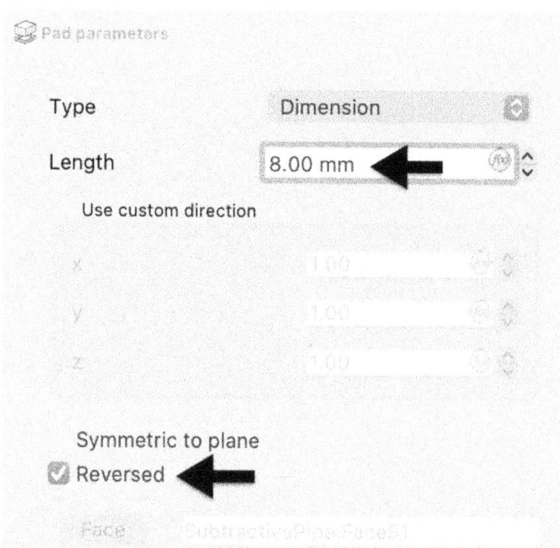

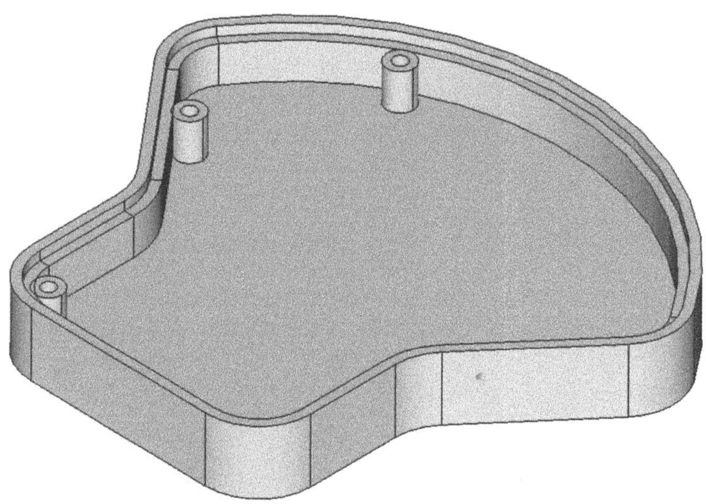

14. Click the Model tab on the **Combo View** panel and click the Pad001 feature.
15. On the toolbar, click the **Mirror** command.
16. Select **Base YZ plane** from the **Plane** drop-down, and click **OK**.

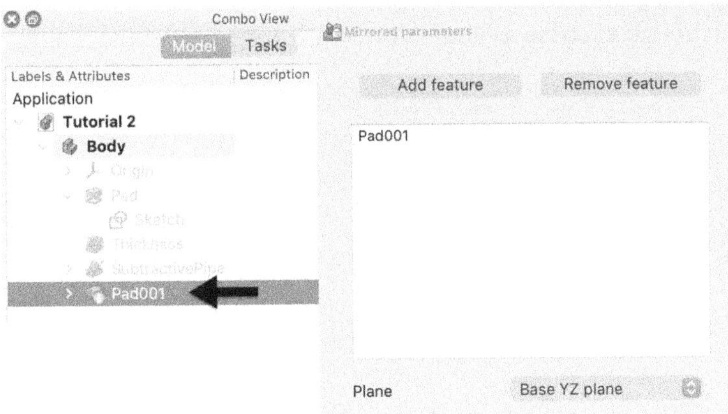

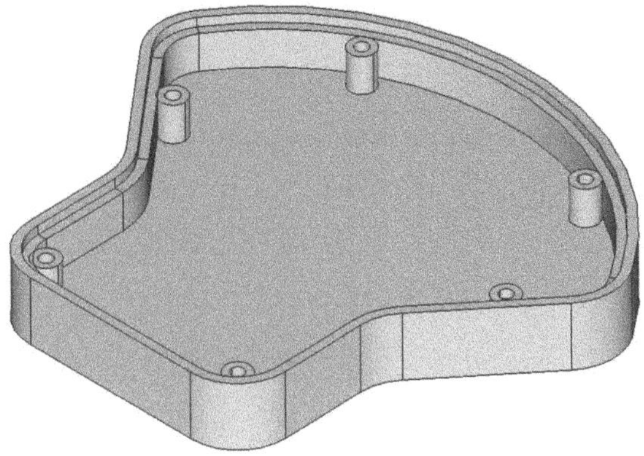

17. Click **View > Draw Style > Wireframe** on the Menu bar.
18. Zoom-in to anyone of the pad features. Next, press and hold the Ctrl key and select the edges where the pad feature meets the wall of the geometry.
19. Click the **Create fillet** icon on the **Part Design Modeling** toolbar. Next, type 1 in the **Radius** box and click **OK**.

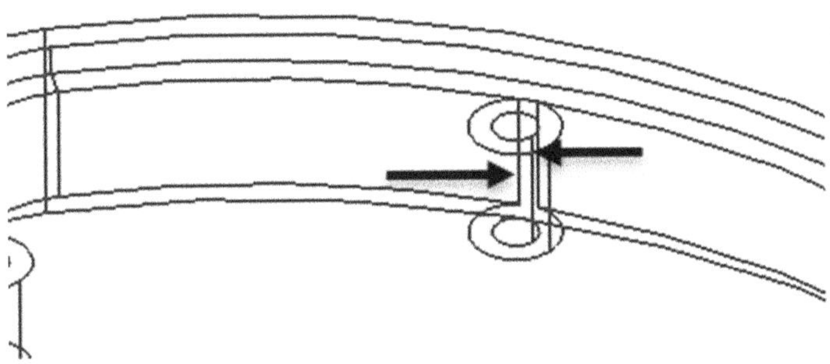

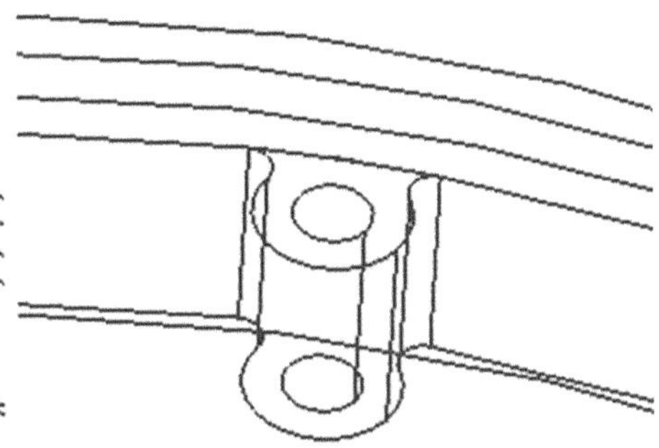

20. Likewise, create fillets for the remaining pad features.
21. Click **View > Draw style > Flat lines** on the menu bar.

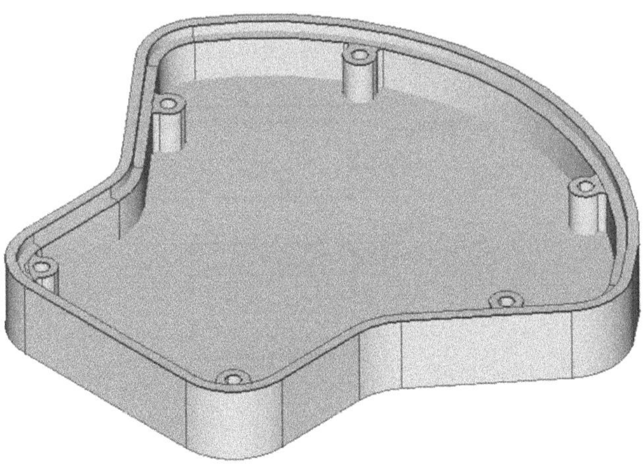

Creating Ribs

1. Select the bottom face of the shell feature and click **Create sketch** on the **Part Design Helper** toolbar.

2. Click the **External geometry** icon on the **Sketcher geometries** toolbar. Next, select the edges of the model, as shown,

3. Click the **Center and rim point** command on the **Sketcher geometries** toolbar and create two circles, as shown.

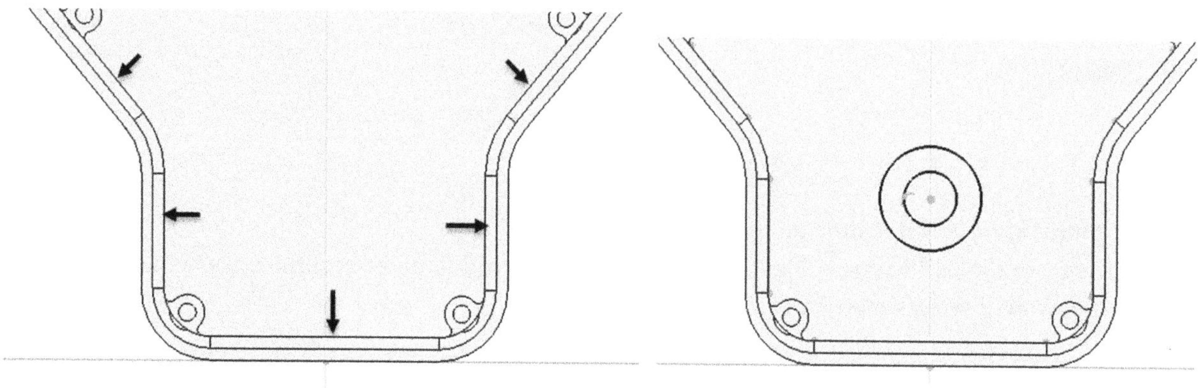

4. Click the **Polyline** command on the **Sketcher geometries** toolbar and create the lines, as shown.

5. On the **Sketcher constraints** toolbar, click **Constrain point onto object** . Next, constrain the endpoints of the lines onto the projected edges, as shown.

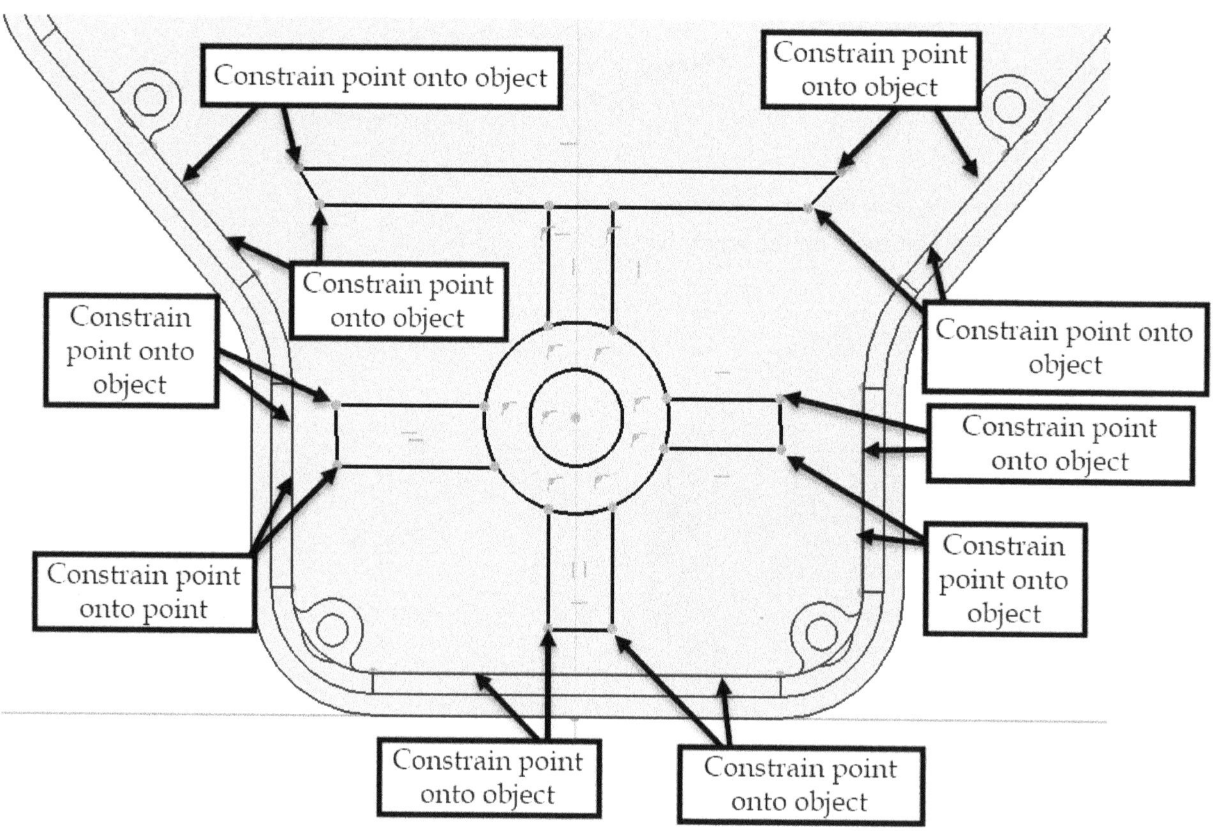

6. Click the **Trim edge** command on the **Sketcher geometries** toolbar and trim the portions of the sketch, as shown.
7. Create the diameter, radius, vertical distance, and horizontal distance constraints, as shown.

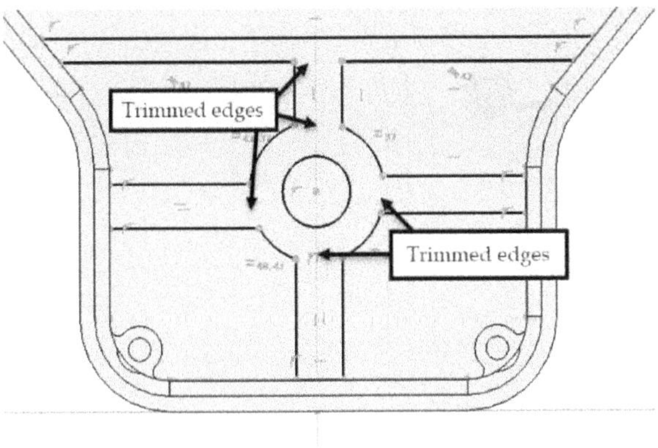

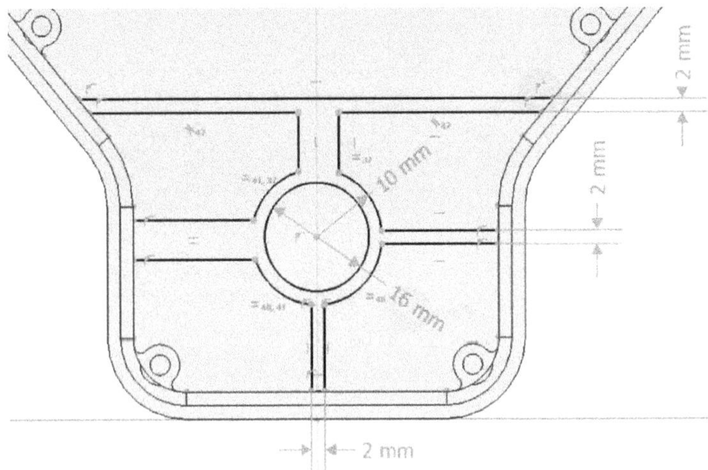

8. Apply the horizontal and vertical constraints between the endpoints of the lines, as shown.

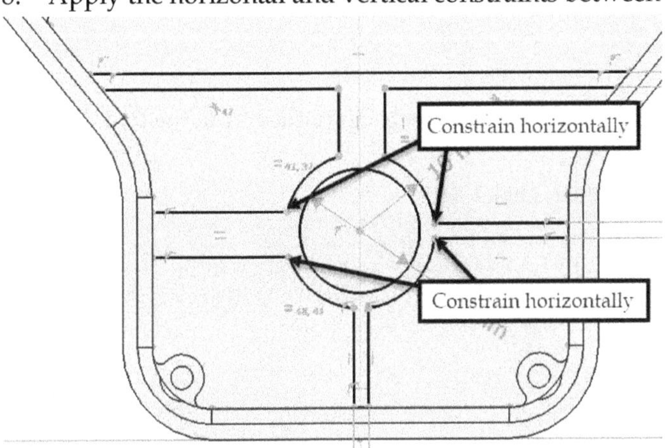

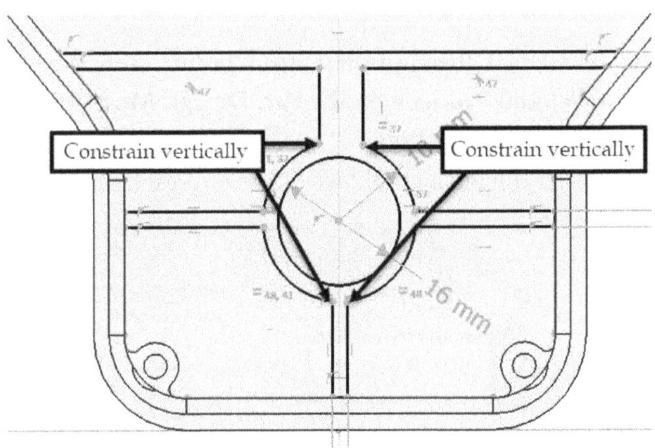

9. Create vertical distance and horizontal distance constraints, as shown.

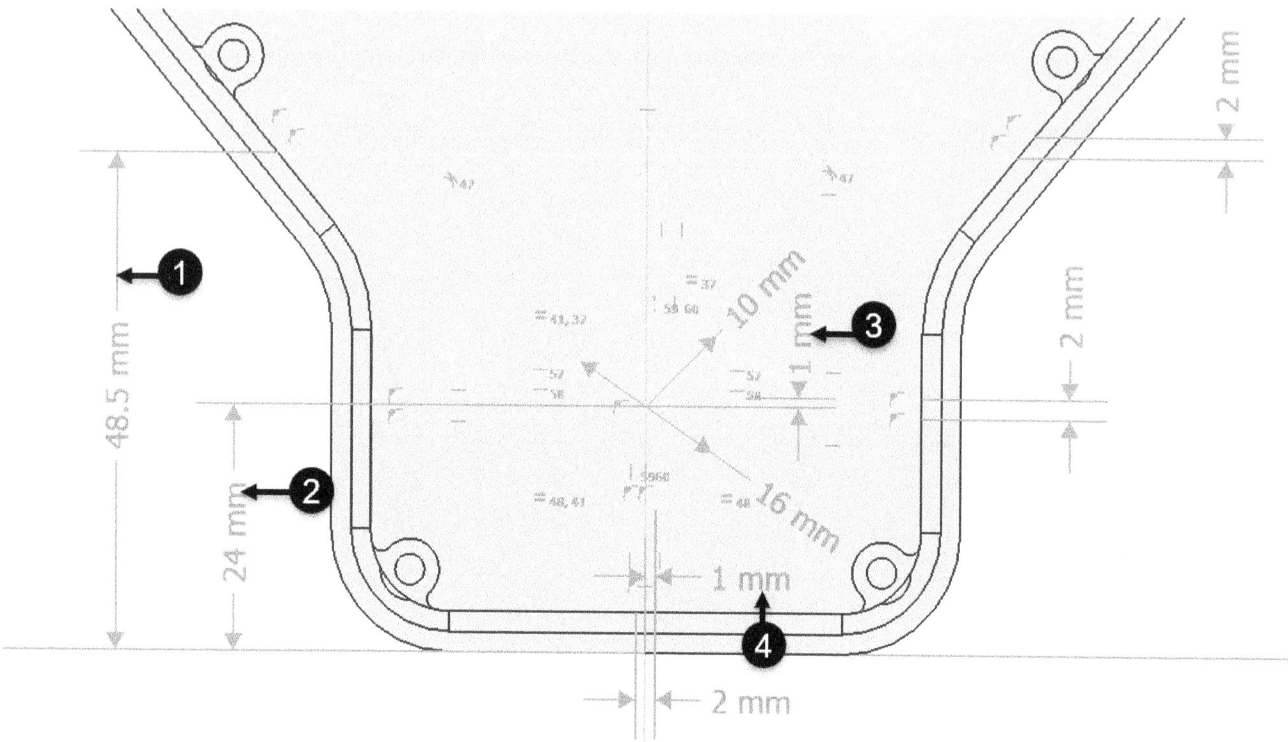

10. Click the **Close** button on the **Combo View** panel.
11. Click the **Pad** icon on the **Part Design Modeling** toolbar. Next, select **Type > Up to face** from the **Pad parameters** section.
12. Select the horizontal face of the Subtractive pipe feature. Next, click **OK**.

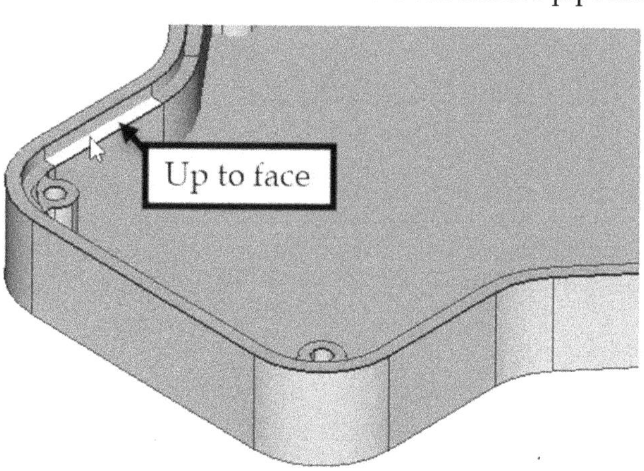

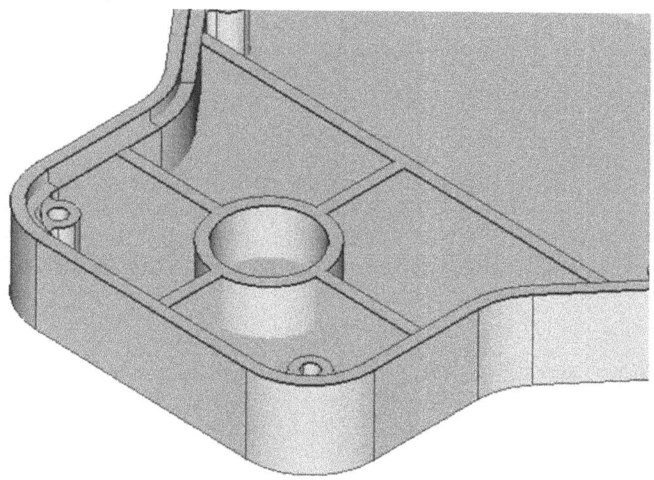

13. Click **File > Save** on the menu bar. Next, type **C7_example1** in the **File name** box, and click **Save**.
14. Click **File > Close** on the menu bar.

Questions

1. Describe the procedure to create a mirrored feature.
2. List any two pattern types.
3. Describe the procedure to create a multi transform feature.

4. How to define spacing in a linear pattern.

Exercises

Exercise 1

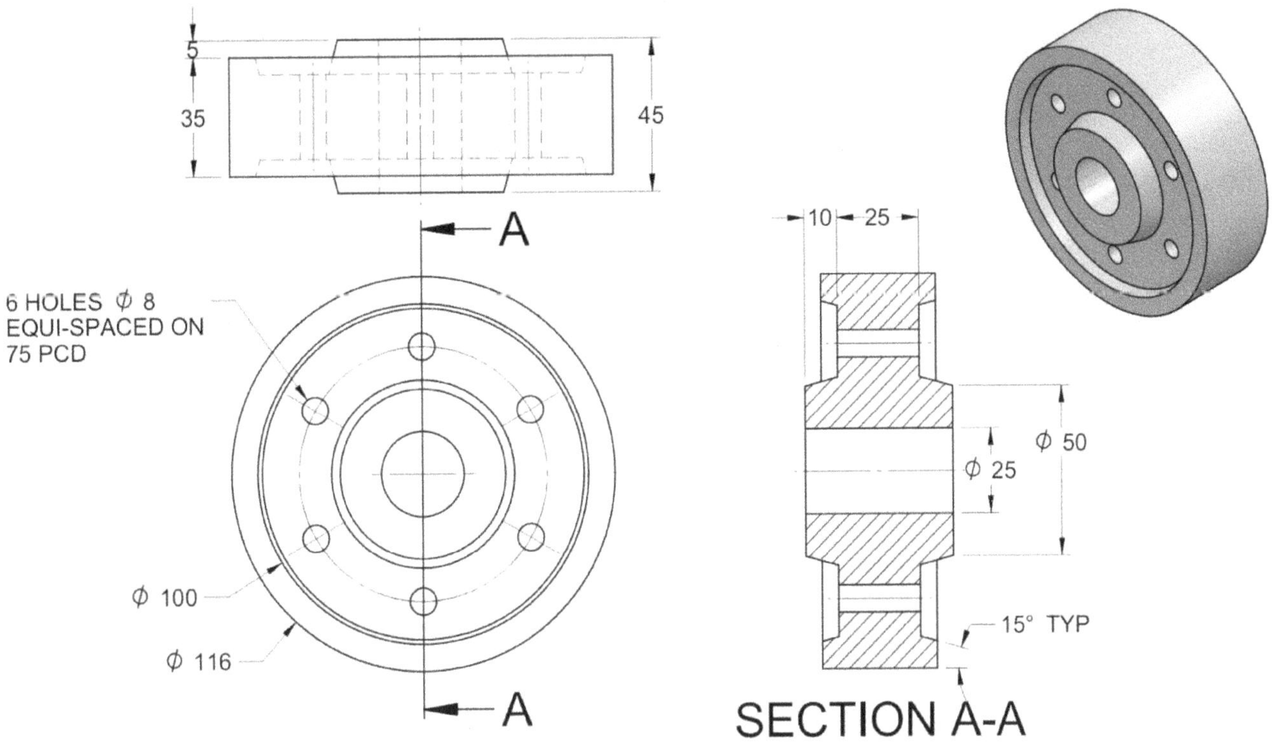

Chapter 6: Sweep Features

The **Additive Pipe** command is one of the basic commands available in FreeCAD 0.19 that allow you to generate solid geometry. It can be used to create simple geometry as well as complex shapes. A sweep is composed of two items: a profile and a path. The profile controls the shape of the sweep while the path controls its direction. For example, take a look at the angled cylinder shown in the figure. This is created using a simple sweep with the circle as the profile and an angled line as the path.

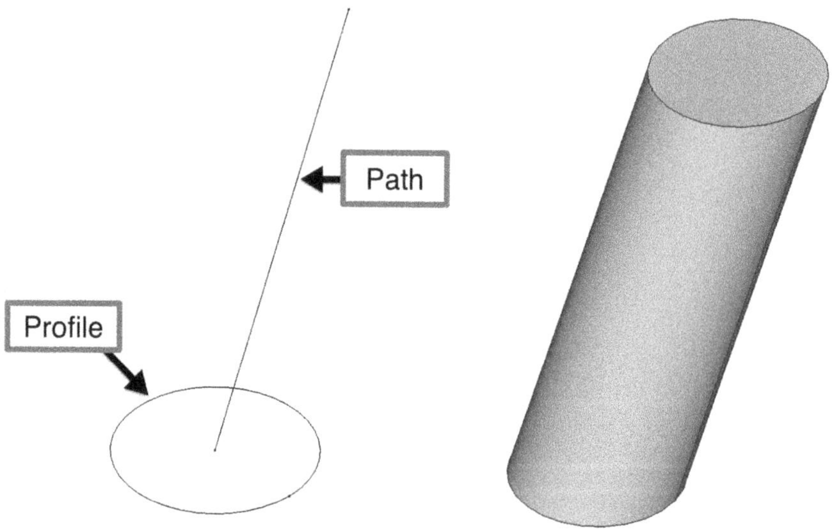

By making the path a bit more complex, you can see that a sweep allows you to create shapes you would not be able to create using commands such as **Extrude** or **Revolve**.

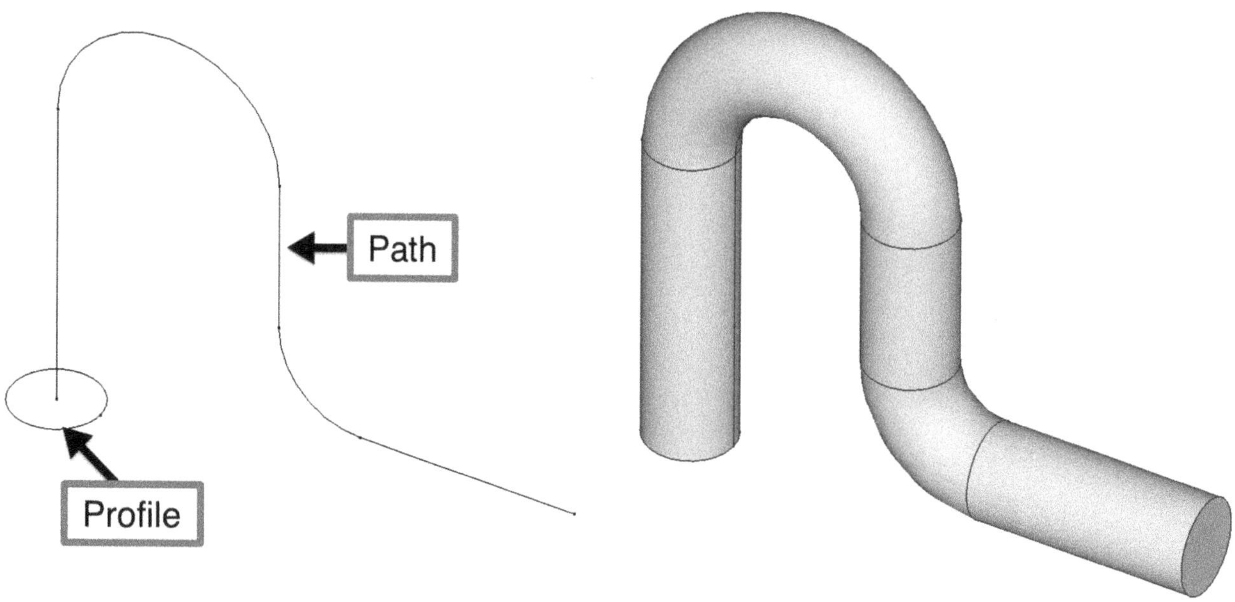

To take the sweep feature to the next level of complexity, you can add another profile and section orientation curve. By doing so, the shape of the geometry is controlled by section orientation curve. For example, the circular profile in the figure varies in size along the path because a section orientation curve controls it.

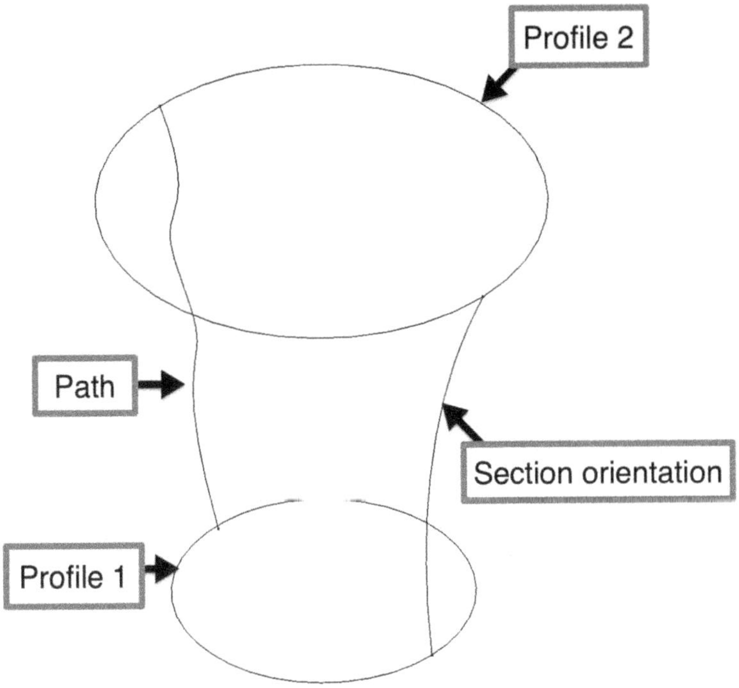

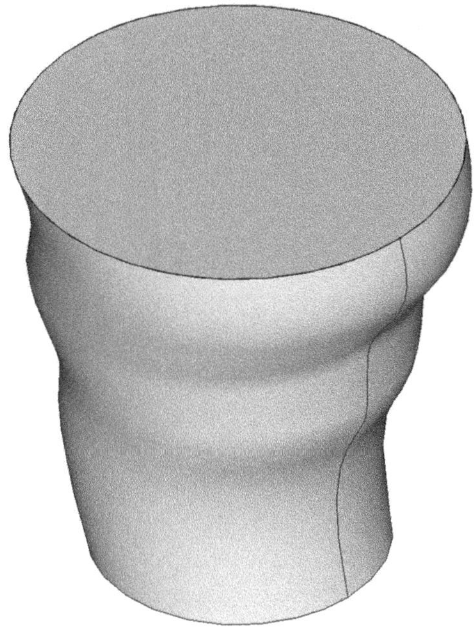

The topics covered in this chapter are:

- *Additive Pipe*
- *Subtractive Pipe*
- *Helix*

Additive Pipe

This feature requires two elements: a path and profile. The profile defines the shape of the sweep along the path. A path is used to control the direction of the profile. A path can be a sketch or an edge. To create a sweep, you must first create a path and a profile. Create a path by drawing a sketch. It can be an open or closed sketch. Next, click the **Datum plane** icon on the **Part Design Helper** toolbar, and then click on an element of the path. Next, select the Normal to edge option from the Attachment mode section of the Plane parameters panel. Click **OK** to create a plane normal to the path. Sketch the profile on the plane normal to the path.

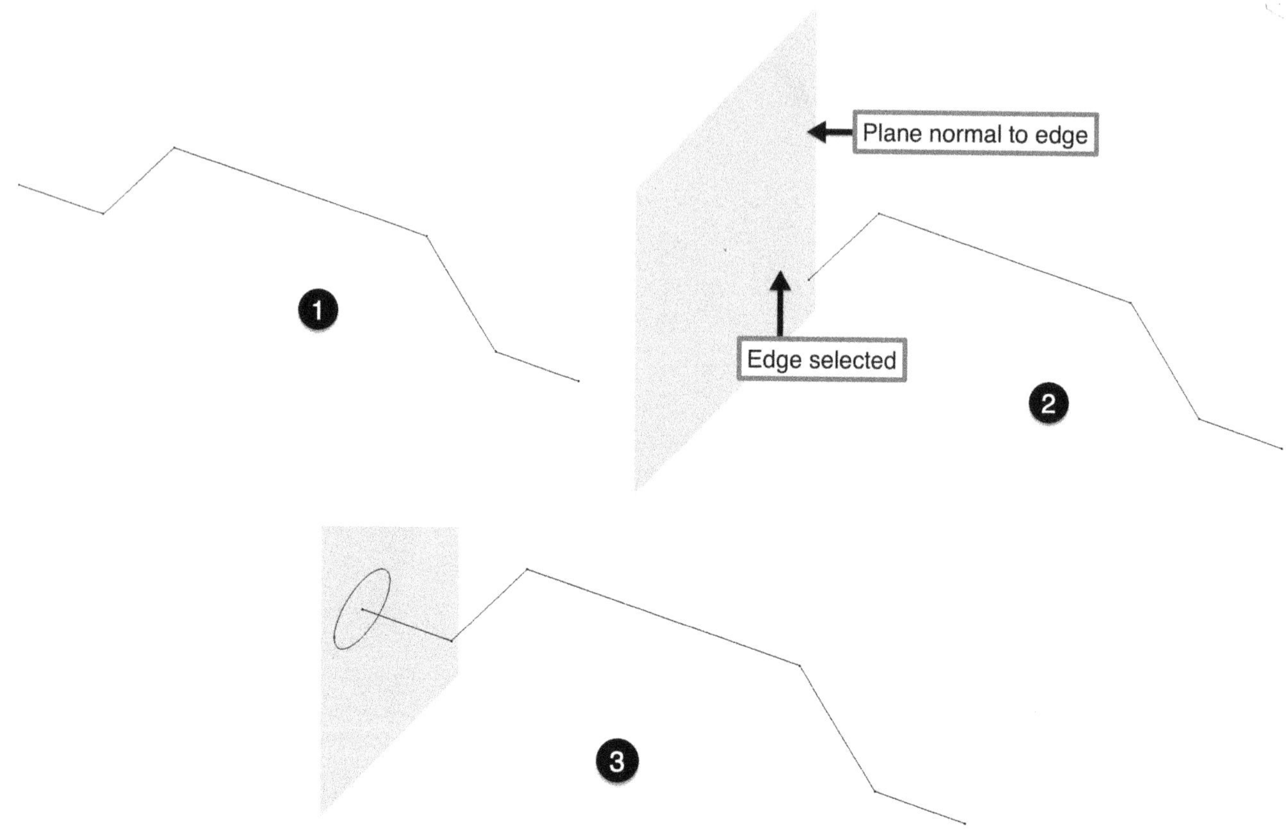

Click the **Additive Pipe**  on the **Part Design Modeling** toolbar (or) click **Part Design > Create an additive feature > Additive Pipe**. Select the profile from the graphics window and click **OK**. Next, click the **Object** button in the **Path to sweep along** section and select the path.

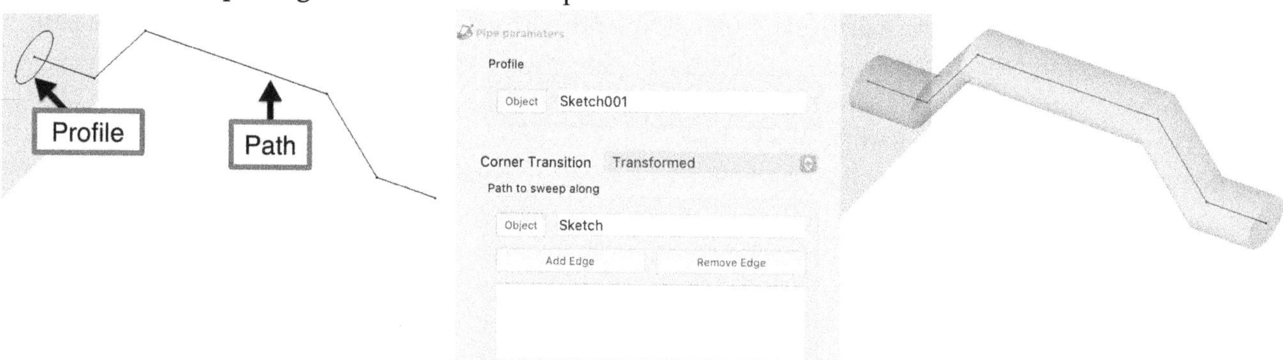

Next, specify the **Corner Transition** option. There are three **Corner Transition** options: **Transformed, Right Corner**, and **Round Corner**. The **Transformed** option creates corners that are parallel to the sketch plane of the profile. The **Right Corner** option creates the corners that are perpendicular to the path. The **Round Corner** option creates the rounded corners.

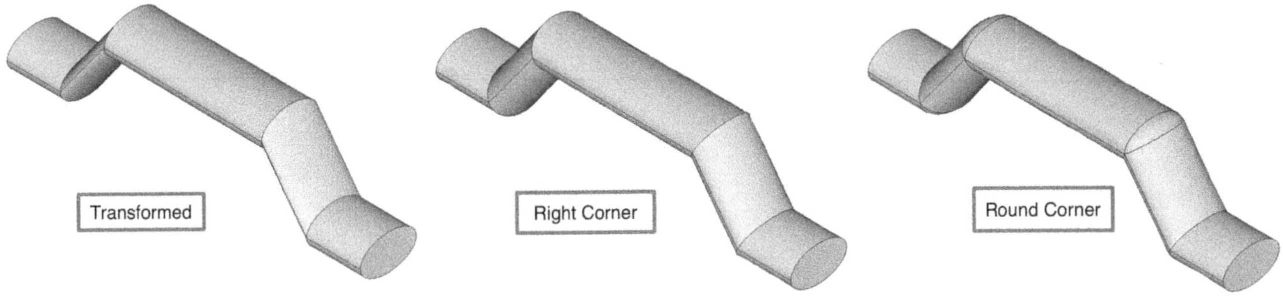

A sweeping profile must be created as a sketch. However, a path can be a sketch or an edge. The following illustrations show types of paths and resultant sweep features.

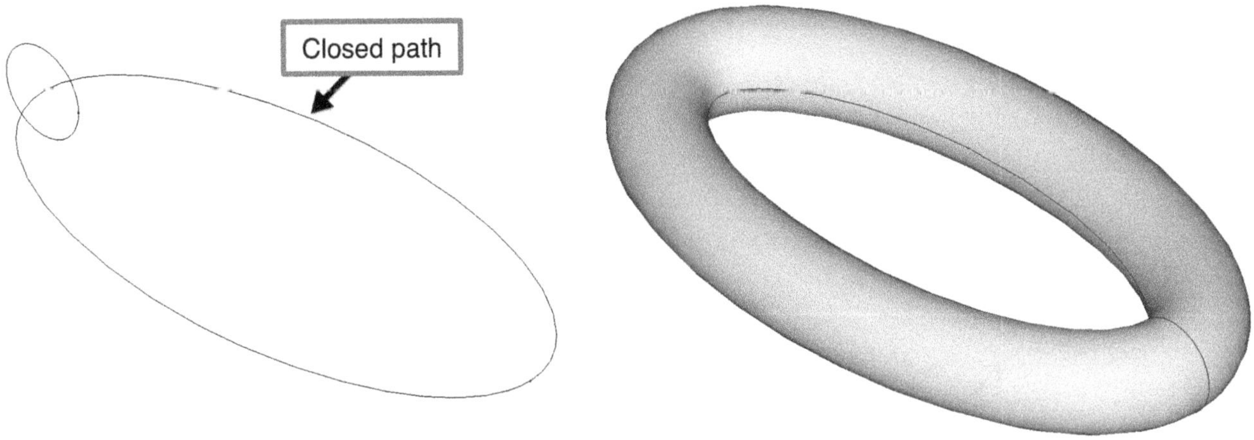

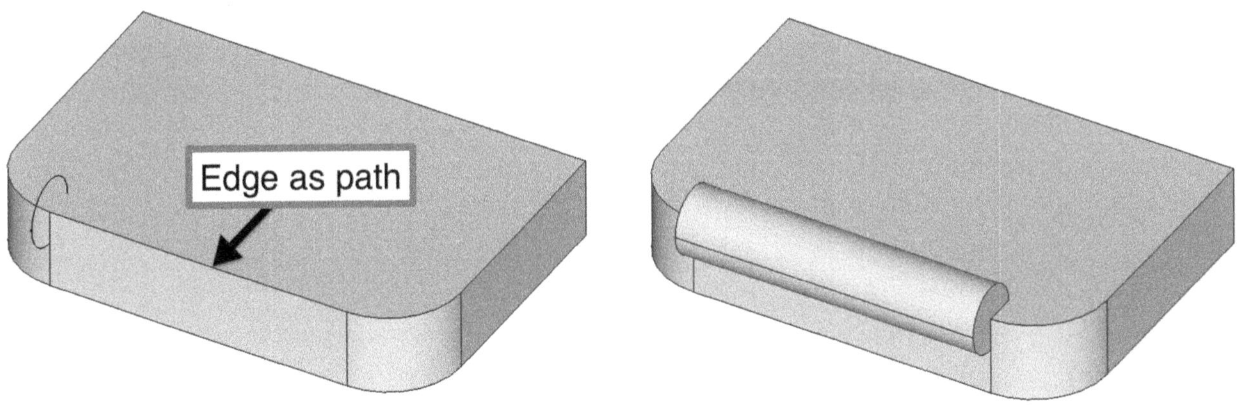

Section Orientation

The **Section Orientation** options define the orientation of the resulting geometry. The **Standard** option sweeps the cross-section in the direction perpendicular to the path. The **Fixed** option sweeps the cross-section in the direction parallel to itself.

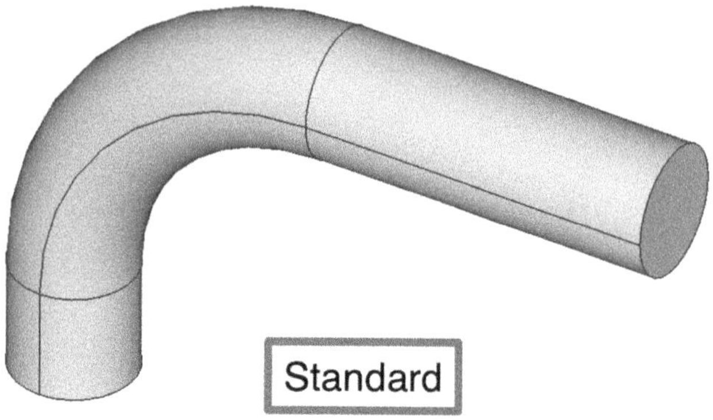

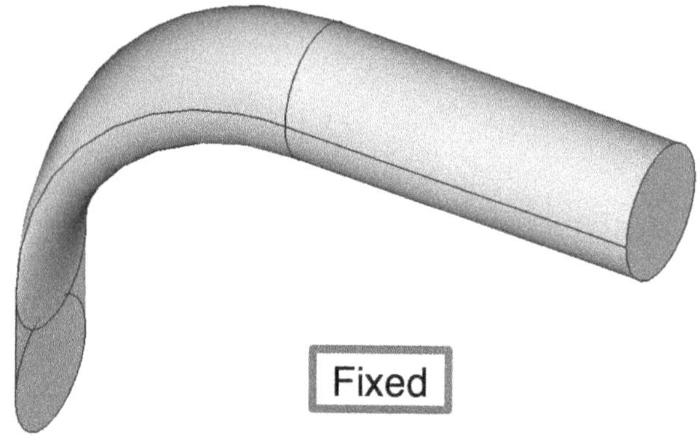

The **Auxiliary** option orients the cross-section perpendicular to a selected auxiliary element.

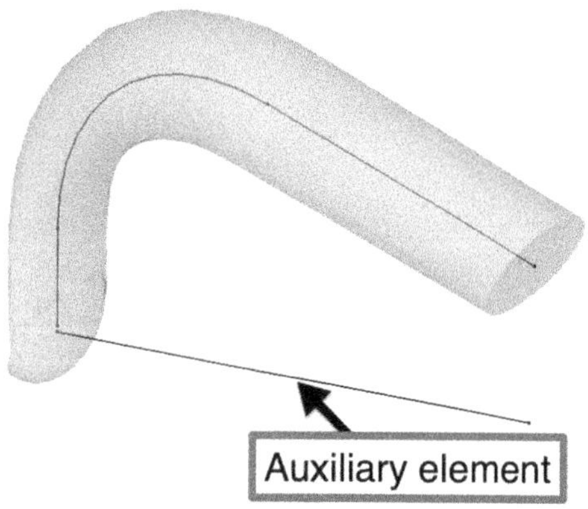

Creating an Additive Pipe between two Cross-sections

Create two cross-sections, as shown. Next, create a path connecting the two cross-sections. For example, start a sketch on the plane intersecting with the cross-sections, and then create a spline.

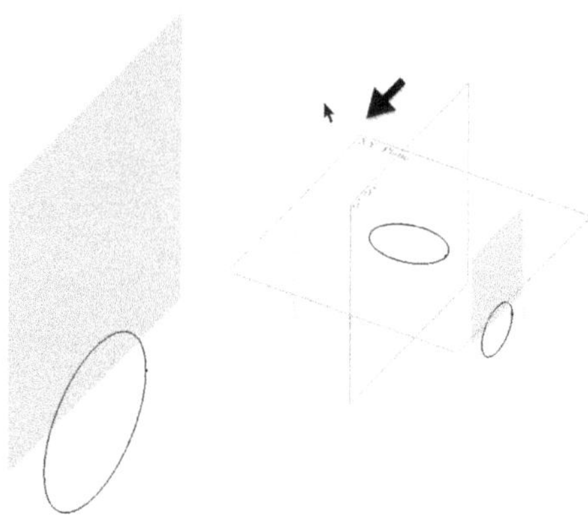

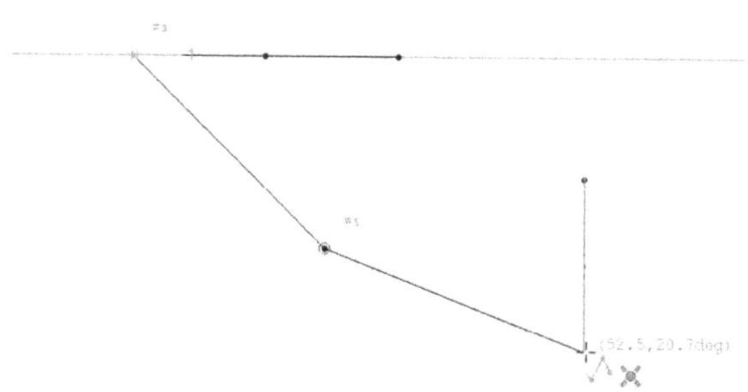

Next, you need to connect the endpoints of the spline with the profiles. Click the **External geometry** icon on the toolbar. Next, select the two existing cross-section. Next, make the endpoints of the spline coincident with the external geometry. Click **Close** on the **Tasks** tab of the **Combo View** panel.

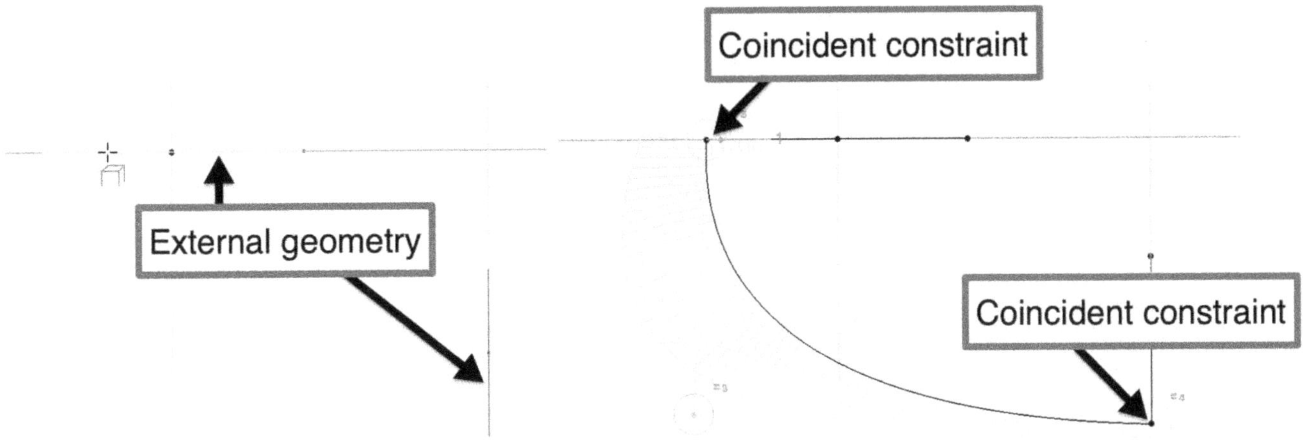

Activate the **Additive Pipe** command and select the first cross-section, and then click **OK**. Click the **Object button** in the **Path to sweep along** section. Next, select the path. You can add more paths by clicking the **Add edge** button in the **Path to sweep along** section and selecting the path; the preview updates. Next, select **Transform mode > Multisection** from the **Section Transformation** section. Click the **Add section** button and select the second cross-section. Select **Orientation mode > Fixed** and click **OK**.

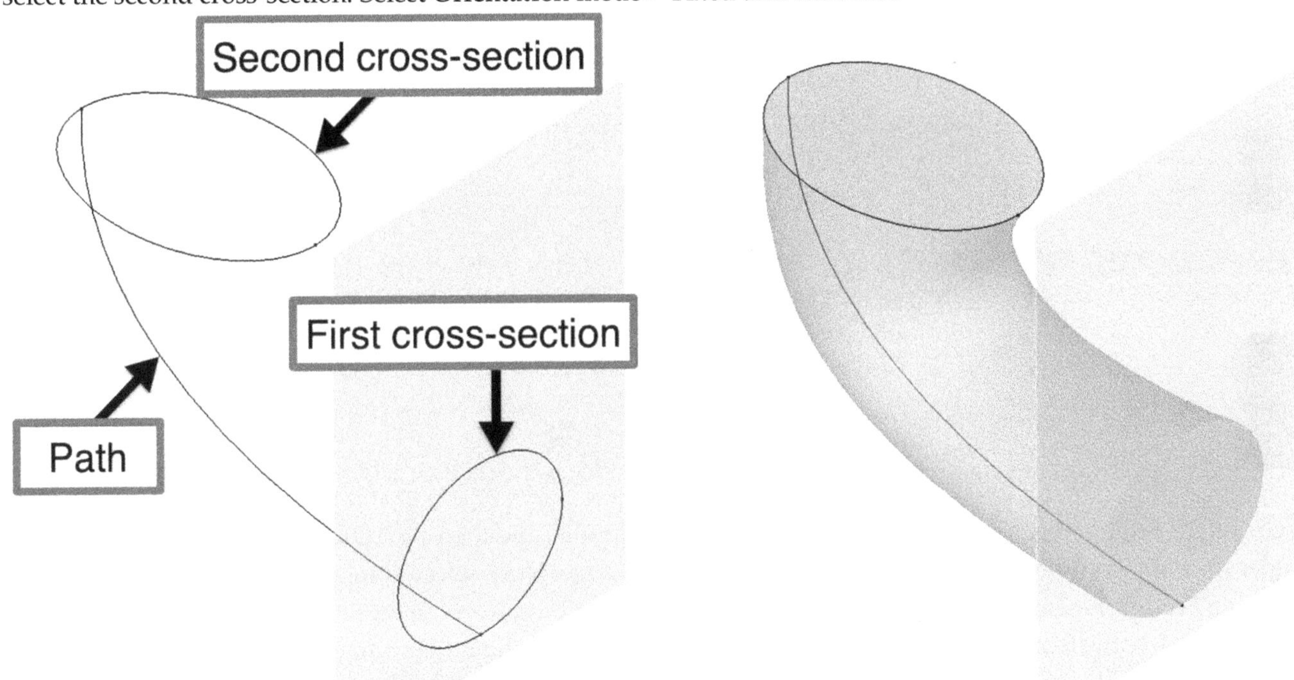

Subtractive pipe

In addition to adding swept features, FreeCAD 0.19 allows you to remove geometry using the **Subtractive Pipe** command. Click the **Subtractive pipe** icon on the **Part Design Modeling** toolbar (or) click **Part Design > Create a subtractive feature > Subtractive pipe.** Next, select the profile and click **OK**. Next, click the **Add edge** button and select the path. Click **OK** to create the subtractive pipe.

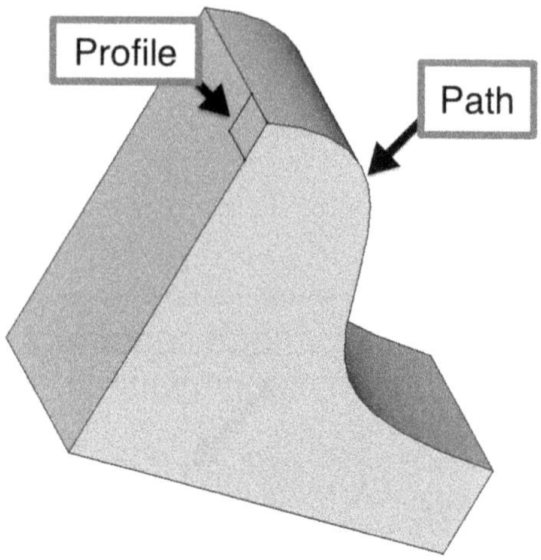

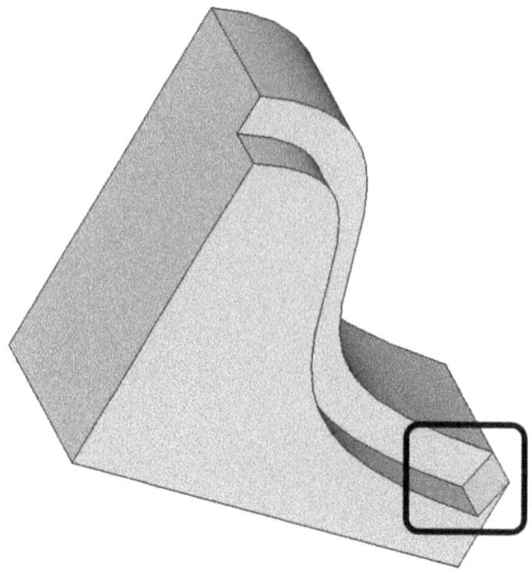

You will notice that the swept cutout is not created throughout the geometry. This is because the profile is swept only up to the endpoints of the path. You can solve this problem by adding another profile at the end face of the model. To do this, right-click on the **Subtractivepipe** feature and select **Delete**. Next, create a sketch on the end face similar in size and shape of the first profile.

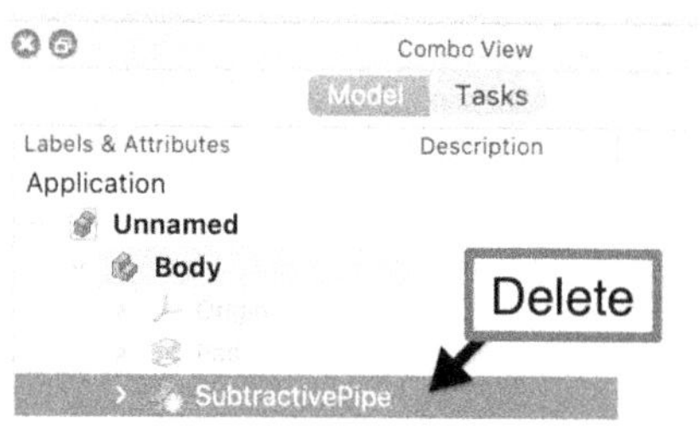

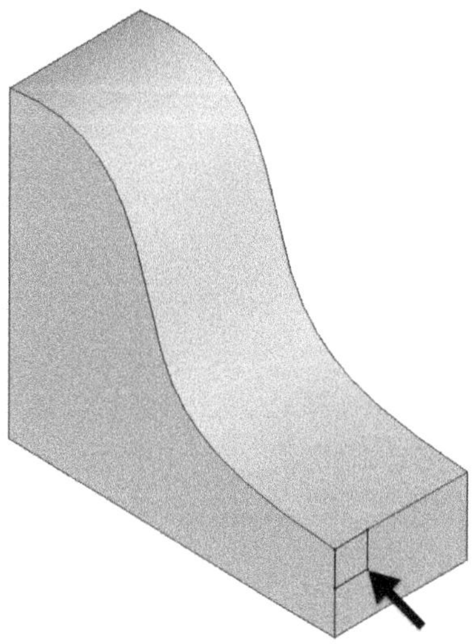

Activate the **Subtractive pipe** command and select the first profile, and then click **OK**. Click the **Add Edge** button under the **Path to sweep along** section and select the curved edge. Next, select **Transform mode > Multisection** from the **Section Transformation** section. Click the **Add Section** button and select the newly created sketch. Click **OK** to complete the feature. The resultant swept cutout will be throughout the geometry.

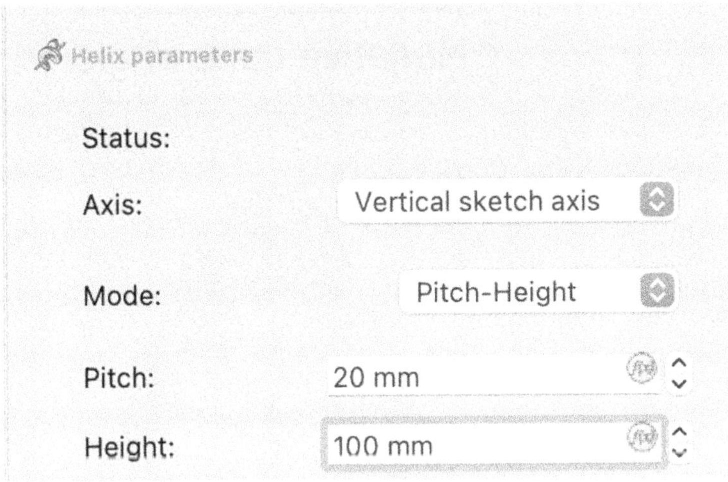

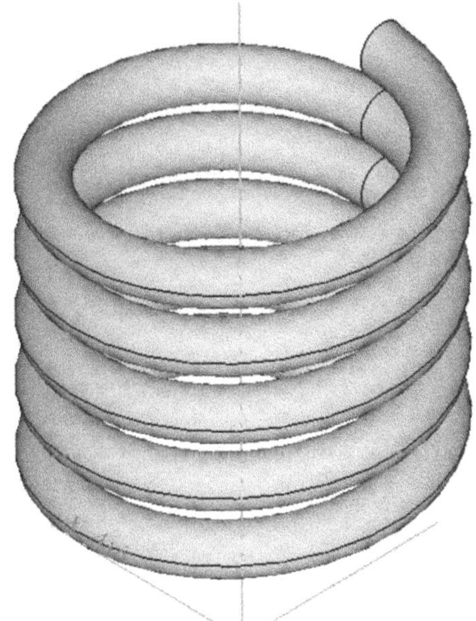

The default rotation direction of the helix is clockwise (i.e., Right-hand side). Check the **Left handed** option to change the rotation direction.

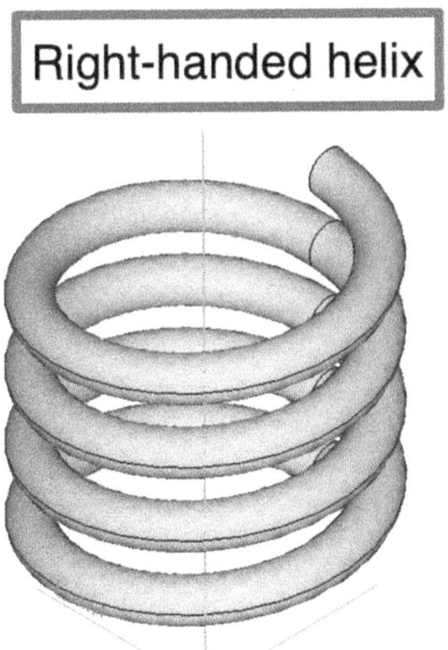

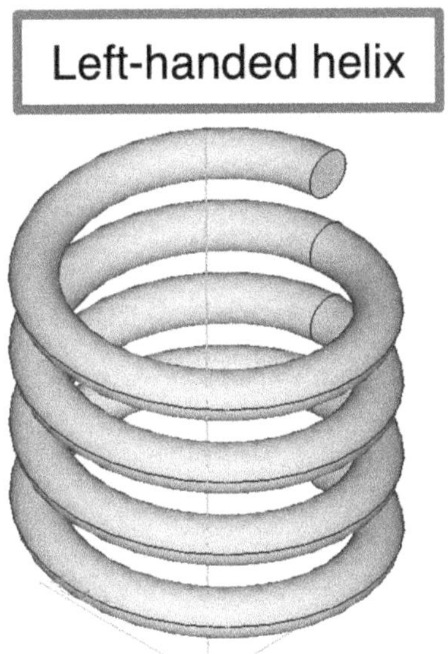

The **Cone angle** box helps you to apply taper to the helix. You can apply taper to the helix by entering a value in the **Cone angle** box. The negative angle reverses the taper direction. Click **OK** to create the helix.

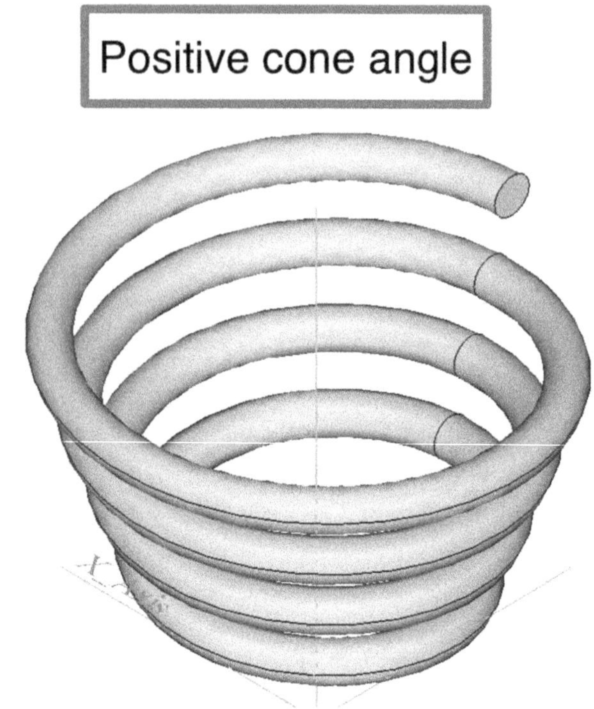

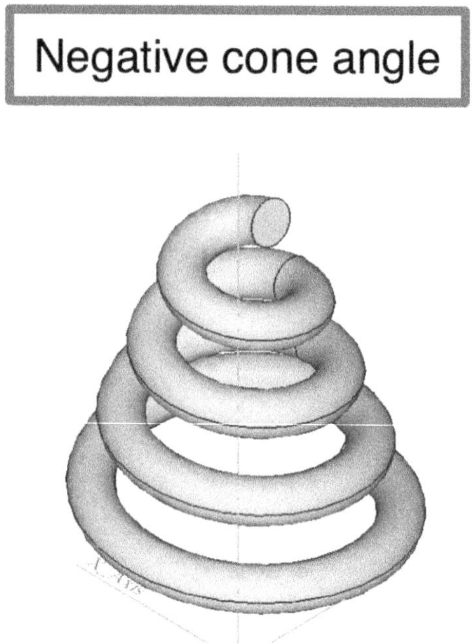

Subtractive helix

The **Subtractive helix** command can be used to remove material from the part geometry by creating a helical feature. To create this feature, first, you must have an existing geometry and sketch of the cross-section. click the **Subtractive helix** icon on the **Part Design Modeling** toolbar (or) click **Part Design > Create a subtractive feature > Subtractive helix**; the cross-section is selected automatically. Define the number of turns and pitch using anyone of the **Modes** described in the previous topic. Next, click **OK** to create the subtractive helix.

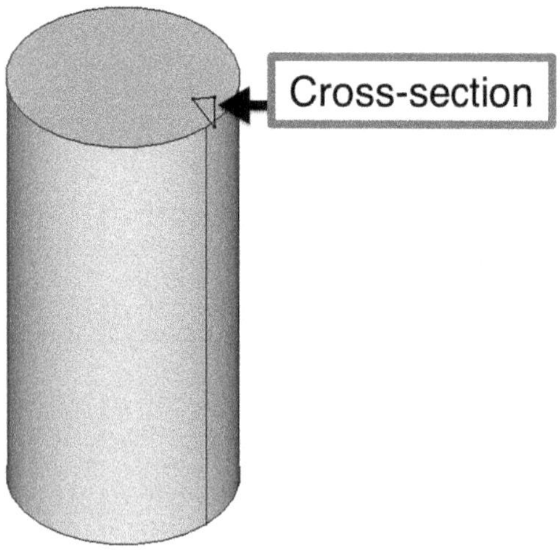

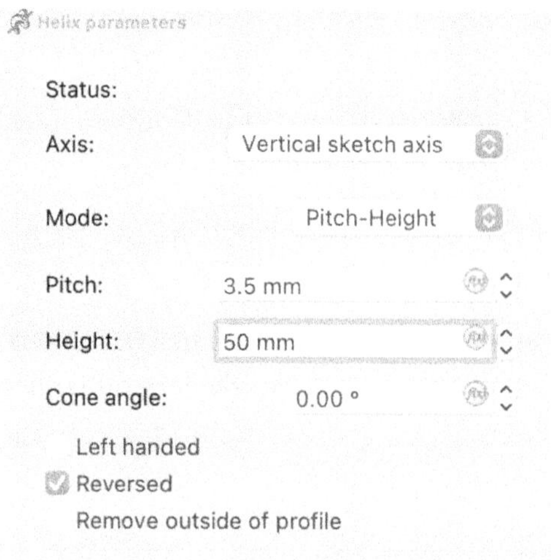

Tutorial 1

In this example, you create the part shown below.

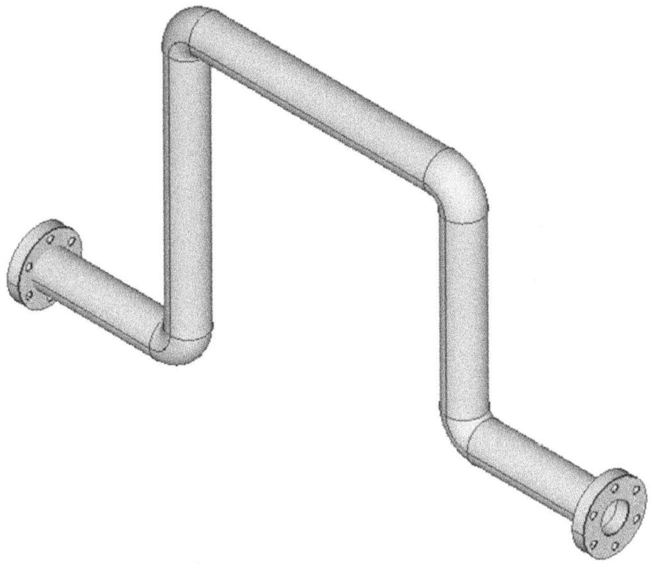

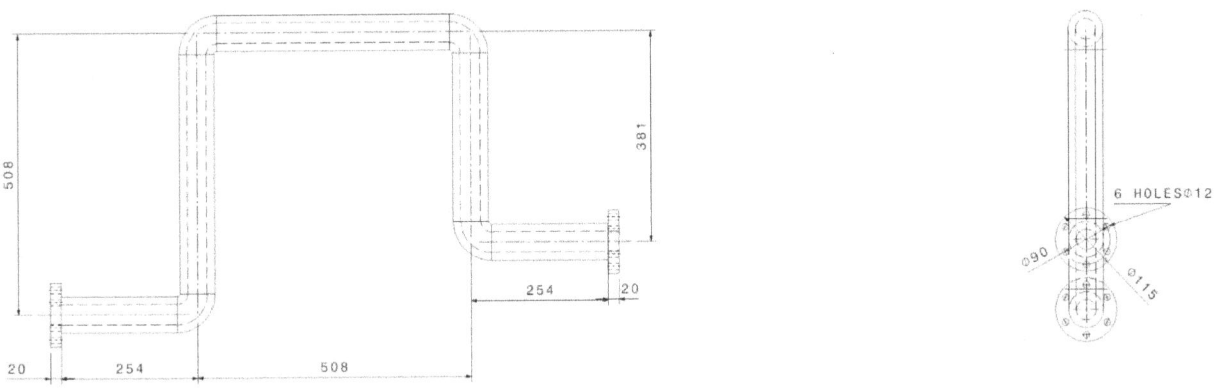

PIPE I.D. 51

PIPE O.D. 65

Creating a New document

1. Click **FreeCAD** on the desktop.
2. On the Home page, click **Documents > Create New**; it creates a new document.
3. On the **Workbench** toolbar, select **Workbench** drop-down **> Part Design**.
4. Click **Edit > Preferences** on the **Menu** bar; the **Preferences** dialog appears on the screen.
5. Click **Units** tab and select **Unit System > Standard (mm/kg/s/degree)**.
6. Select **Number of decimals > 2** and click **OK** on the **Preferences** dialog.

Creating the Sweep Feature

1. Click **Create Sketch** command on the **Part Design Helper** toolbar and then select **XZ_plane**.
2. Click **OK** on the **Combo View** panel.

3. Click the **Polyline** icon on the **Sketcher geometries** toolbar. Next, create the lines, as shown.

4. On the **Sketcher geometries** toolbar, click the **Create fillet**. Select the corner points of the sketch.

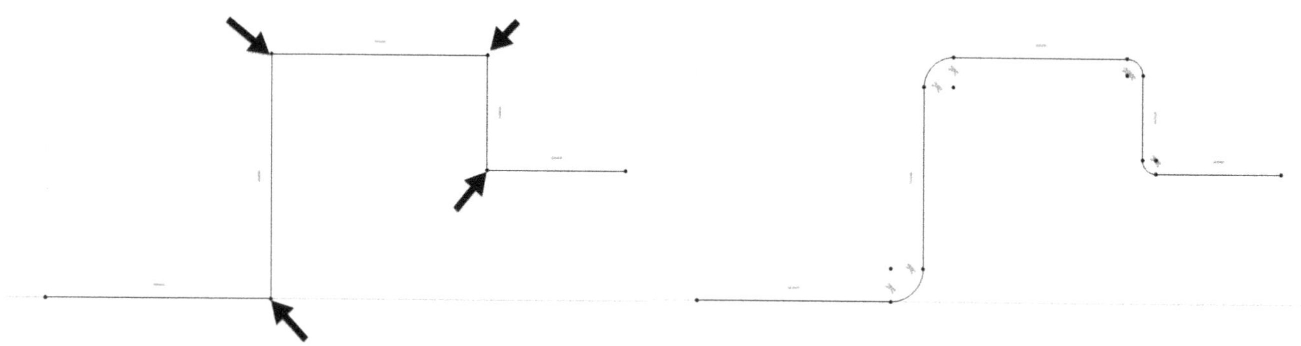

5. Click the **Constrain horizontal distance** icon on the **Sketcher constraints** toolbar.
6. Select the lower horizontal line of the sketch. Click **OK** on the **Insert Length** dialog.
7. Likewise, create the horizontal distance constraints, as shown.

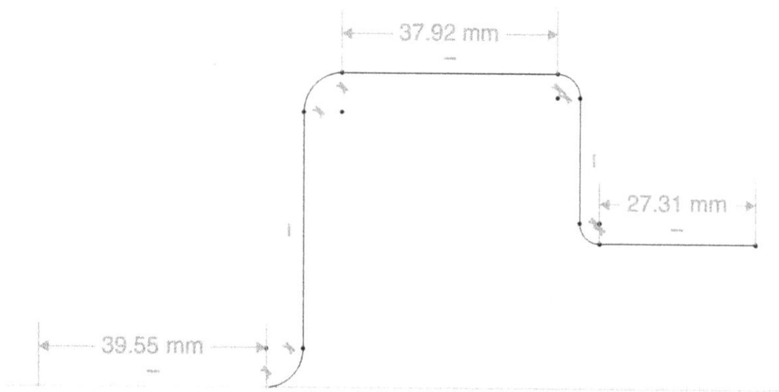

8. Click the **Constrain vertical distance** icon on the **Sketcher constraints** toolbar.
9. Select the left vertical line and click **OK** on the **Insert Length** dialog.
10. Likewise, select the right vertical line and click **OK** on the **Insert Length** dialog.

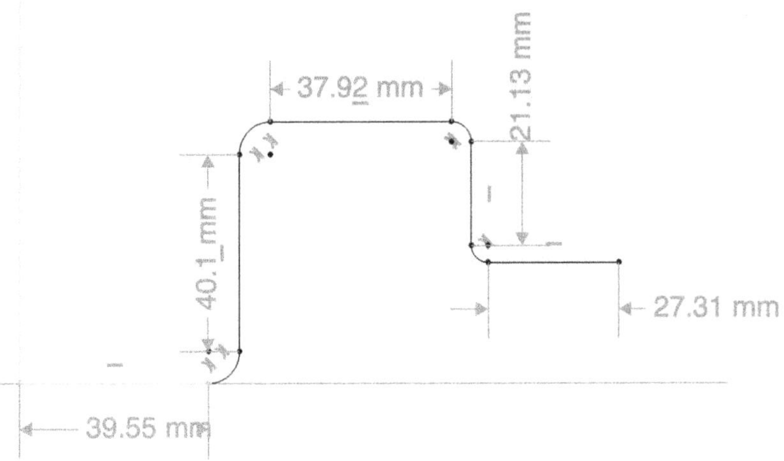

11. Click the **Constrain equal** icon on the **Sketcher constraints** toolbar. Select all the fillets.
12. Again, select the first and last fillets.

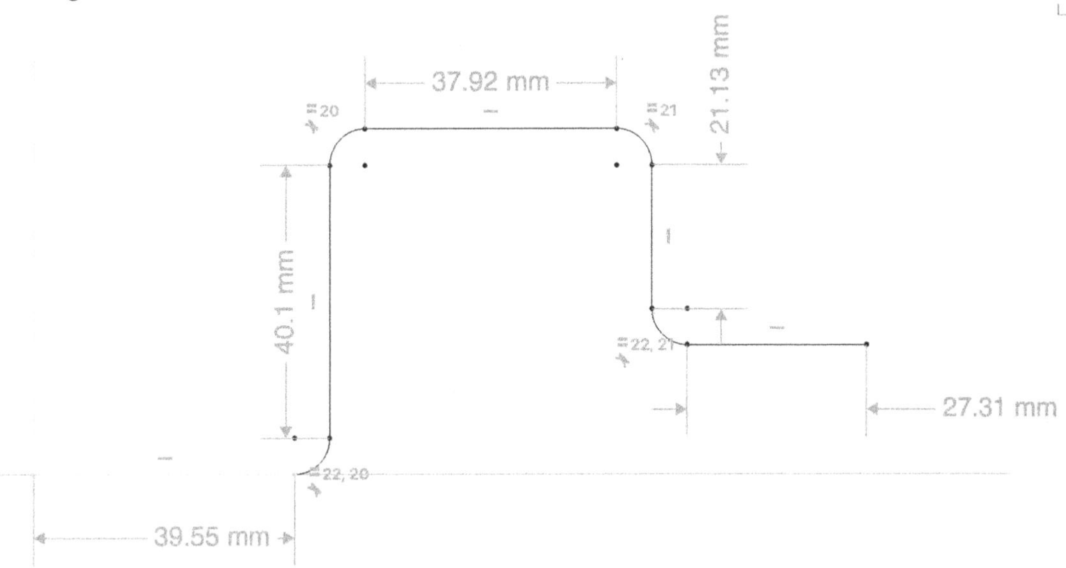

13. Click the **Constraint radius** icon on the **Sketcher constraints** toolbar.
14. Select any one fillet and click **OK** on the **Insert Length** dialog.

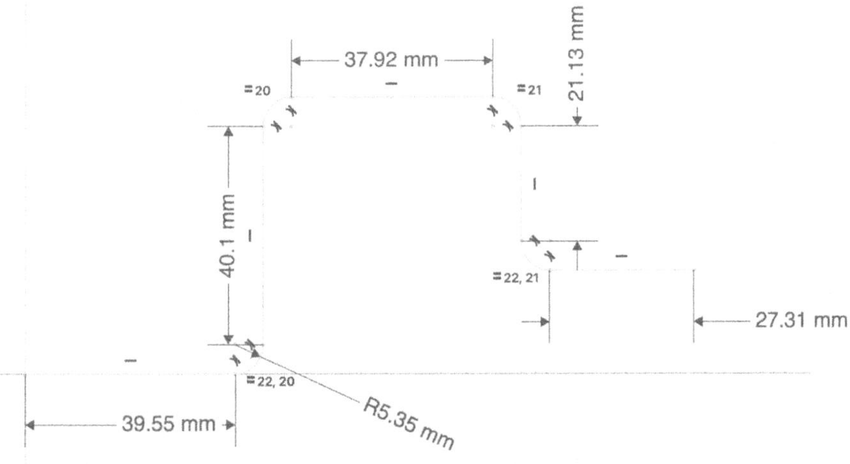

15. Double-click on the left vertical dimension and type-in **432** in the **Length** box on the **Insert Length** dialog.
16. Click **OK** to edit the dimension.
17. Likewise, edit all the values of the dimensions, as shown in the figure.

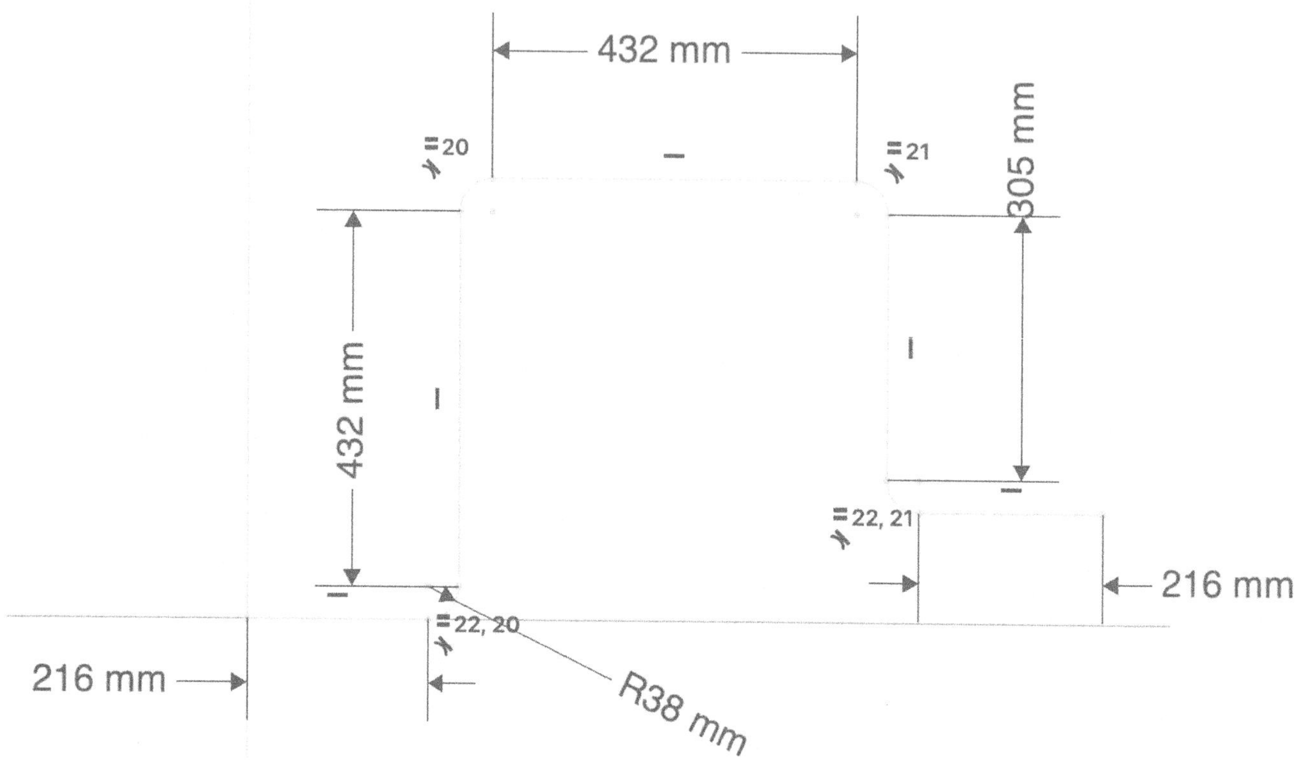

18. Click the **Close** button on the **Combo View** panel.

19. On the **Part Design Helper** toolbar, click the **Create a datum plane** 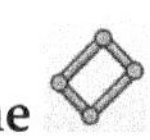 icon.
20. Select the right horizontal line of the sketch to define the first reference.
21. Select the endpoint of the horizontal line to define the second reference.
22. Select **Normal to edge** from the **Attachment mode** section and click **OK**.

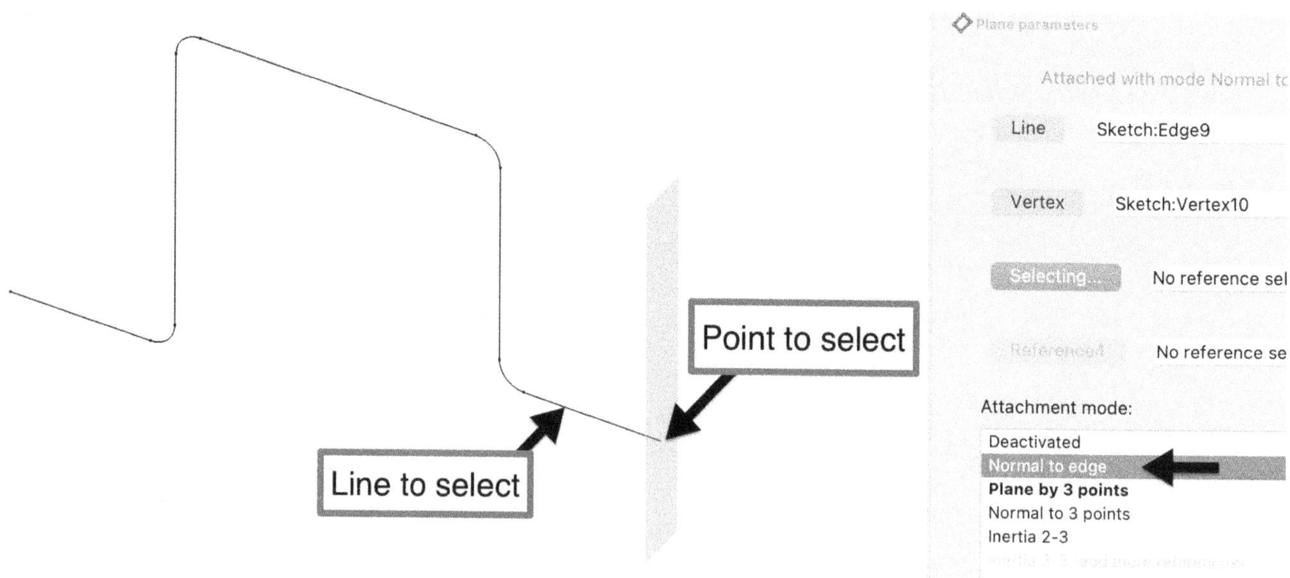

23. Click the **Create Sketch**  icon on the **Part Design Helper** toolbar. Next, select the newly created datum plane and click **OK**.
24. Click the **Create circle**  tool on the **Sketcher geometries** toolbar.
25. Change the view orientation to the Isometric view. (on the graphics area, click the **Camera and render options > Isometric** from the right-side).
26. Click on the sketch origin to define the center point of the circle.
27. Move the pointer outward and click to create a circle.
28. Again, click on the sketch origin to define the center point of the circle.
29. Move the pointer and click to create another circle. Make sure that the center points of the circles are coincident with each other.

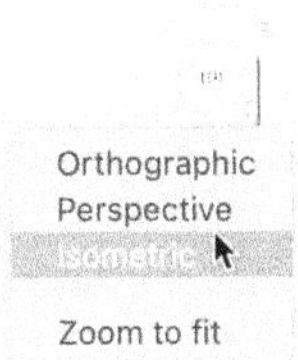

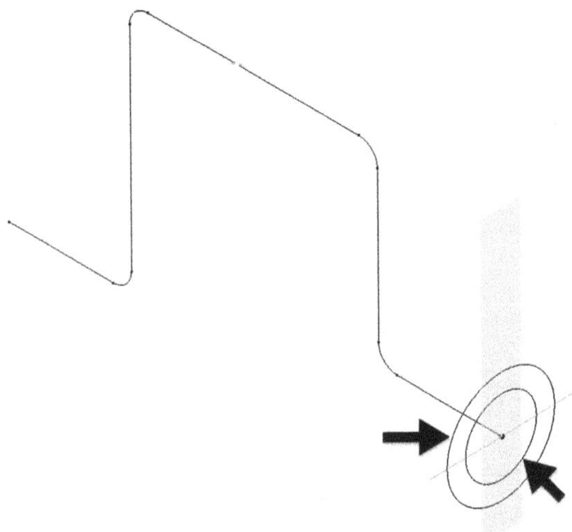

30. On the **Sketcher constraints** toolbar, click the **Constrain an arc or circle** drop-down and select **Constrain diameter**.
31. Select the inner circle. Next, type-in **51** in the **Length** box and click **OK** on the **Insert Length** dialog.
32. Select the outer circle. Next, type-in **65** in the **Length** box and click **OK** on the **Insert Length** dialog.

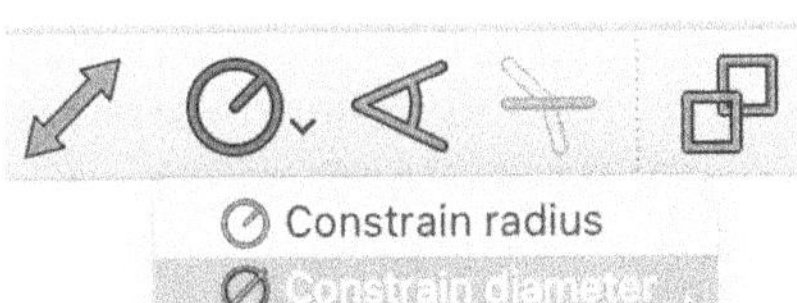

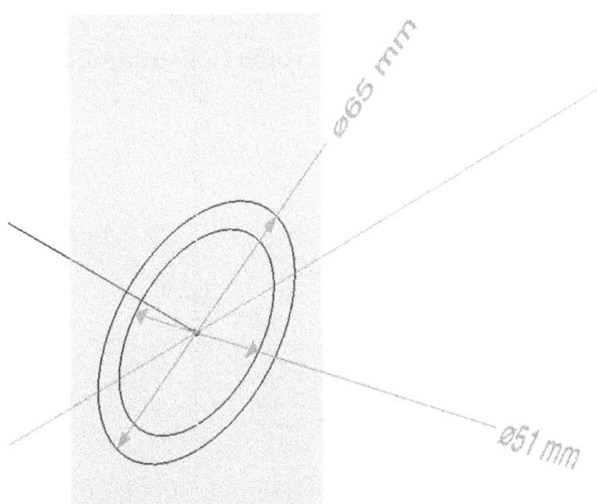

33. Click the **Close** button on the **Combo View** panel.

34. On the **Part Design Modeling** toolbar, click the **Additive pipe** icon. Next, select the **Sketch001** from the **Select feature** section on the **Combo View** panel.
35. Click **OK** on the **Combo View** panel.
36. Under the **Pipe parameters** section on the **Combo View** panel, click the **Object** button under the **Path to sweep along** section.
37. Select the first sketch from the graphics area and click **OK** to create the *Sweep* feature.

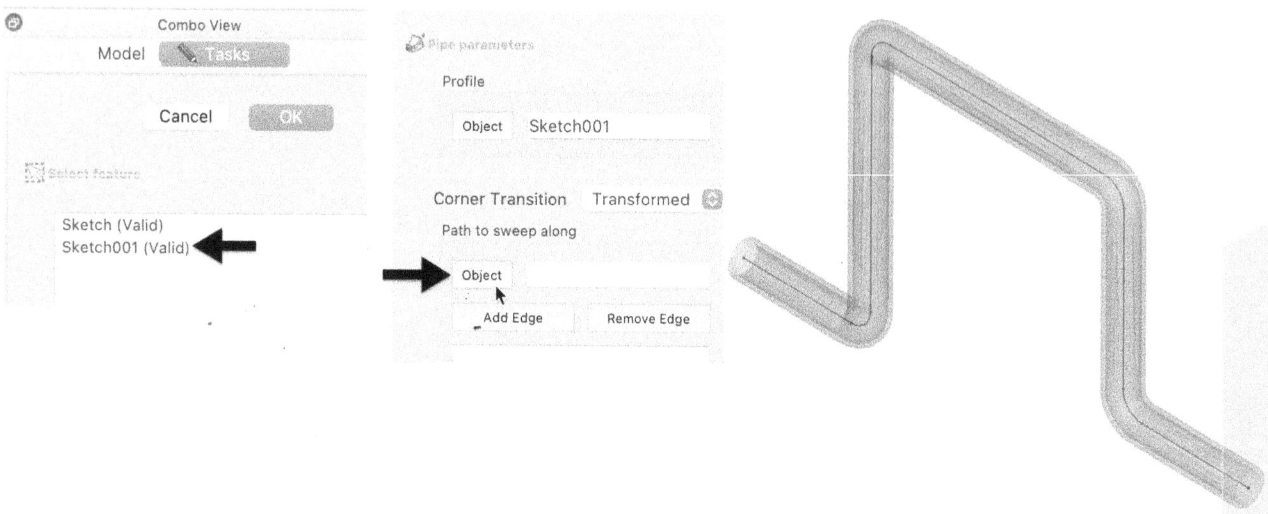

Adding the Padded Feature

1. Click the **Create Sketch** command on the **Sketcher geometries** Toolbar.
2. Click on the datum plane displayed on the front-end face of the *Sweep* feature.
3. Change the view orientation to the Isometric view. (on the graphics area, click the **Camera and render options > Isometric** from the right-side).
4. Create two circles that are concentric with each other.
5. Click the **Constrain coincident** icon on the **Sketcher constraints** toolbar. Next, select the centerpoint of the circles and the sketch origin.
6. Click the **Constrain radius** drop-down and select the **Constrain diameter** icon on the **Sketcher constraints** toolbar.

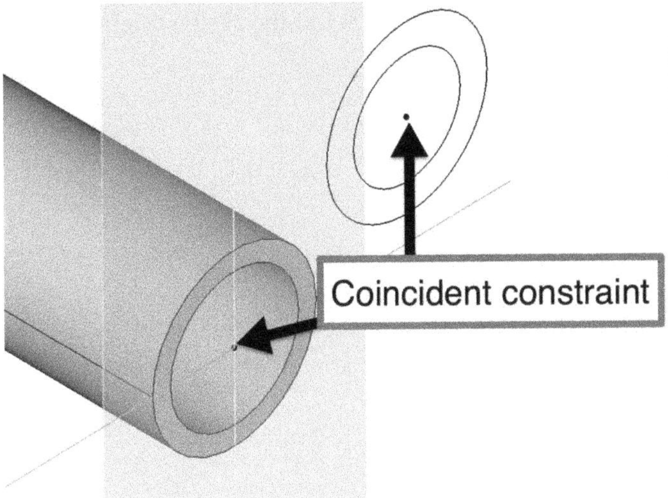

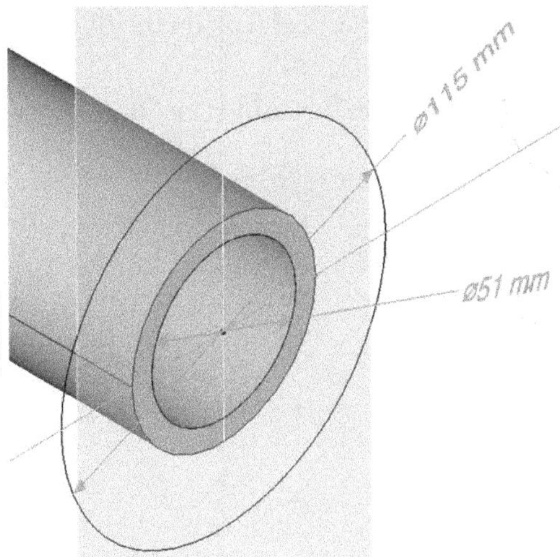

7. Click the **Close** button on the **Combo View** panel to confirm the sketch.
8. Activate the **Pad** command; the sketch is selected automatically.
9. Type-in **20** in the **Length** box under the **Pad parameters** section on the **Combo View** panel. Next, check the **Reversed** option.

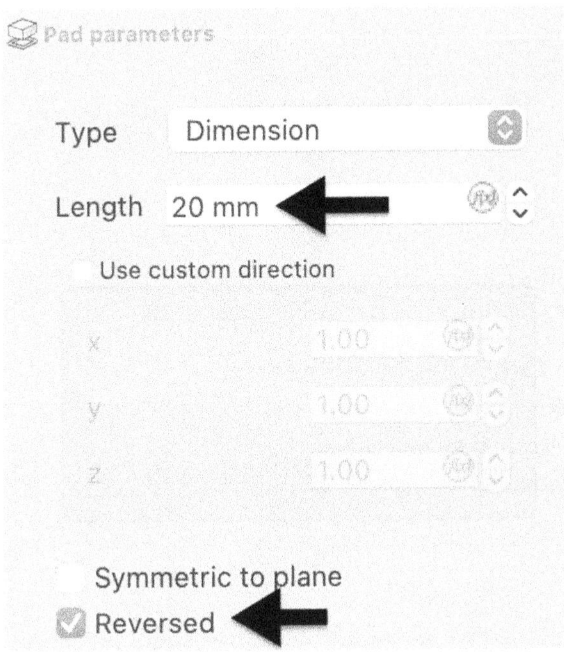

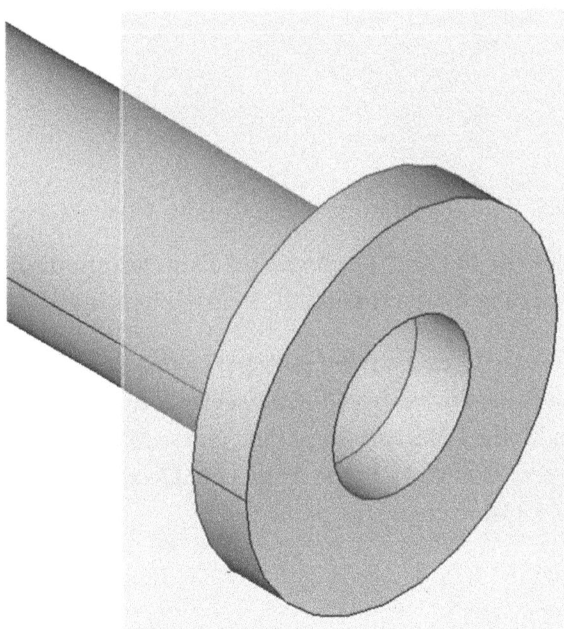

10. Click **OK** on the **Combo View** panel to complete the *Pad* feature.

Creating the Polar pattern

1. Select the face of the pad feature and activate the **Create Sketch** command.
2. Click the **Create circle**  icon on the **Sketcher geometries** toolbar. Next, click the vertical sketch axis and move the pointer outward. Again, click to create a circle.
3. On the **Sketcher constraints** toolbar, click **Constrain radius** drop-down > **Constrain diameter**.
4. Select the circle and enter 12 in the **Diameter** box. Next, click **OK** to add a diameter constraint.

5. Click the **Constrain vertical** icon on the **Sketcher constraints** toolbar. Next, select the sketch origin and the centerpoint of the circle.
6. Type **45** in the **Length** box and click **OK**.

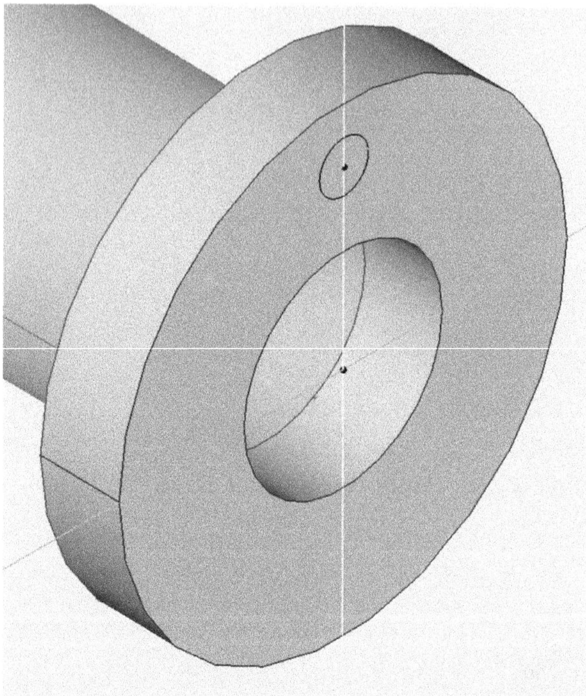

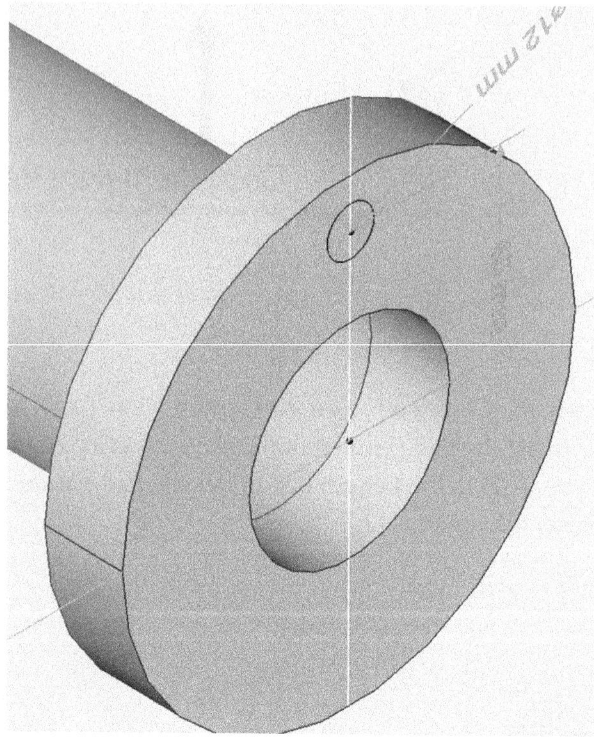

7. Click the **Close** button on the **Combo View** panel.
8. Click the **Pocket** icon on the **Part Design Modeling** toolbar.
9. Select **Type > To first** on the **Pocket parameters** section of the **Combo View** panel. Next, click **OK**.
10. Click the **Create a polar pattern feature** command on the **Part Design Modeling** toolbar.
11. Select the **Pocket** from the **Select feature** panel. Next, click **OK**.
12. Under the **PolarPattern parameters** section and select **Axis > Normal sketch axis.**
13. Type-in **360** and **6** in the **Angle** and **Occurrences** boxes, respectively.
14. Click **OK** to create the pattern.

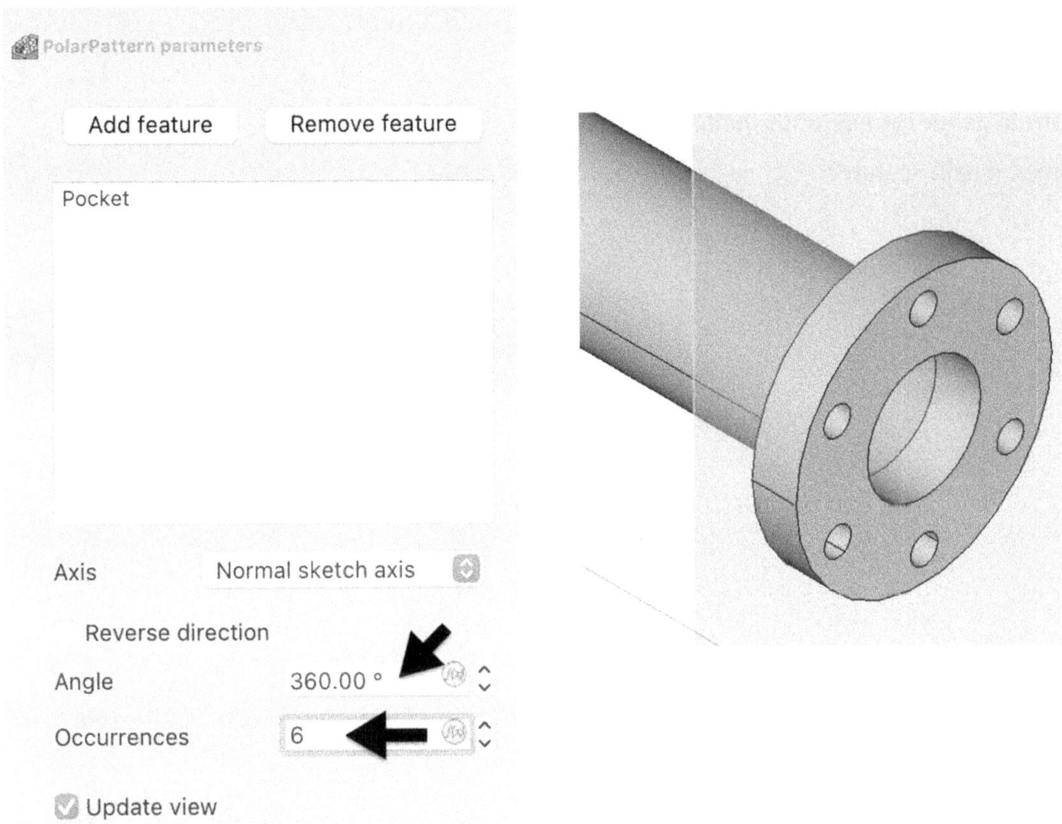

15. Create the *Pad, Pocket,* and *Polar pattern* features on the other end of the model.

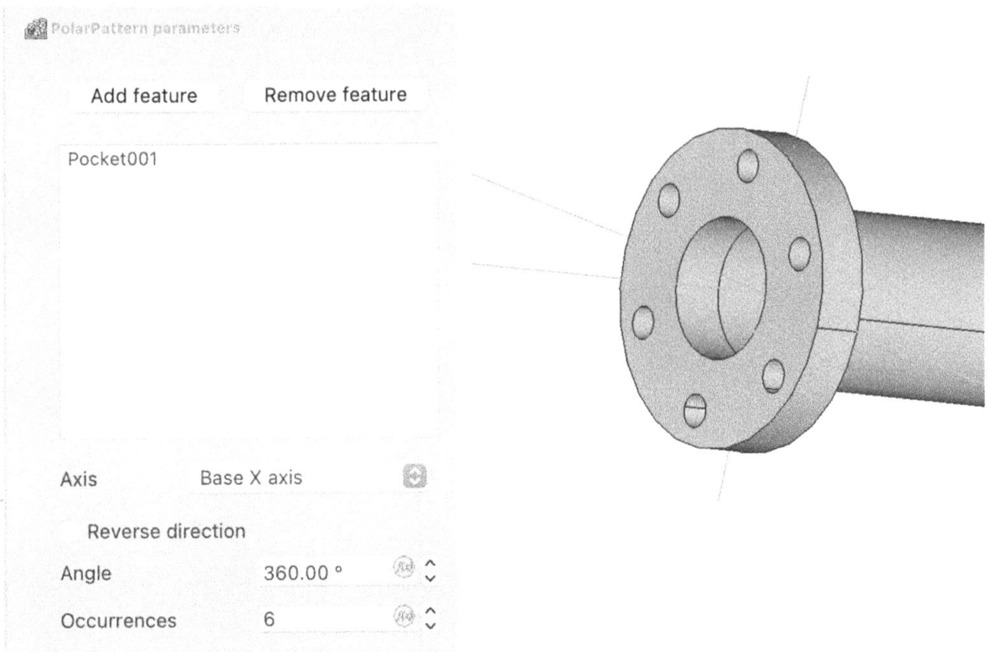

16. Click **File > Save** on the menu bar. Next, type C6_example1 in the **File name** box and click the **Save** button.
17. Click **File > Close** to close the document.

Questions

1. What are the section orientation options?
2. List any two options to define the size of the helix features.
3. How to create a tapered helix feature?

Exercises

Exercise1

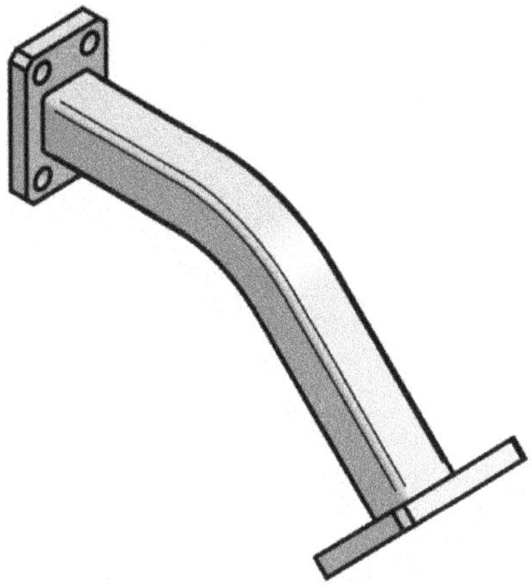

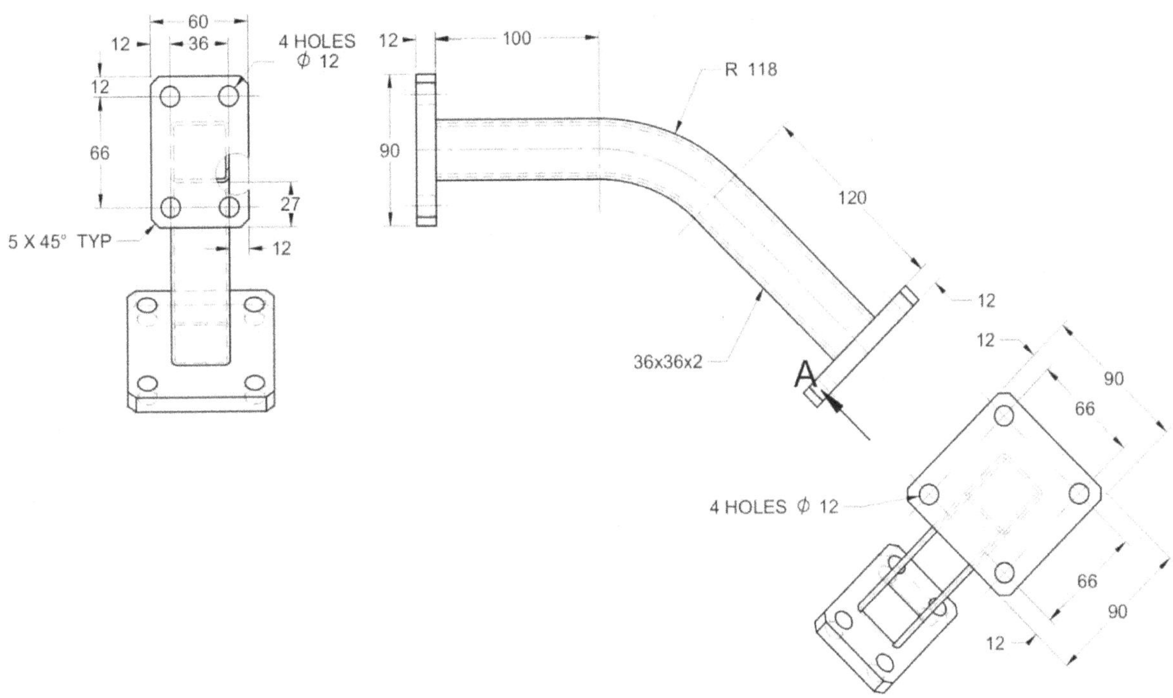

Chapter 7: Loft Features

The **Loft** command is one of the advanced commands available in FreeCAD 0.19 that allows you to create simple as well as complex shapes. A basic loft is created by defining two profiles and joining them together. For example, if you create a loft feature between a circle and a square, you can easily change the cross-sectional shape of the solid. This ability is what separates the loft feature from the sweep feature.

The topics covered in this chapter are:

- *Additive Loft*
- *Subtractive Loft*

Additive Loft

This command creates a loft feature between different profiles. To create a loft, first, create two or more profiles on different planes. The planes can be parallel or perpendicular to each other. Click the **Additive Loft** icon on the **Part Design Modeling** toolbar (or) click **Part Design > Create an additive feature > Additive Loft**. Select the first cross-section and click **OK**. Next, select the second cross-section from the graphics window and click **OK** to create the loft.

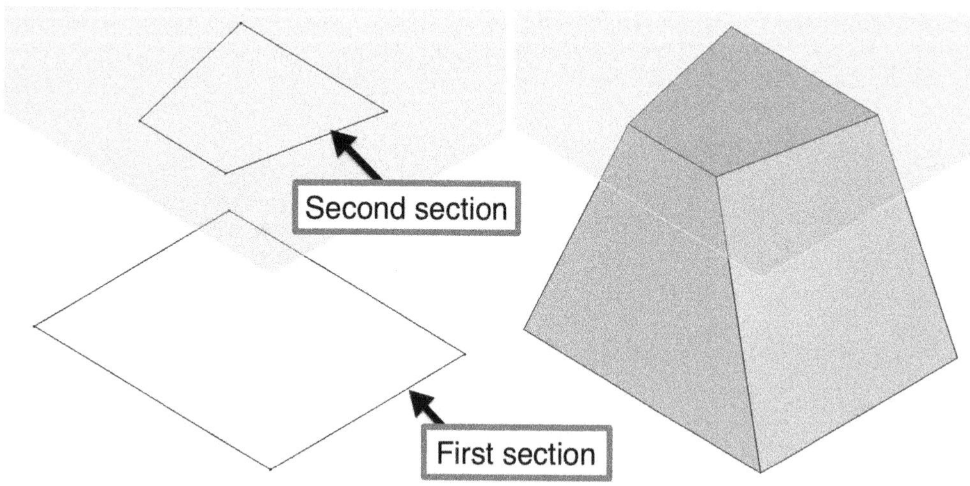

Ruled Surface

This option creates a straight transition between the selected cross-sections. Note that you need to select more than two cross-sections to view the effect of this option.

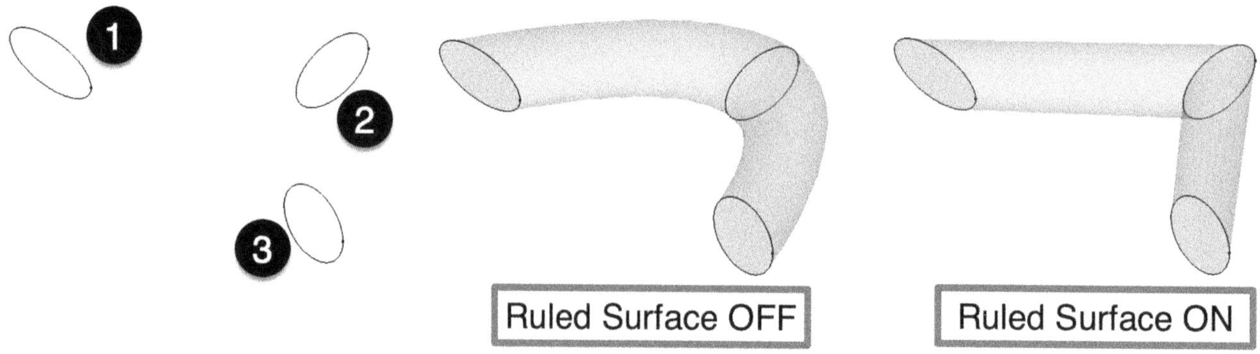

Subtractive Loft

Like other standard features such as extrude, revolve and sweep, the loft feature can be used to add or remove material. You can remove material by using the **Subtractive Loft** command. To do this, click the **Subtractive Loft** icon on the **Part Design Modeling** toolbar (or) click **Part Design > Create a subtractive feature > Subtractive Loft** on the menu bar. Next, select the first cross-section and click **OK**. Click the **Add section** button and select the second cross-section. Likewise use the **Add section** button to select the remaining cross-sections. Click **OK** to create the subtractive loft.

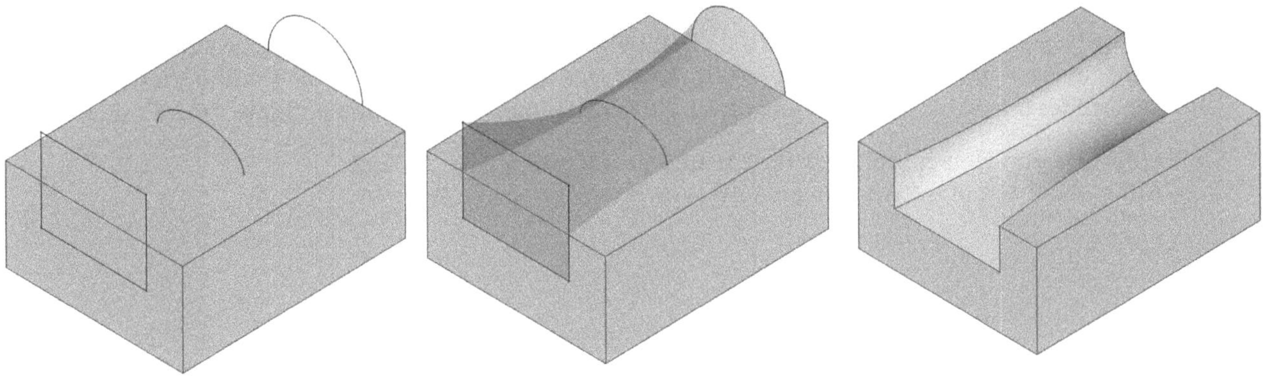

Tutorial 1

In this example, you create the part shown below.

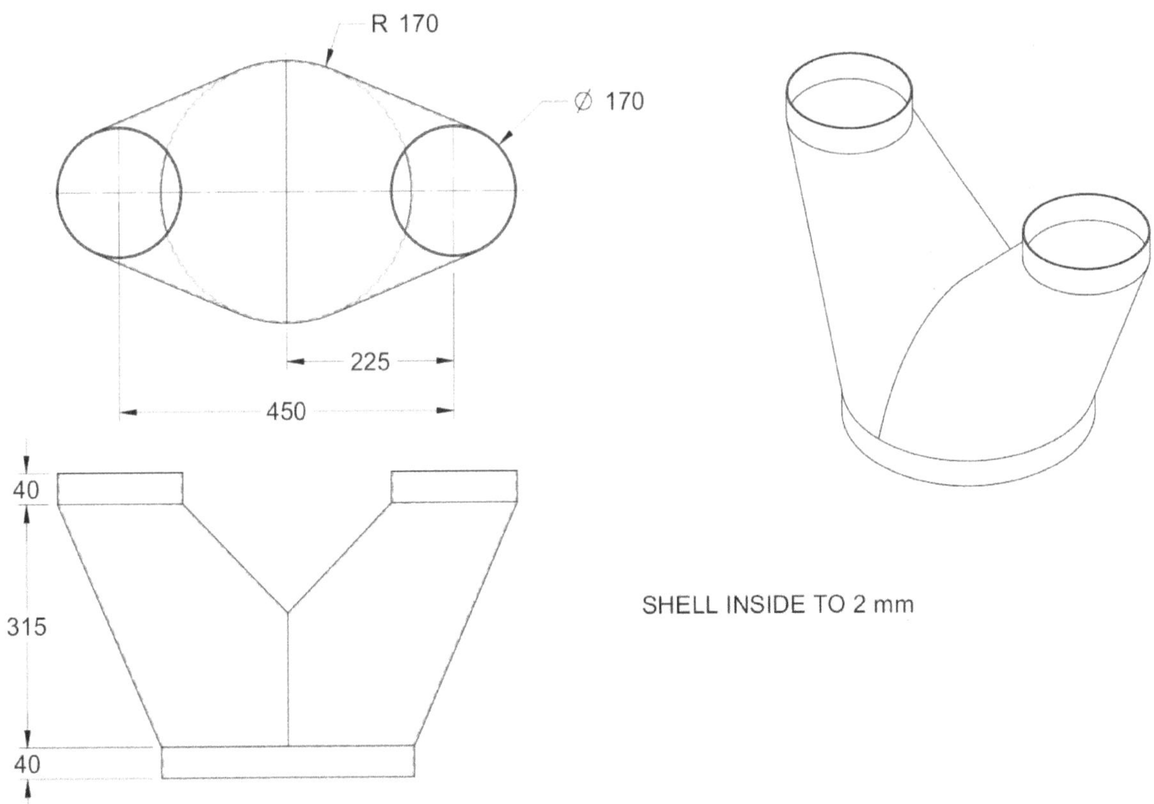

Creating a New document

1. Click **FreeCAD** on the desktop.
2. On the menu bar, click **File > New**; it creates a new document.
3. On the **Workbench** toolbar, select **Workbench** drop-down > **Part Design**.
4. Click **Edit > Preferences** on the **Menu** bar; the **Preferences** dialog appears on the screen.
5. Click **Units** tab and select **Unit System > Standard (mm/kg/s/degree)**.
6. Select **Number of decimals > 2** and click **OK** on the **Preferences** dialog.

Creating a Loft Feature

1. On the **Part Design Helper** toolbar, click the **Create Sketch** icon. Select the XY_Plane and click **OK**.
2. Draw two circles of **338** and **340** mm diameters. Next, click the **Close** button on the **Combo View** panel.
3. Click the **Model** tab on the **Combo View** panel. Next, expand the **Origin** node and select the **XY_Plane**.
4. On the **Part Design Helper** tool, click the **Create a new datum plane** tool.
5. Type **315** in the **Z** box of the **Attachment Offset** section. Next, click **OK**.

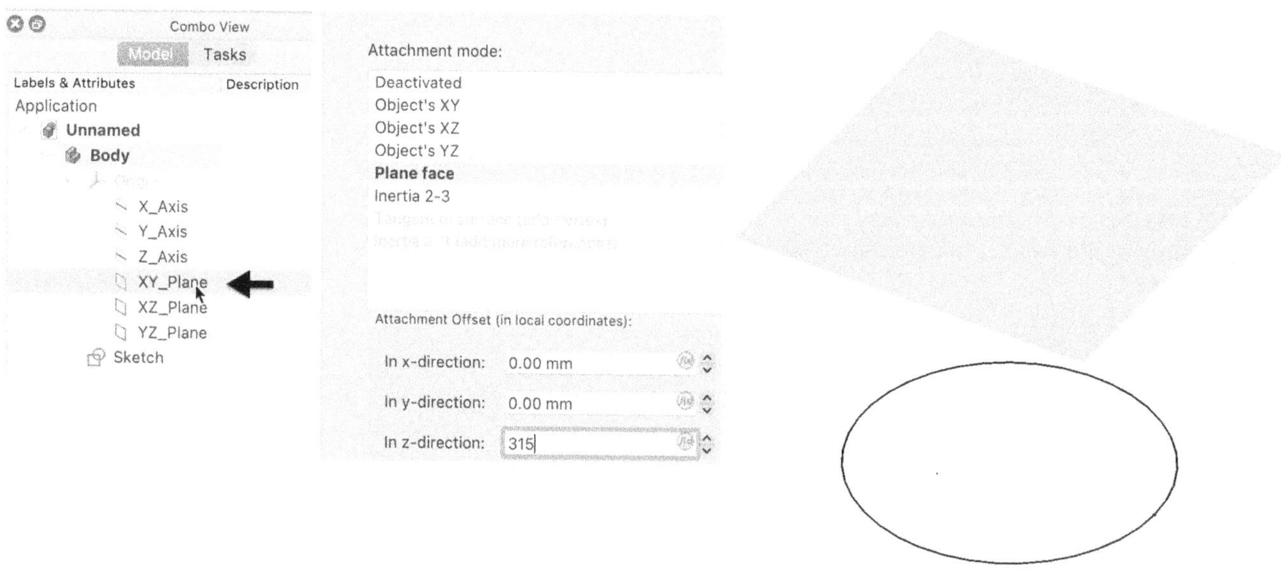

6. Select the newly created datum plane and click the **Create Sketch** icon on the **Part Design Helper** toolbar.
7. Draw two circles of **168** and **170** mm diameters and add constraints to them, as shown.
8. Click the **Close** button on the **Combo View** panel. Click in the graphics window to deselect the circle.

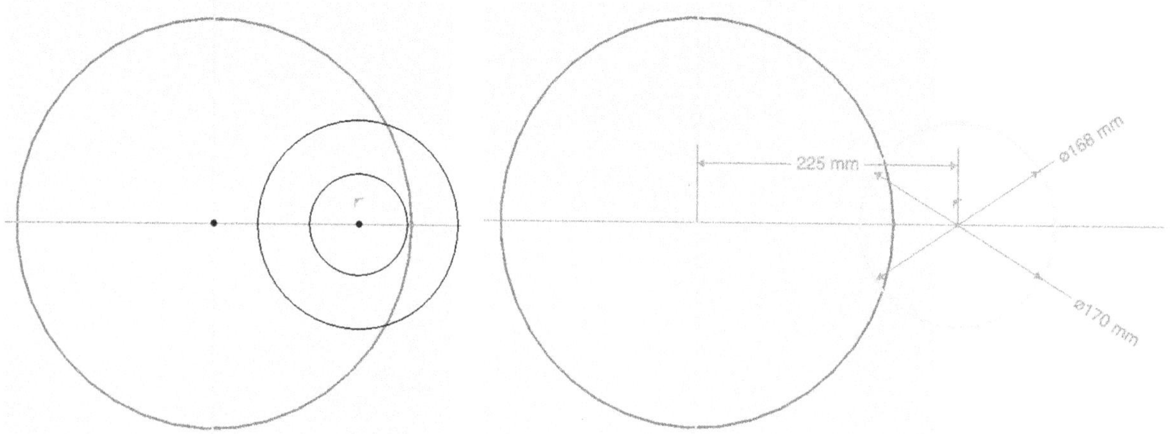

9. On the **Part Design Modeling** toolbar, click the **Loft** icon. Select **Sketch** from the **Select feature** section of the **Combo View** panel. Next, click **OK**.
10. Click the **Add section** button on the **Loft parameters** section and select the circle from the graphics window.
11. Click **OK**.

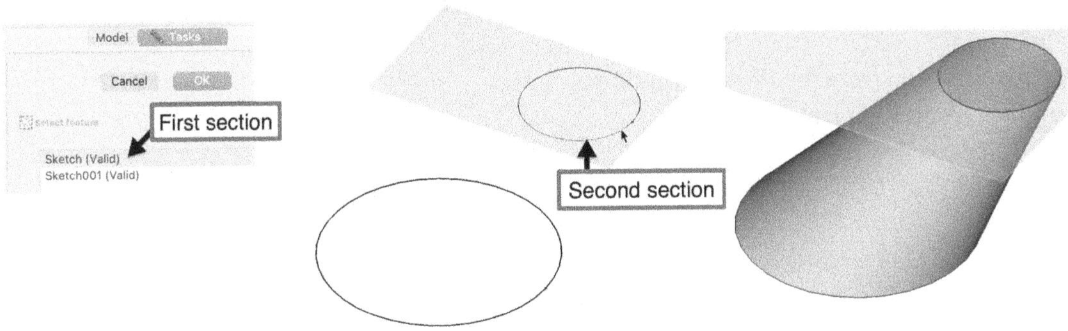

12. In the **Combo View** panel, expand the **AdditiveLoft** feature and select the first sketch.
13. Click the **Pad** icon on the **Part Design Modeling** toolbar. Next, check the **Reversed** option.
14. Enter **40** in the **Length** box and click **OK** to create a pad feature.

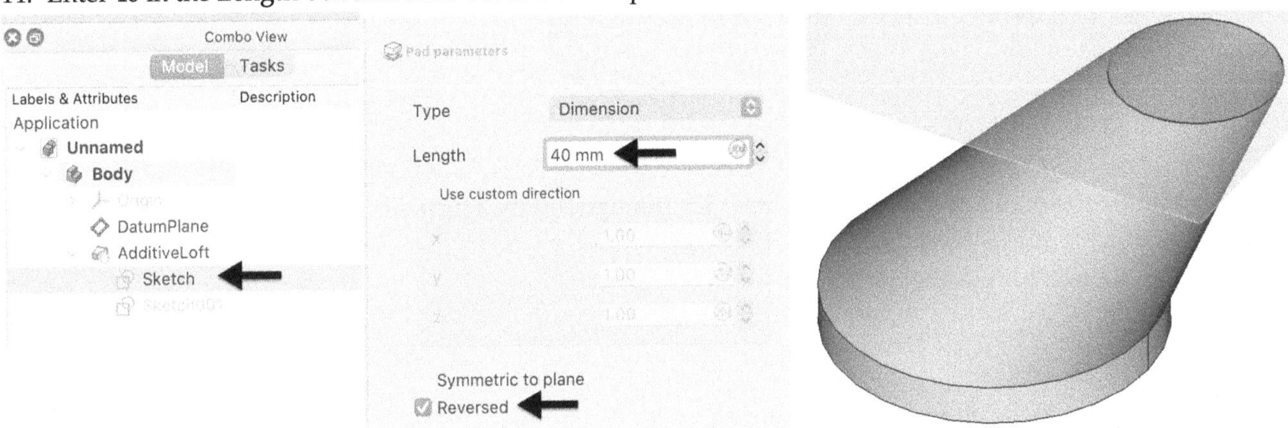

15. In the **Combo View** panel, expand the **AdditiveLoft** feature and select the second sketch.
16. Click the **Pad** icon on the **Part Design Modeling** toolbar. Next, enter **40** in the **Length** box and click **OK**.

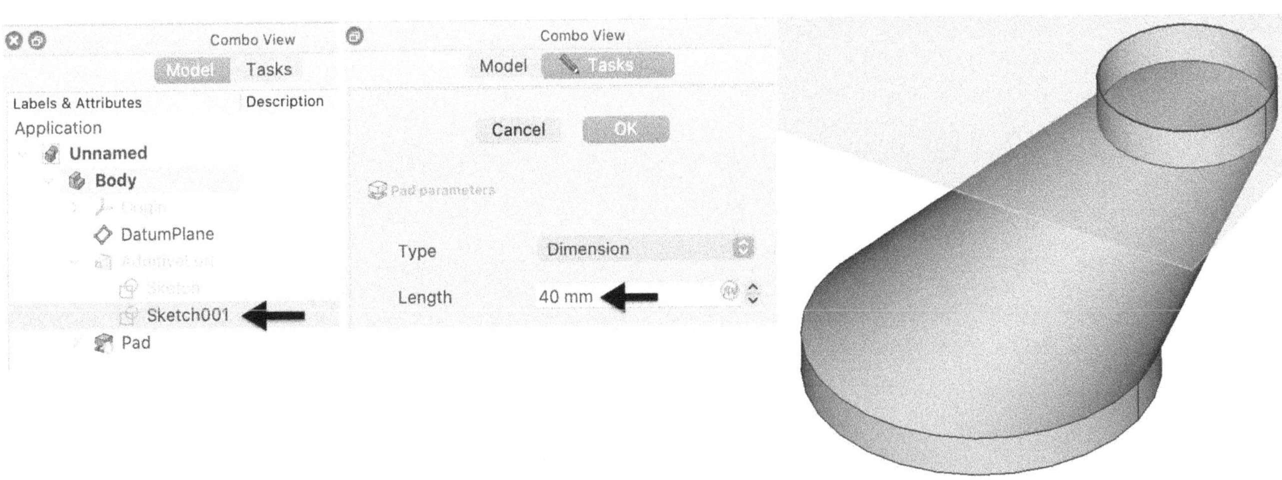

Mirroring the Loft and Pad features

1. On the **Part Design Modeling** toolbar, click the **Mirrored**  icon. Next, select the **AdditiveLoft** feature from the **Select feature** section, and then click **OK**.
2. Select the *Pad* feature from the graphics window. Next, click the **Add feature** button on the **Mirrored parameters** section.
3. Select the **Base YZ plane** option from the **Plane** drop-down
4. Click **OK** to mirror the *Loft* and *Pad* features.

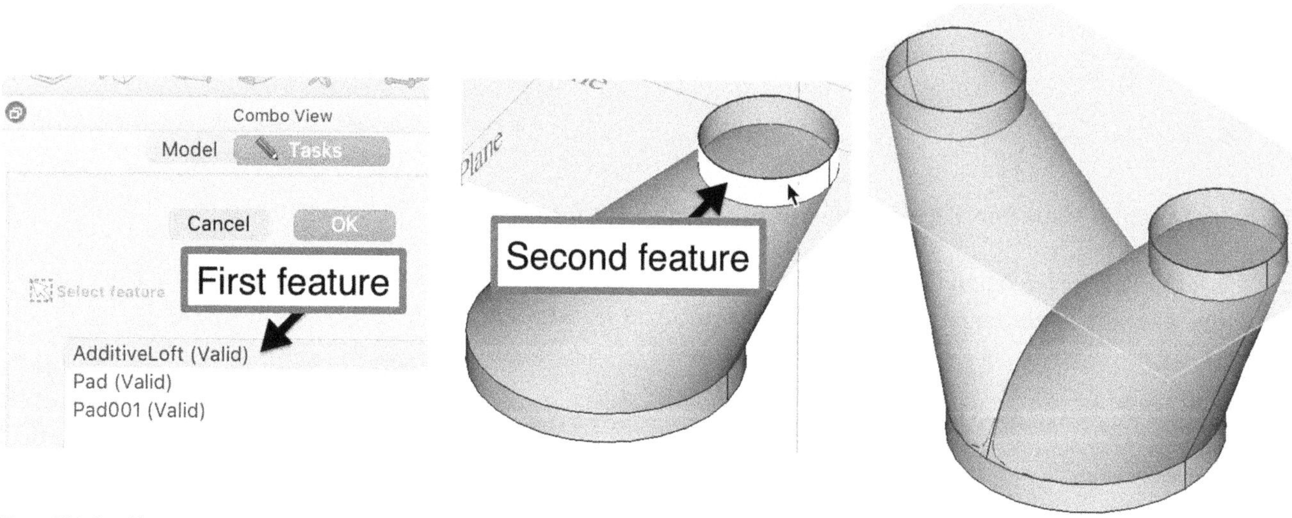

5. Click **File > Save** on the Menu bar. Next, enter C6_example2 in the **File name** box**.** Click **Save** to save the document
6. Click **File > Close** to close the document.

Questions

1. Describe the procedure to create a *Loft* feature.
2. What is the use of the **Ruled Surface** option?

Exercises

Exercise 1

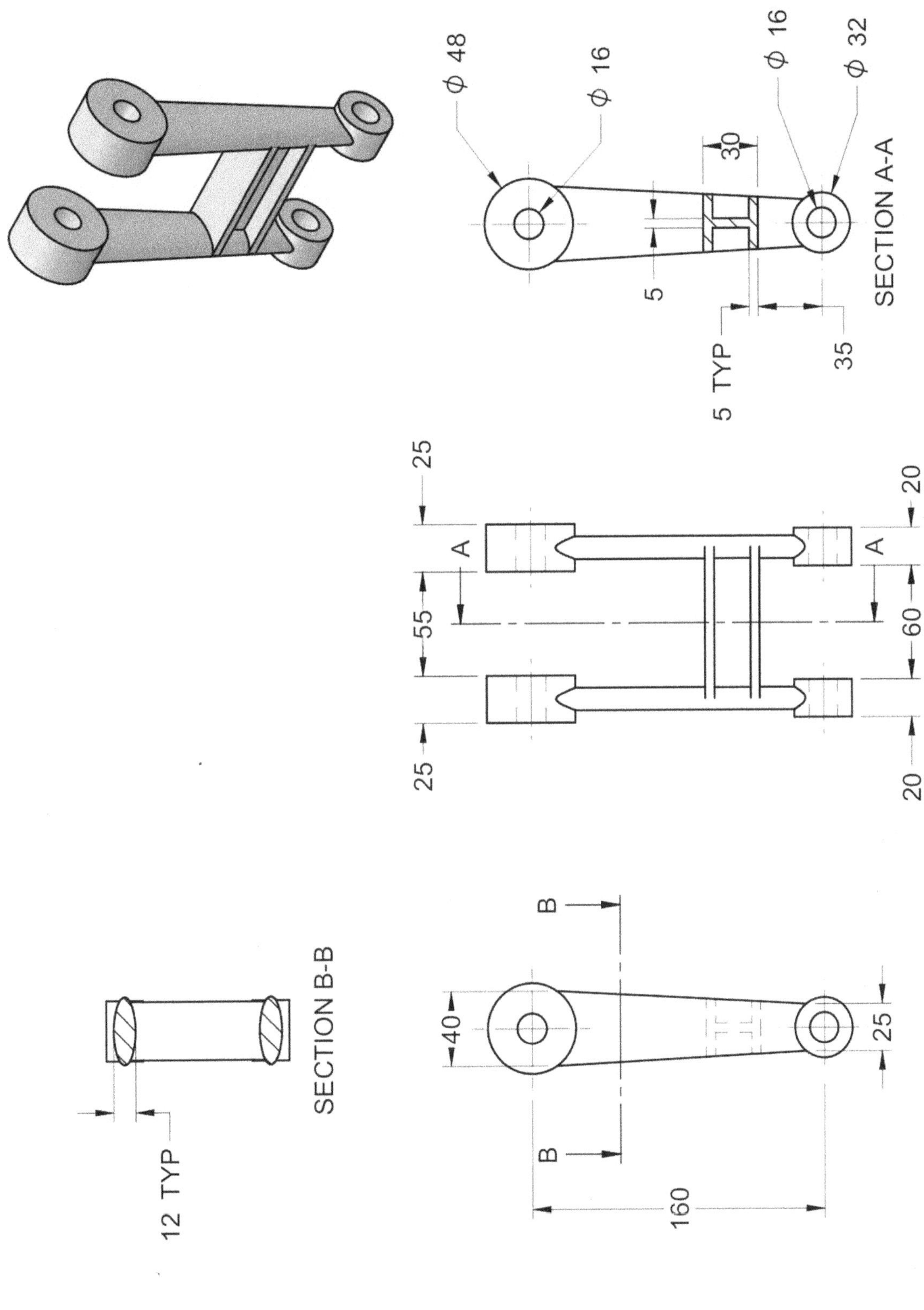

ϕ 48
ϕ 16
ϕ 16
ϕ 32
30
5
5 TYP
35
SECTION A-A
25
55
25
A
A
20
60
20
B
B
40
25
160
12 TYP
SECTION B-B

Exercise 2

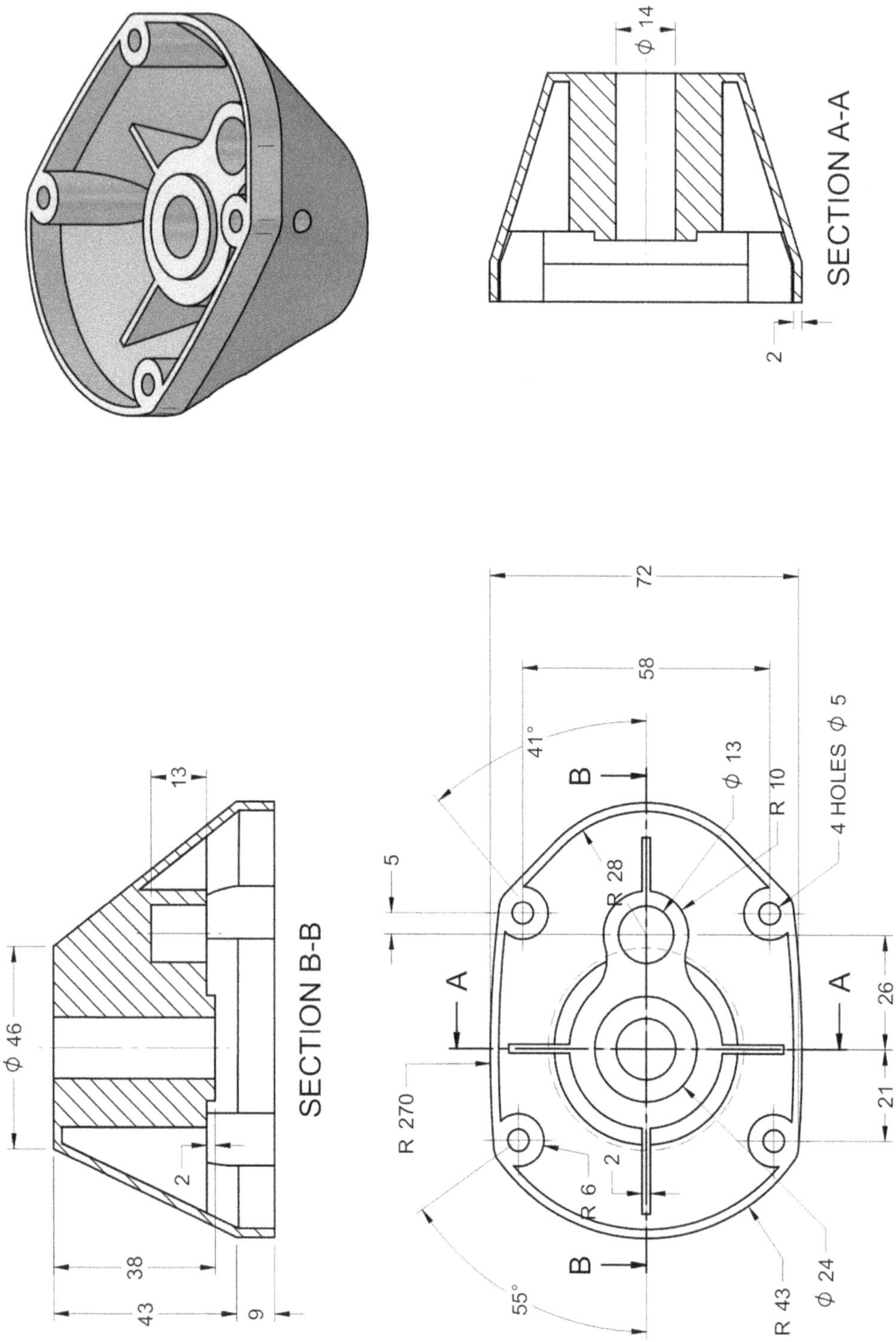

Chapter 8: Modifying Parts

In the design process, it is not required to achieve the final model in the first attempt. There is always a need to modify the existing parts to get the desired part geometry. In this chapter, you will learn various commands and techniques to make changes to a part.

The topics covered in this chapter are:

- *Edit Sketches*
- *Edit Features*

Edit Sketches

Sketches form the base of a 3D geometry. They control the size and shape of the geometry. If you want to modify the 3D geometry, most of the time, you are required to edit sketches. To do this, expand the feature in the **Model** tab of the **Combo View** panel. Next, right click on the sketch and select **Edit Sketch**. Now, modify the sketch and click **Close** on the **Combo View** panel. You will notice that the part geometry updates immediately.

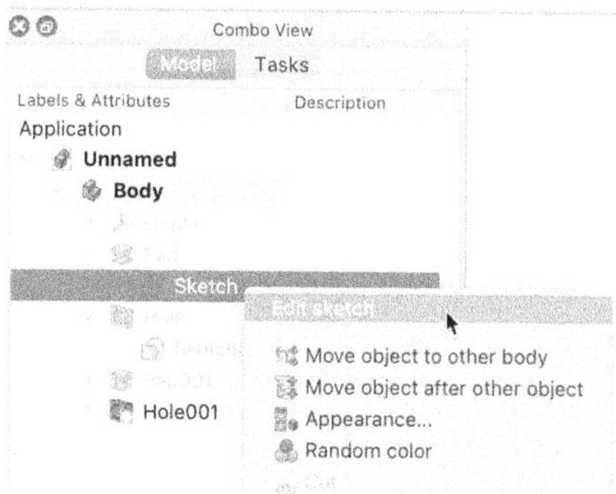

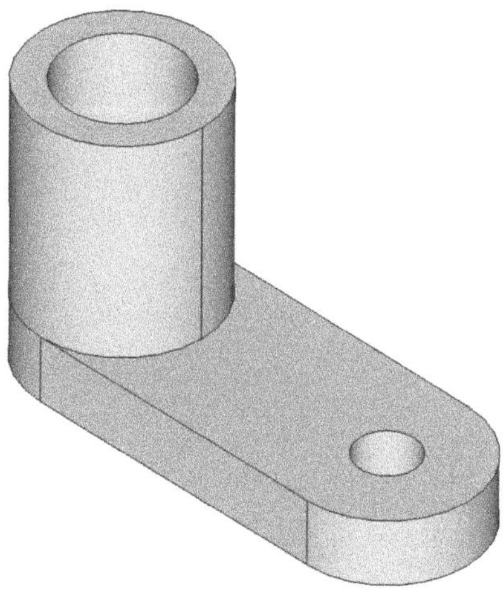

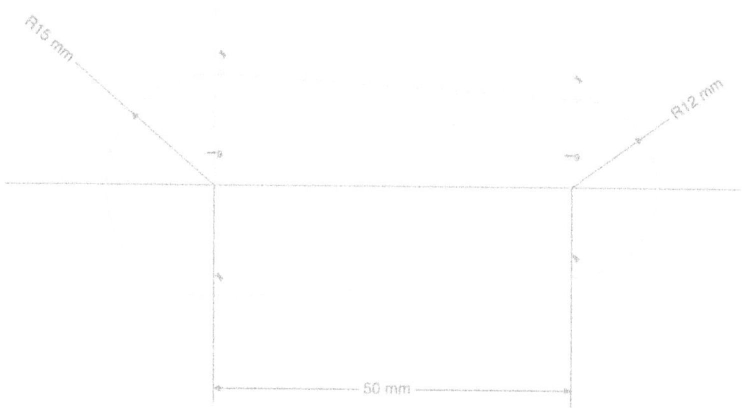

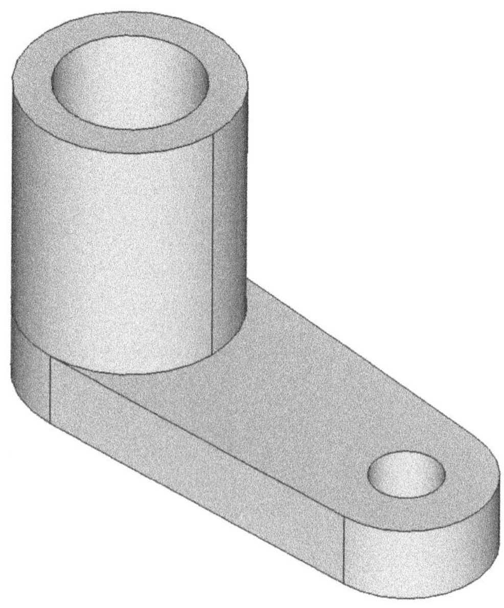

Edit Feature

Features are the building blocks of model geometry. To modify a feature, click the right mouse button on it and select **Edit Feature**. The parameters panel related to the feature appears. Next, modify the parameters of the feature and click **OK**. The changes take place immediately.

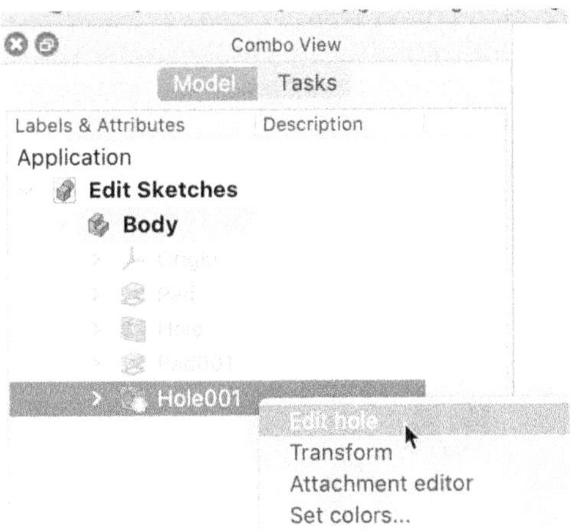

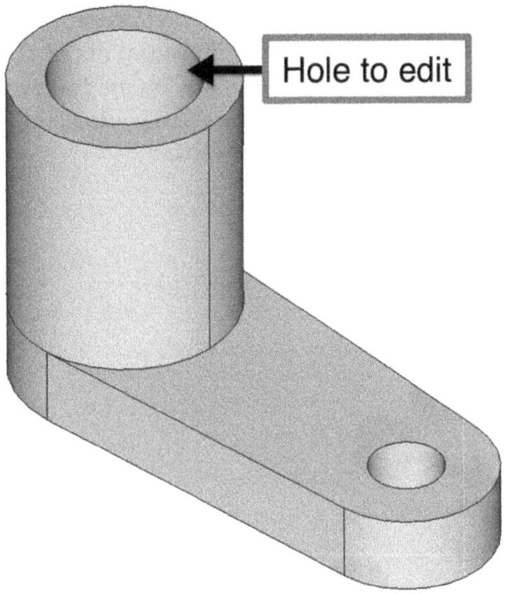

Tutorial 1 (Inches)

In this example, you create the part shown below and then modify it.

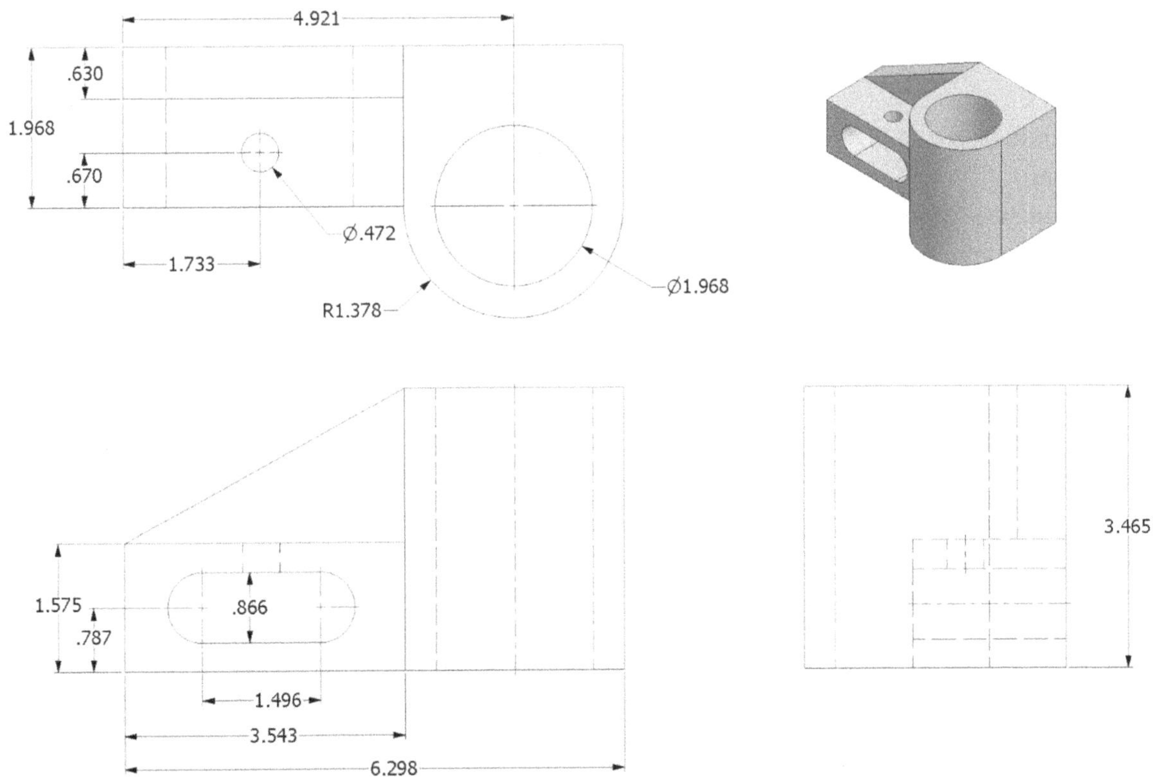

Creating a New document

1. Click **FreeCAD 0.19** on the desktop.
2. On the menu bar, click **File > New**; it creates a new document.
3. On the **Workbench** toolbar, select **Workbench** drop-down **> Part Design**.
4. Click **Edit > Preferences** on the **Menu** bar; the **Preferences** dialog appears on the screen.
5. Click **Units** tab and select **Unit system > Imperial decimal**.
6. Select **Number of decimals > 3** and click **OK** on the **Preferences** dialog.
7. Create the part using the tools and commands available in FreeCAD, as shown in the figure. You can also download this file from the companion website.

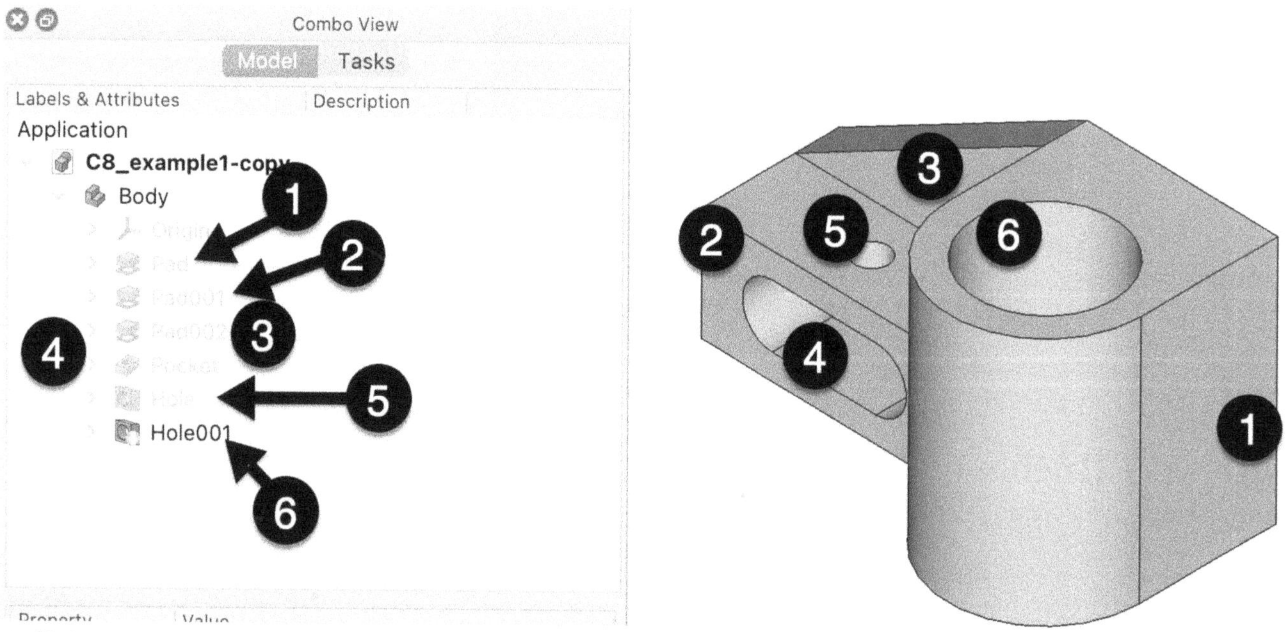

Editing a Feature

1. Click the right mouse button on the **Hole001** in the **Model** tab of the **Combo View** panel. Next, select **Edit hole** from the menu.
2. On the **Hole parameters** section, enter **1.378** in the **Diameter** box.
3. Select **Type > Counterbore** from the **Hole cut** section. Next, enter **1.968** and **0.787** in the **Diameter** and **Depth** boxes available under the **Hole cut** section, respectively.
4. Click **OK**.

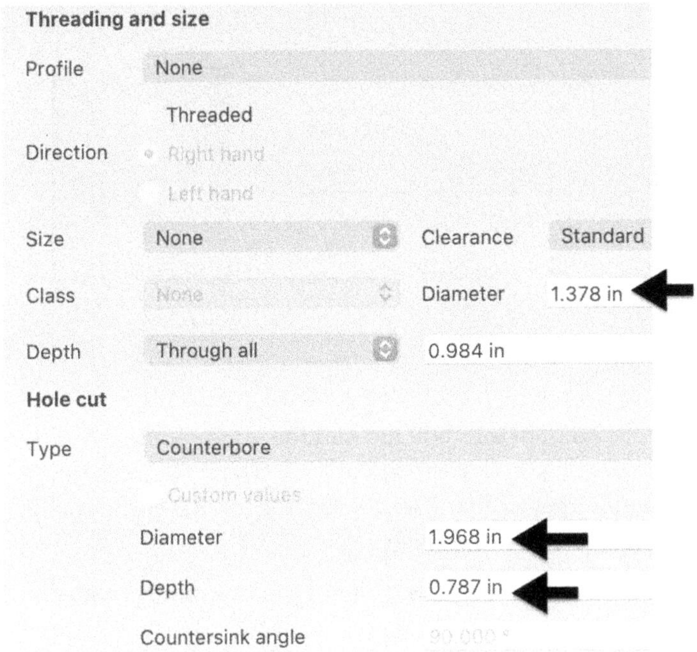

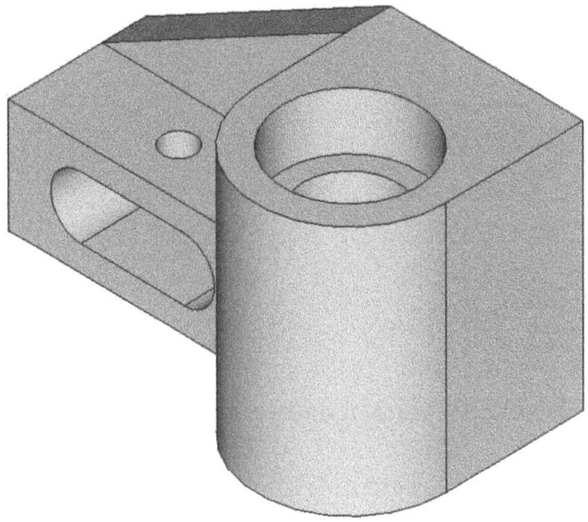

Editing Sketches

1. Expand the **Pad001** feature in the **Model** tab of the **Combo View** panel. Next, right-click on the *Sketch001* in and select **Edit sketch**. Modify the sketch, as shown. Click **Close** on the **Combo View** panel.

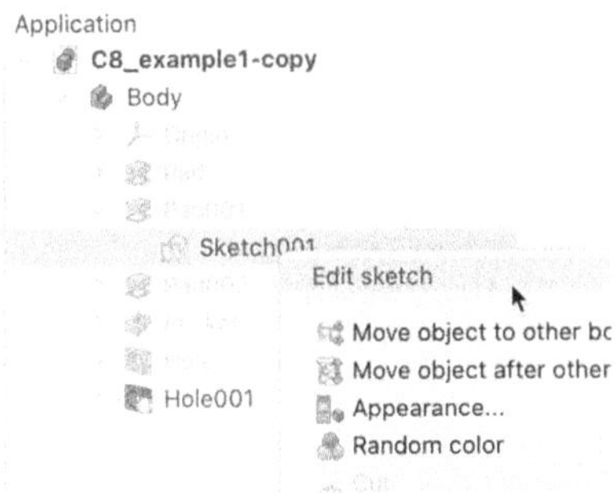

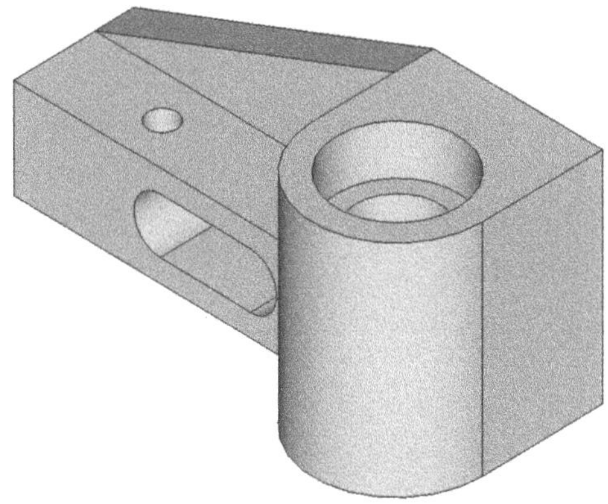

2. Expand the **Pocket** feature and click the right mouse button on *Sketch003*. Next, select **Edit sketch**.
3. Click the **External geometry**  icon on the **Sketcher geometries** toolbar. Next, select the left vertical edge.
4. Delete the horizontal distance constraint of the slot, as shown. Next, add a horizontal distance constraint between the center point of the left arc.

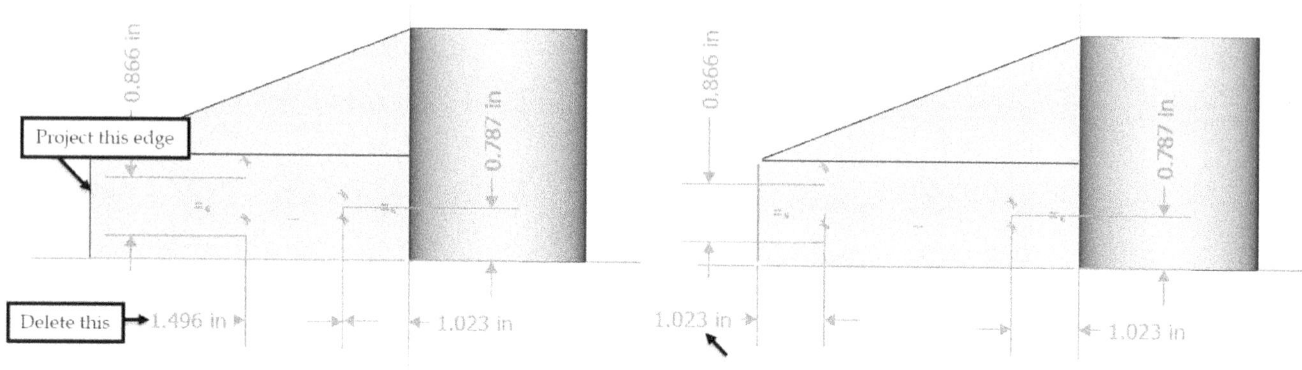

5. Delete the vertical distance constraint between the centerpoint of the right arc and the origin point.
6. Click the **Create polyline** icon on the **Sketcher geometries** toolbar. Next, select the top left corner point.
7. Select the centerpoint of the left arc. Next, select the bottom left corner point.
8. Click the **Constrain equal** icon on the **Sketcher constraints** toolbar. Next, select the two newly created lines.

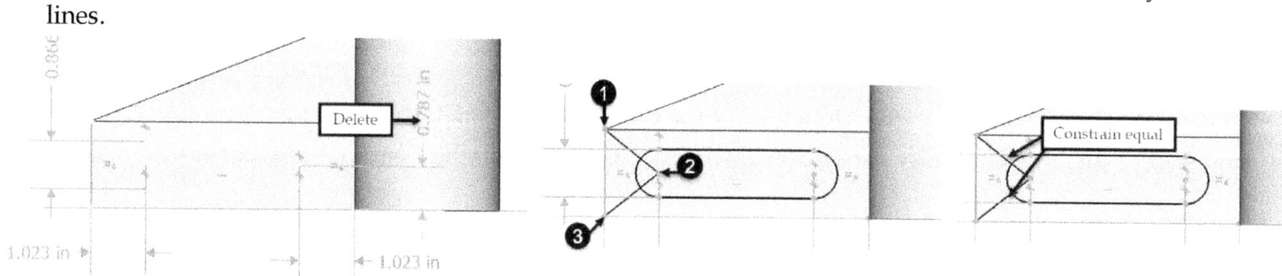

9. Select the two new lines, and click the **Toggle Construction geometry** icon on the **Sketcher geometries** toolbar.
10. Click **Close** on the **Combo View** panel.

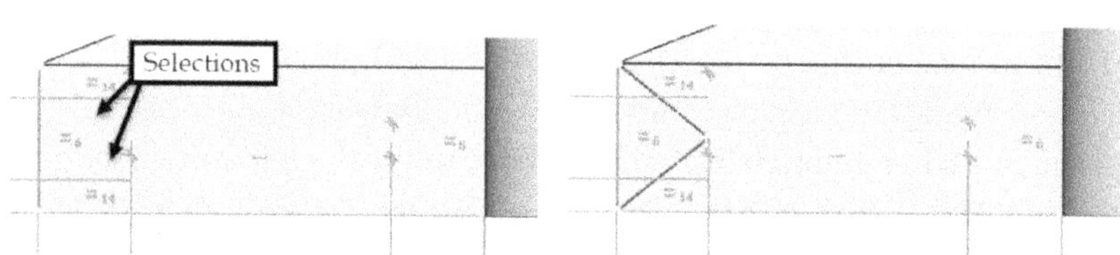

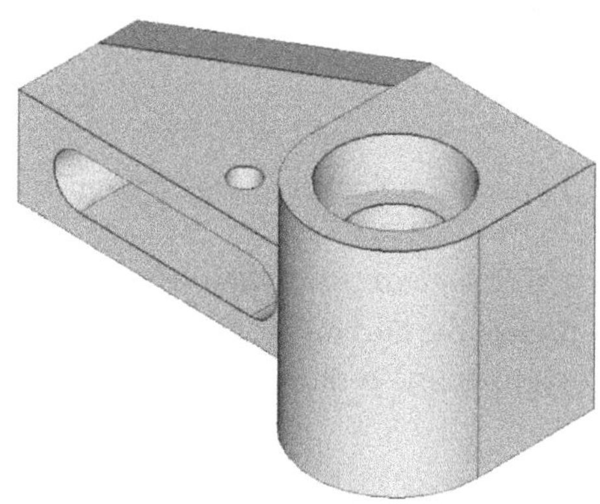

11. In the **Model** tab of the **Combo View** panel, expand the first **Hole** and click the right mouse button on the Sketch. Next, select **Edit sketch** from the menu.
12. Delete the vertical and horizontal distance constraints.
13. Create three lines connecting the centerpoint of the circle and the corner points, as shown.
14. Click the **Constrain equal** icon on the **Sketcher constraints** toolbar. Next, select two lines to make them equal in length. Select the third and second lines to make them equal.

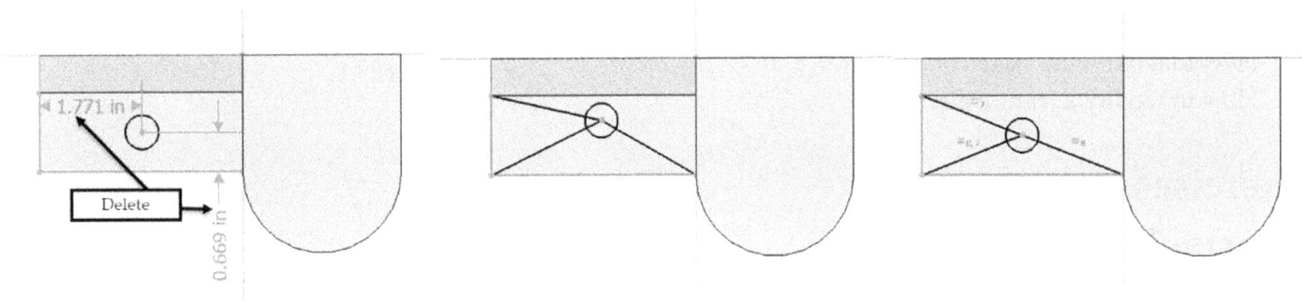

15. Select all the three lines and click the **Toggle construction geometry** icon on the **Sketcher geometries** toolbar.
16. Click **Close** on the **Combo View** panel.

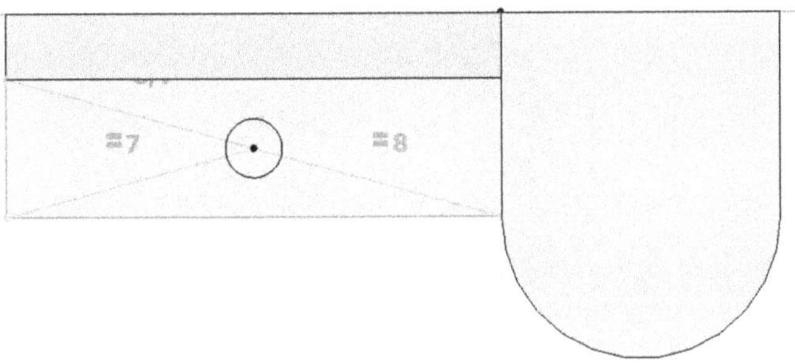

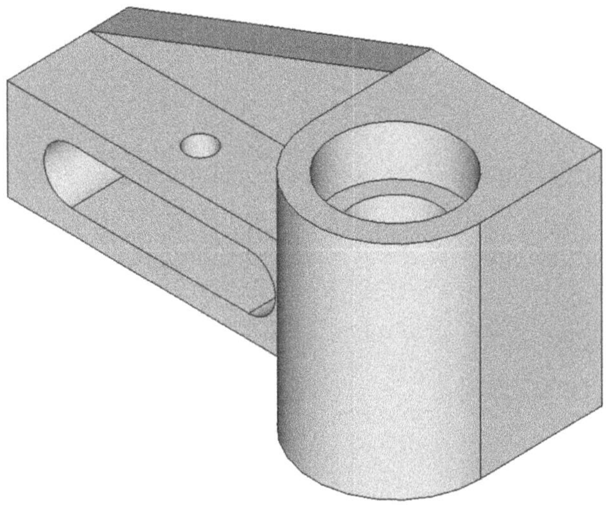

17. Now, change the size of the rectangular pad feature. Notice that the slot and hole are adjusted automatically.

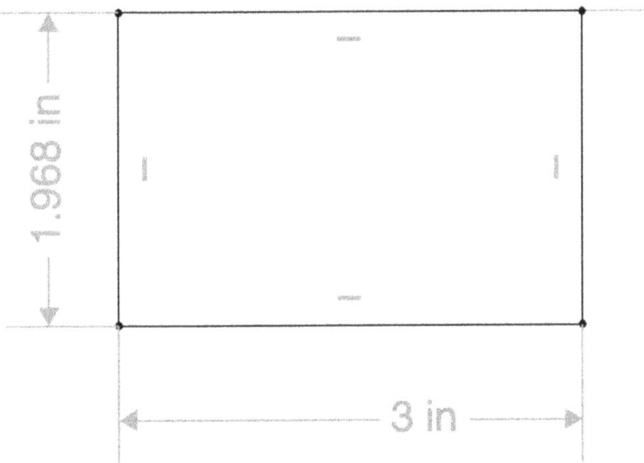

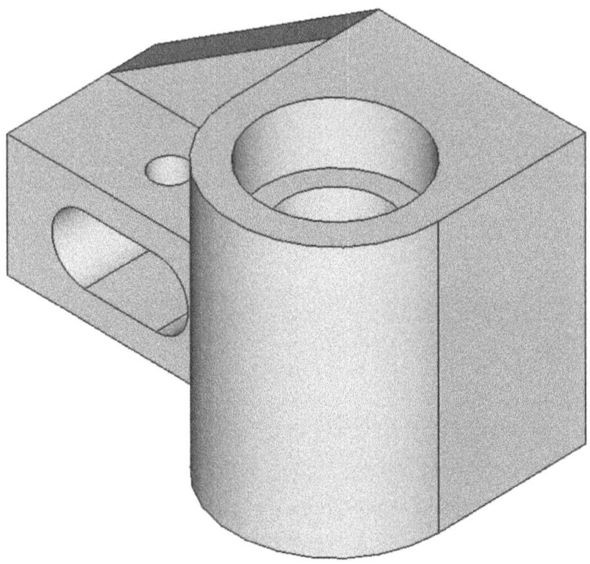

18. Save and close the file.

Questions

1. How to modify the sketch of a feature?
2. How to modify a feature directly?

Exercises

Exercise 1

Initial Part

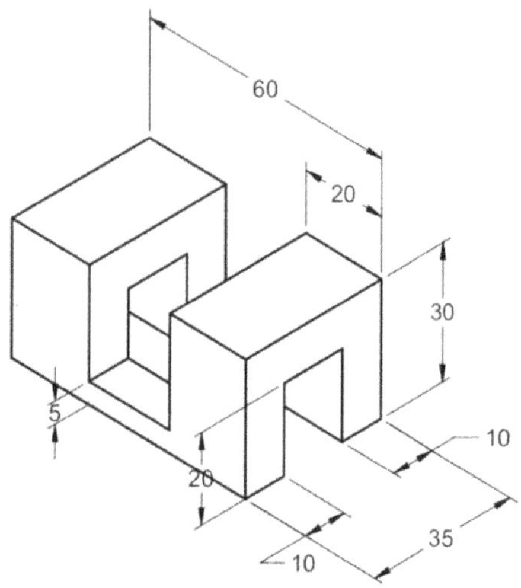

Modified Part

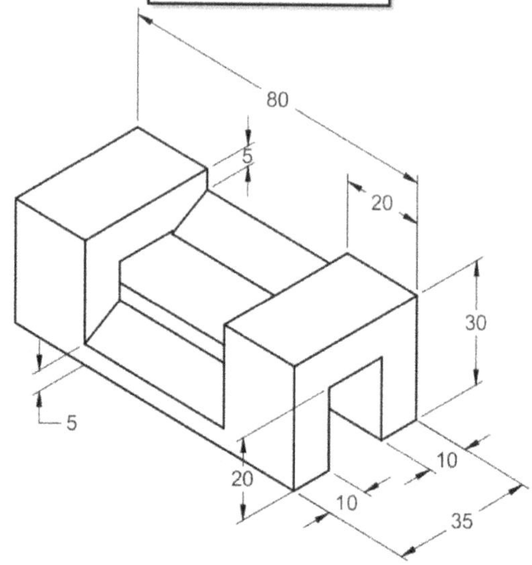

Chapter 9: Assemblies

After creating individual components, you can bring them together into an assembly. By doing so, it is possible to identify incorrect design problems that may not have been noticeable at the component level. In this chapter, you will learn how to bring components together and create real-life movements between them.

The topics covered in this chapter are:

- *Starting an assembly*
- *Inserting Components*
- *Moving and Rotating parts*
- *Adding Constraints*
- *Editing Components*
- *Sub-assemblies*

Starting an Assembly

In FreeCAD 0.19, there is no in-built workbench to create an assembly. You need to install the **A2plus** workbench to create assemblies. To do this, click **Tools > Addon Manager** on the menu bar and click **OK**. Next, select the **A2plus** addon from the **Workbenches** list and click the **Install/update selected** button. Click the **Close** button after installing the selected addon.

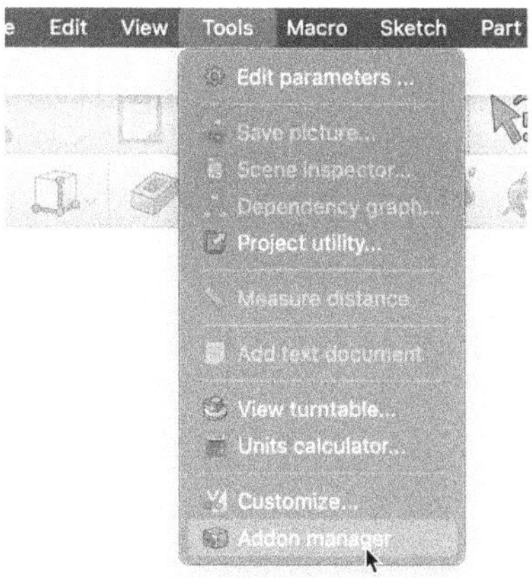

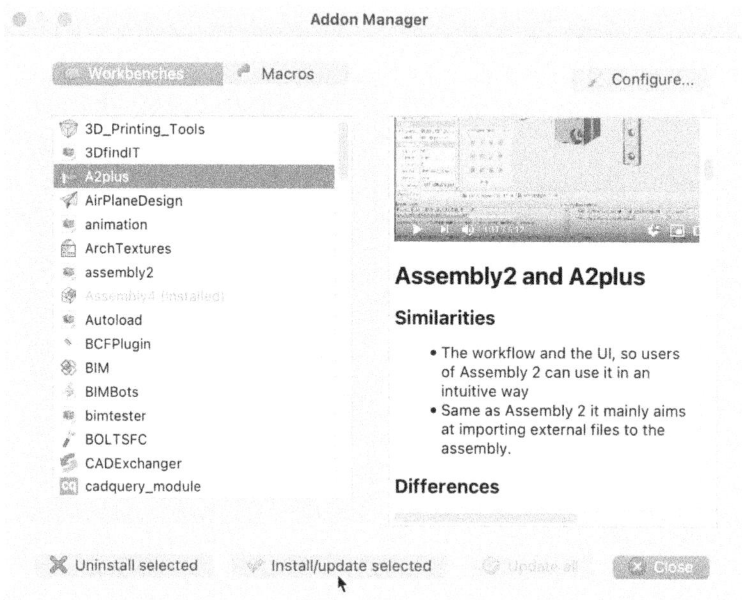

Click **OK** to restart the FreeCAD application. Next, select the A2plus workbench from the **Workbenches** drop-down. Click the **Create a new empty document** icon to open a new document.

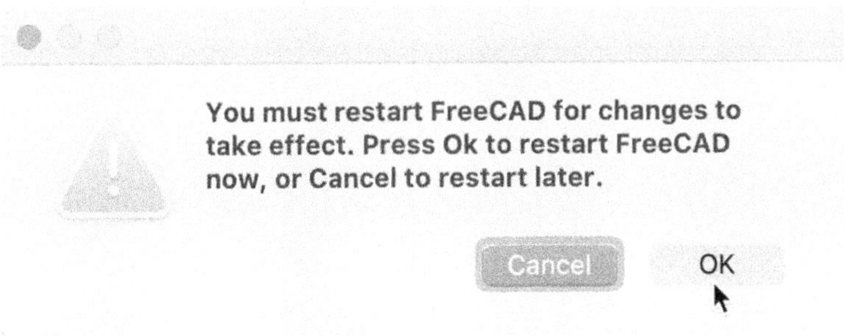

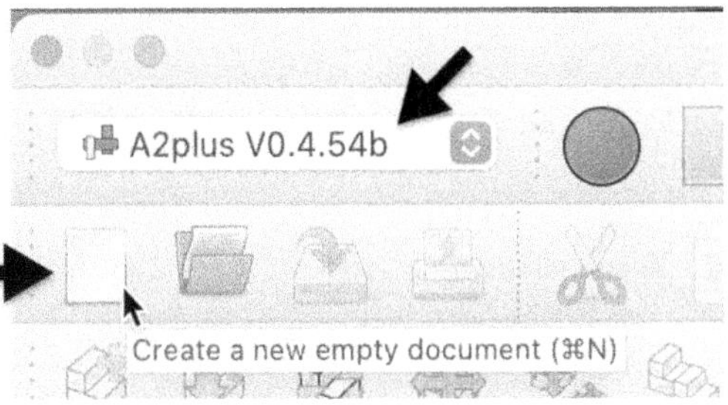

Inserting Components

To insert components into a design file, first, you need to save it. Next, Click the **Add a part from an external file** icon on the **A2p_Part** toolbar (or) click **A2plus > Add a part from an external file** on the menu bar. Go to the location of the parts and select the component. Next, click **Open** or double-click on the part to insert.

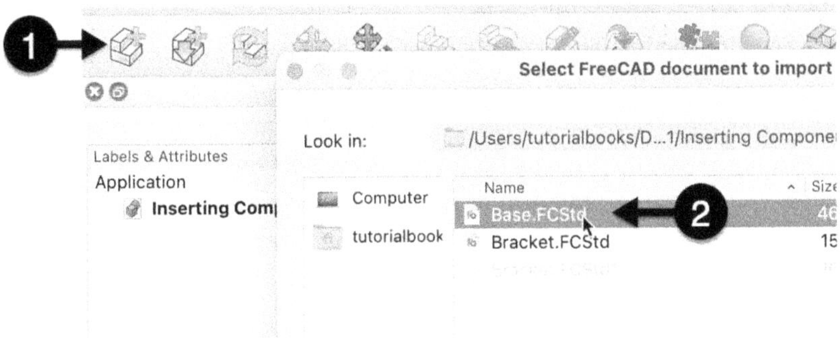

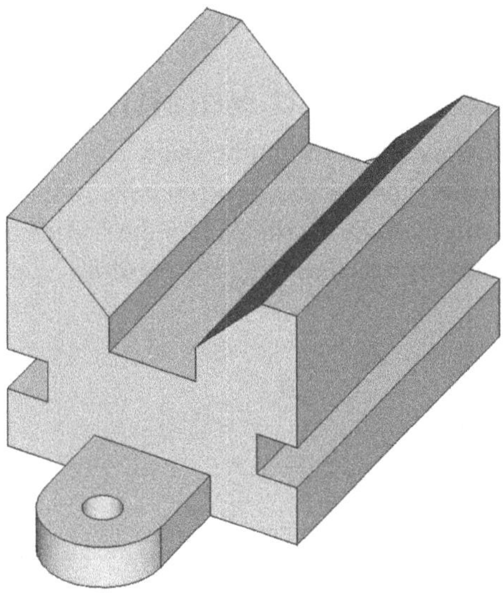

Likewise, insert the second component into the design.

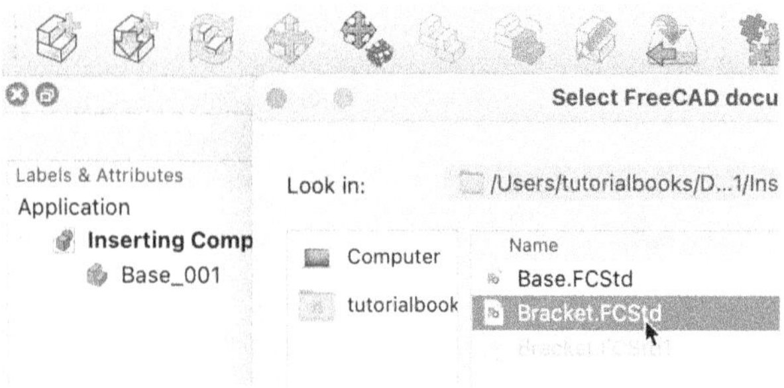

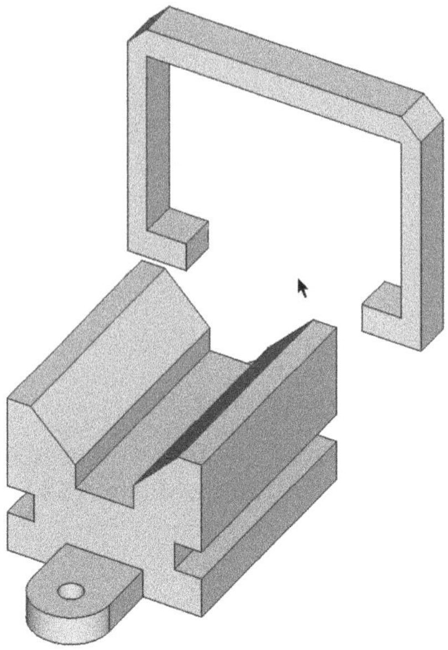

Moving and Rotating Parts

To move the component, select it and click the **Move the selected part** icon on the **A2p_Part** toolbar (or) click **A2plus > Move the selected part** on the menu bar. Now, use the manipulator to move the part. For example, to move the part in the X-direction, select the X-axis handle and move the part (press and hold the left mouse button and drag the pointer).

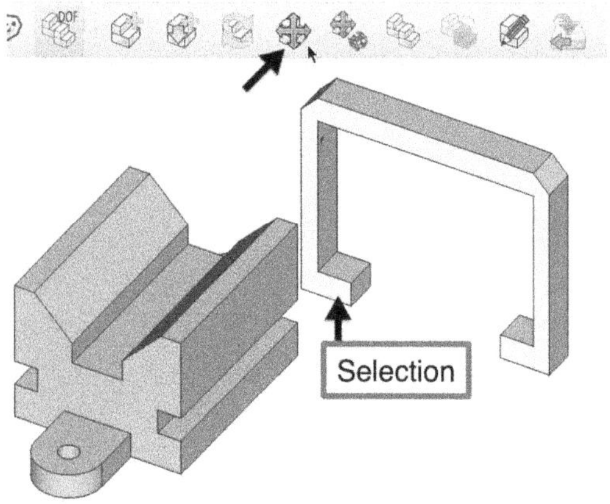

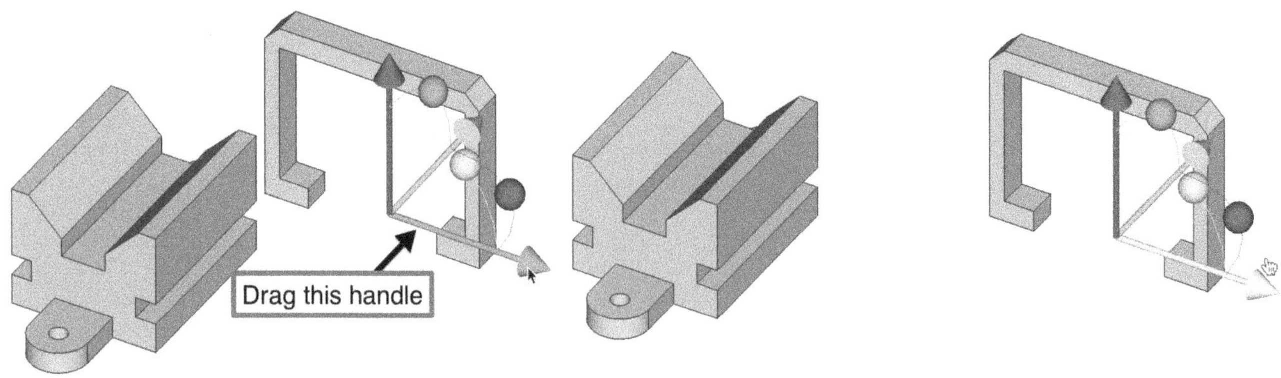

Use the rotate handles on the manipulator to rotate the part about an axis. For example, to rotate the part about the X-axis, select the sphere handle displayed between the Y and Z axes and rotate the part (press and hold the left mouse button and drag the pointer).

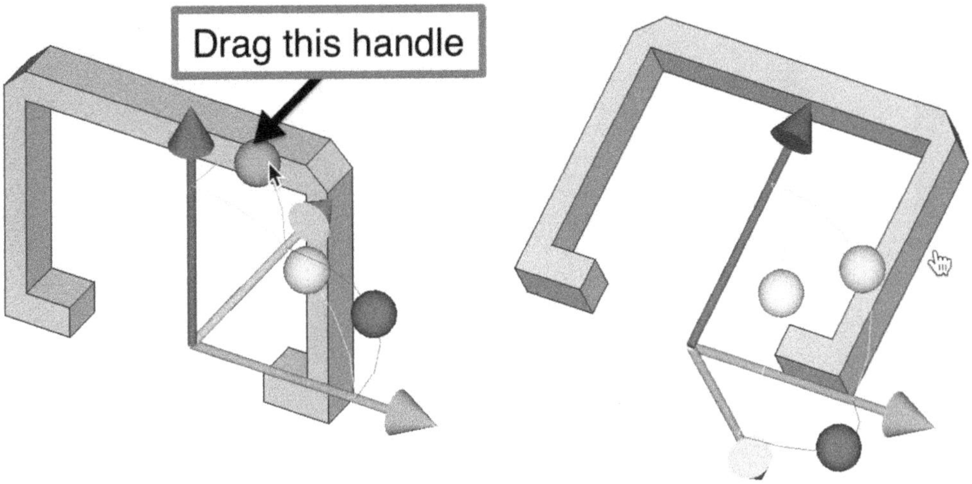

Click **OK** on the **Combo View** panel to complete the move operation.

Applying Constraints

After inserting parts into an assembly, you have to apply constraints between them. By applying constraints, you can make parts to flush with each other or make two round faces concentric with each other. As you add constraints between parts, the degrees of freedom will be removed from them. By default, there are six degrees of freedom for a part (three linear and three rotational). Eliminating the degrees of freedom will make parts attached and interact with each other as in real life. Now, you will learn to add constraints between parts.

Coincident Plane Constraint

The **Coincident Plane** constraint makes two faces coincident with each other. The orientation of the two coincident faces depends on the type of Direction that you select: **Opposed** or **Aligned**.

The Opposed direction

The **Opposed** direction will make the selected faces coincident with each other. In addition to that, the orientation of the faces will be opposite to each other. To apply this solution, first click the **Define Constraints** icon on the **A2p_Constraint** toolbar (or) click **A2plus > Constraint > Define Constraints** on the menu bar. Press and hold the CTRL key (Command key for Mac users) and select the faces of the two parts, as shown. On the **Constraint Tools** dialog, click the **Coincident Plane Constraint** icon. Next, select **Direction > Opposed** from the **Constraint Properties** dialog. The two selected faces will mate with each other.

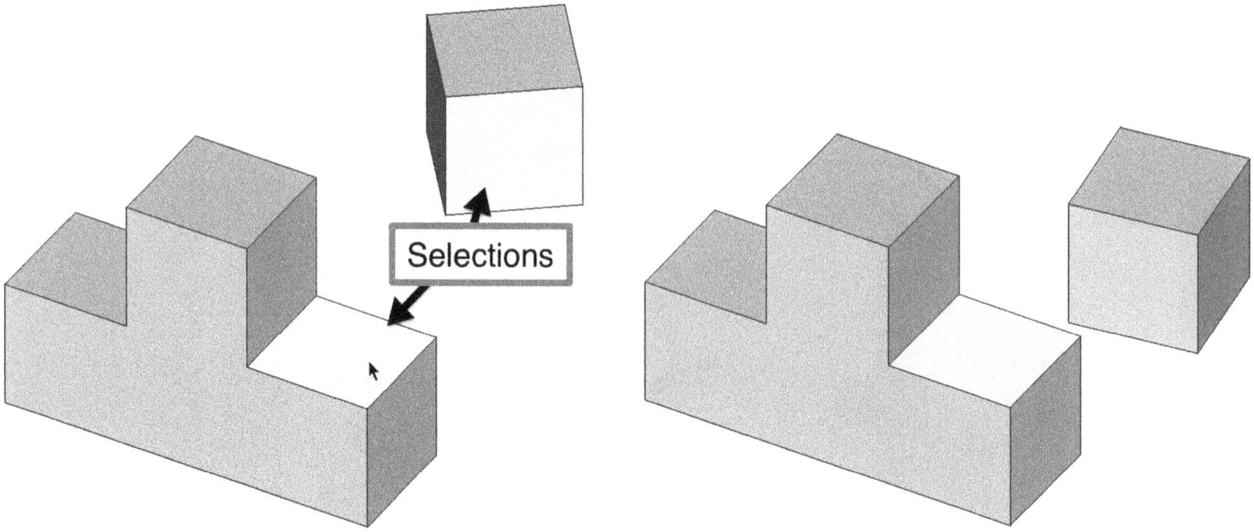

The Aligned Direction

The **Aligned Direction** will make the selected faces coincident with each other and oriented in the same direction. Select **Direction > Aligned** from the **Constraint Properties** dialog; the two faces will be aligned in the same direction.

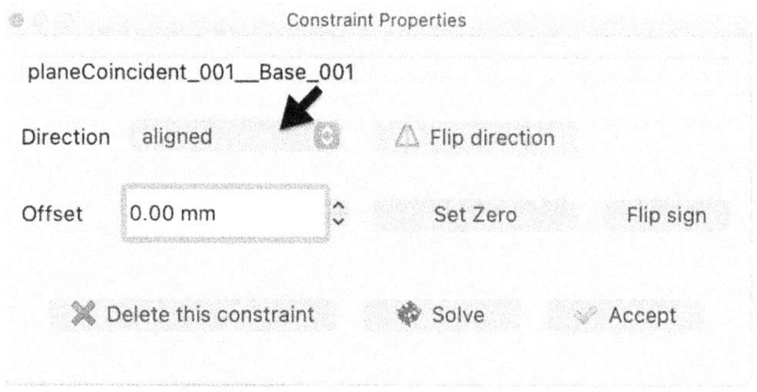

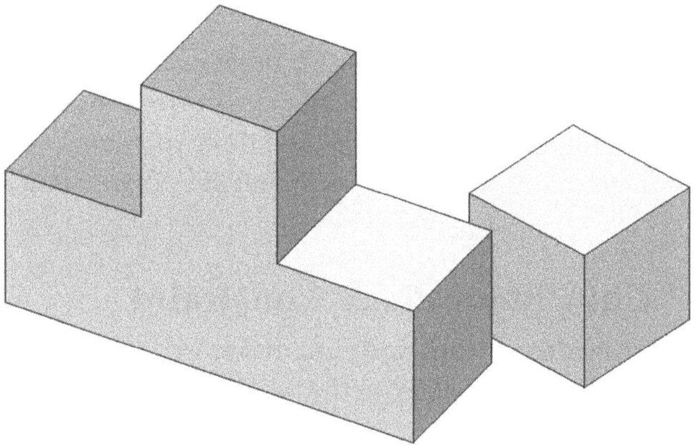

You can also enter an offset value in the **Offset** box on the **Constraint Properties** dialog. The constrained components are offset from each other. Next, click the **Accept** button to create the Coincident Plane Constraint between the two selected faces.

Axis Coincident Constraint

The **Axis Coincident Constraint** makes the axes of two cylindrical faces coincide with each other. Press and hold the CTRL key (Command key for Mac users) click on the round faces of the placement and target parts. Click the **Add axis coincident constraint** icon on the **A2p_Constraint** toolbar (or) click **A2plus > Constraint > Add axis Coincident constraint** on the menu bar; the two cylindrical axes will coincide with each other.

You can notice that there are options in the **Direction** drop-down: **Opposed**, and **Aligned**.

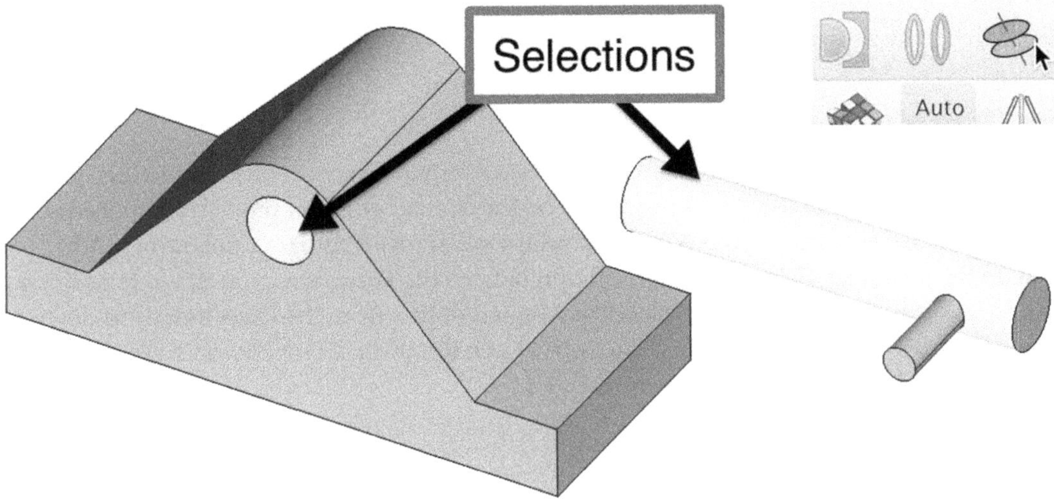

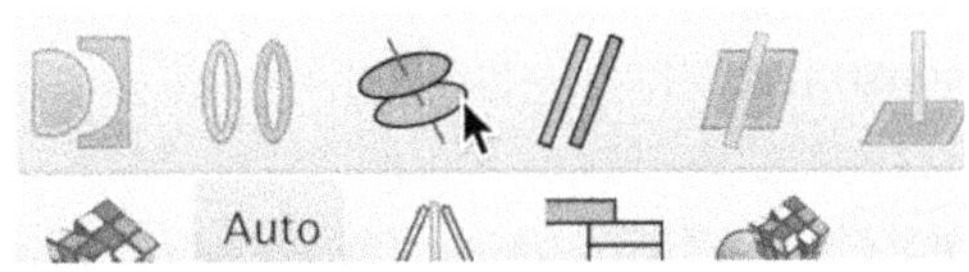

The **Opposed** option positions the two selected axes or edges in the direction opposite to each other.

The **Aligned** option will align the parts' axes in the same direction.

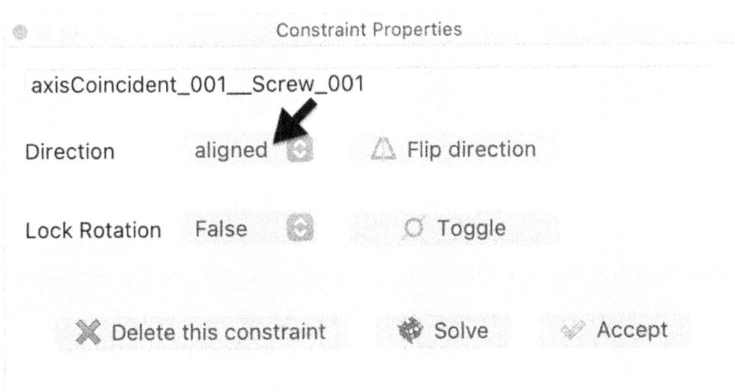

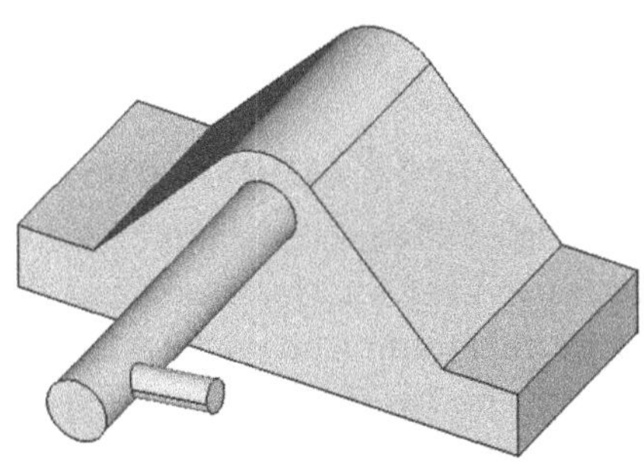

Select **Lock Rotation** > **True**, if you want to lock the rotation of the placement part. Click **Accept** to create the constraint.

Circular Edge Constraint

The **Circular Edge** constraint helps you to make circular edges of two components concentric. To do this, press and hold the CTRL key (Command key for Mac users) and click on the circular edges of the two components. Next, click the **Circular Edge constraint** icon on the **A2p_constraint** toolbar (or) **A2plus > Constraint > Add circularEdge constraint** on the menu bar. Select the **Opposed** option from the **Direction** drop-down to position the faces of the circular edges opposite to each other. Select the **Aligned** option from the **Direction** drop-down to position the faces of the circular edges in the same direction. Select **Lock Rotation** > **True**, if you want to lock the rotation of the placement part. Click **Accept** to create the constraint.

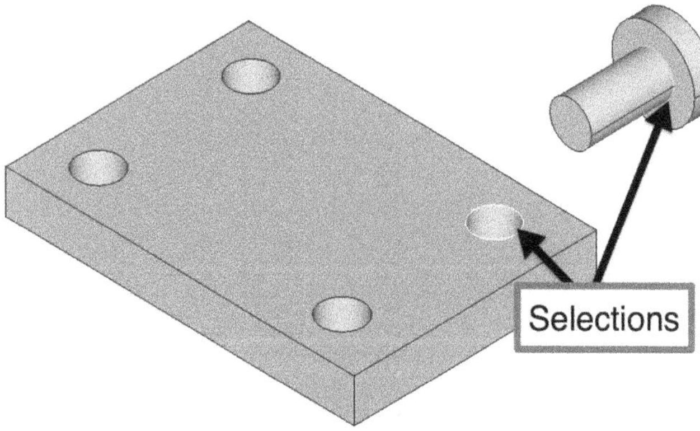

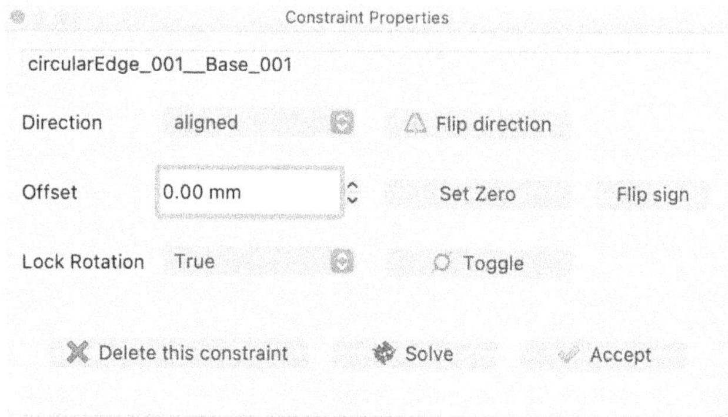

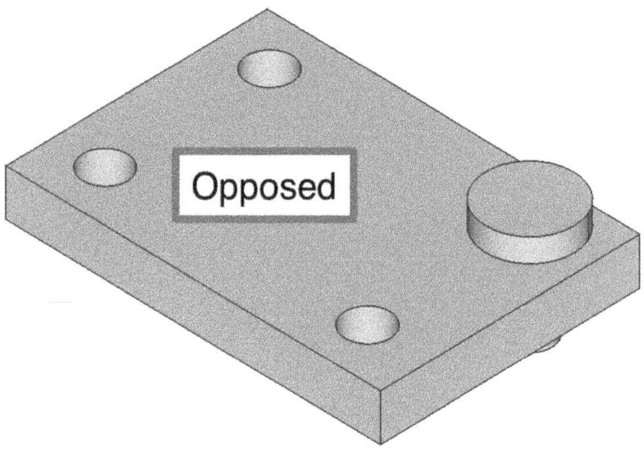

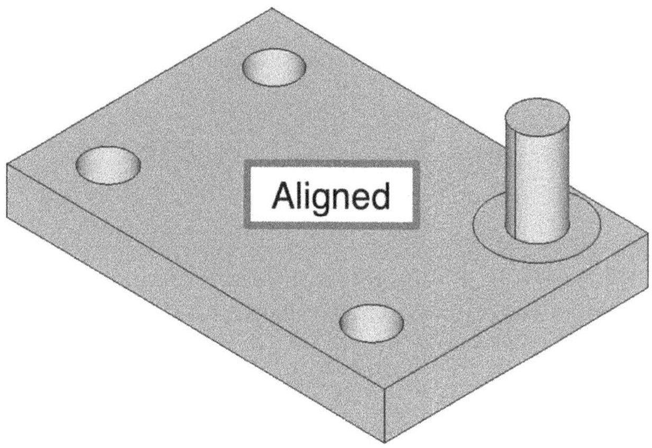

Angled Planes Constraint

The **Angled Planes** constraint is used to position faces at a specified angle. Press and hold the CTRL key (Command key for Mac users) and click on planar faces of the first and second parts. Next, click the **Angled Planes constraint** icon on the **A2p_constraint** toolbar (or) **A2plus > Constraint > Add angledPlanes constraint** on the menu bar. Type in a value in the **Angle** box on the dialog. The first part will be positioned at the specified angle. Click the **Perpendicular** button if you want to add 90 degrees to the angle. Click the **Accept** button to create the constraint.

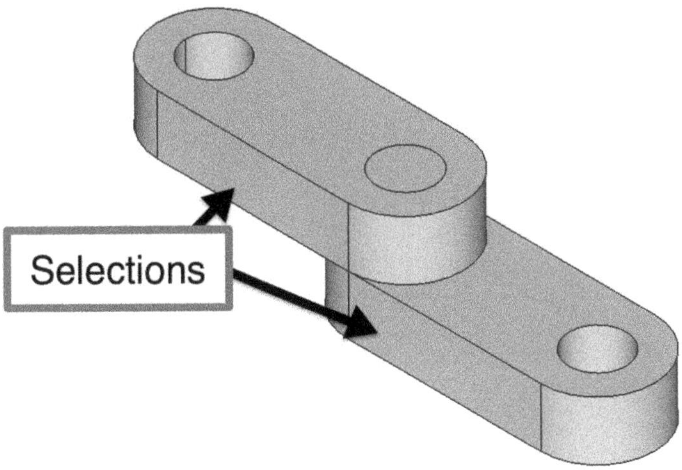

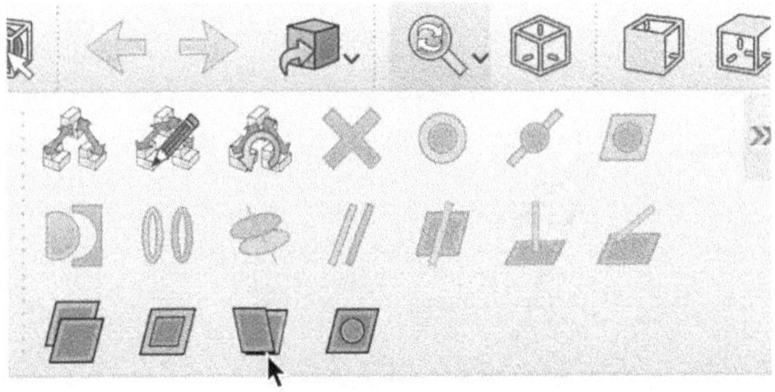

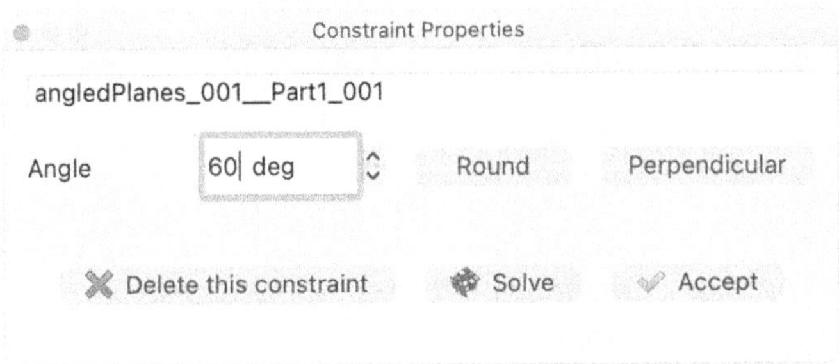

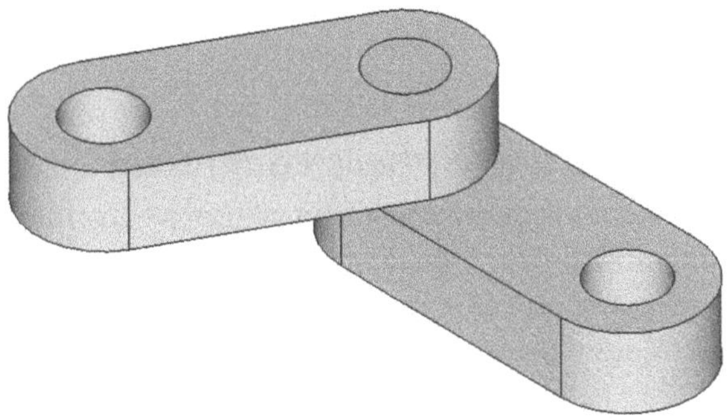

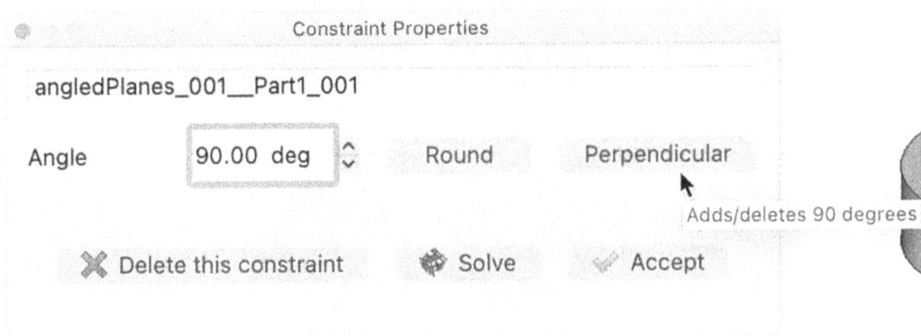

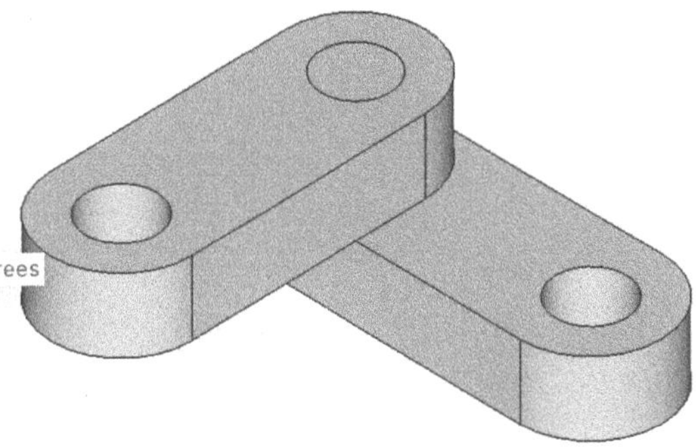

Planes Parallelism Constraint

The **Planes Parallelism constraint** is used to apply a parallel constraint between the two selected faces. Press and hold the CTRL key (Command key for Mac users) and click on planar faces of the first and second parts. Next, click the **Planes Parallelism constraint** icon on the **A2p_constraint** toolbar (or) **A2plus > Constraint > Add planesParallel constraint** on the menu bar. Select the **Opposed** option from the **Direction** drop-down to position the selected faces opposite to each other. Select the **Aligned** option from the **Direction** drop-down to position the selected faces in the same direction. Click the **Accept** button to create the constraint.

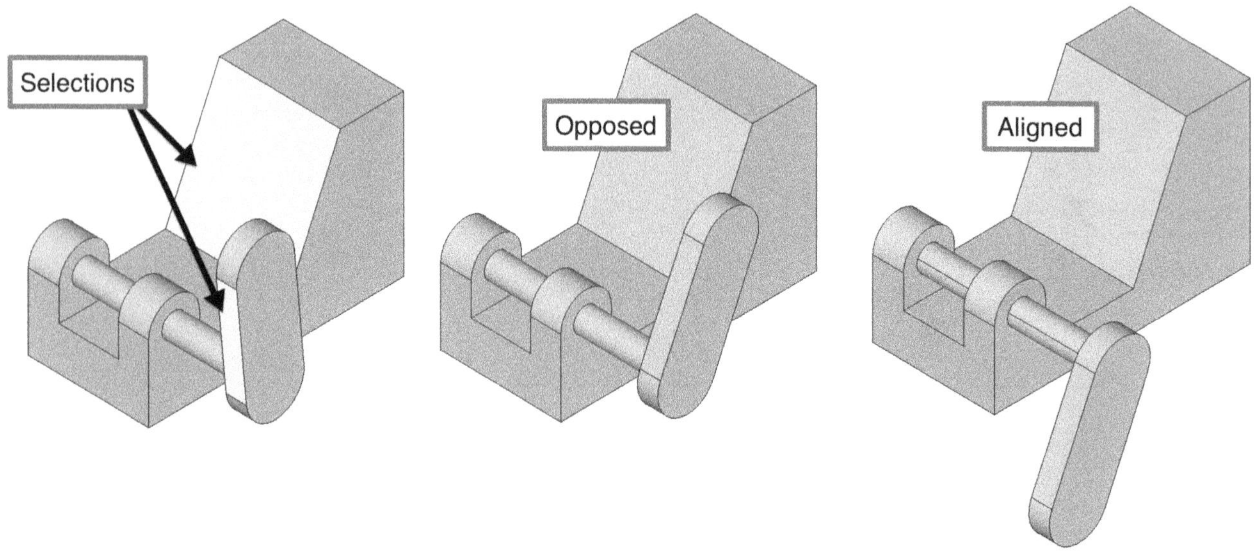

Sphere to Sphere constraint

The **Sphere to sphere constraint** creates a constraint between two spherical faces. Press and hold the CTRL key (Command key for Mac users) and click on spherical faces of the first and second parts. Next, click the **Sphere-to-sphere constraint** icon on the **A2p_constraint** toolbar (or) **A2plus > Constraint > Add sphereCenterIdent constraint** on the menu bar. Click the **Accept** button to create the constraint.

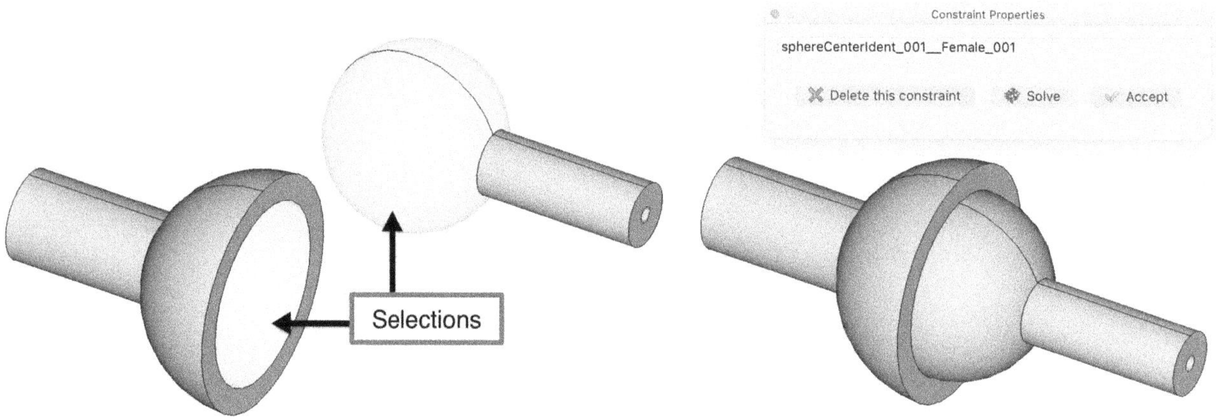

Editing an Imported Part

During the design process, the correct design is not achieved on the first attempt. There is always a need to go back and make modifications. FreeCAD allows you to accomplish this process very quickly. To modify a part in an assembly, click on it and click **Edit an imported part** icon on the **A2p_Part** toolbar; the part model will be opened. Make changes to the part and save and close the part file. Next, switch to the assembly file and click the **Update** parts icon on the **A2p_Part** toolbar; the modified part is updated in the assembly.

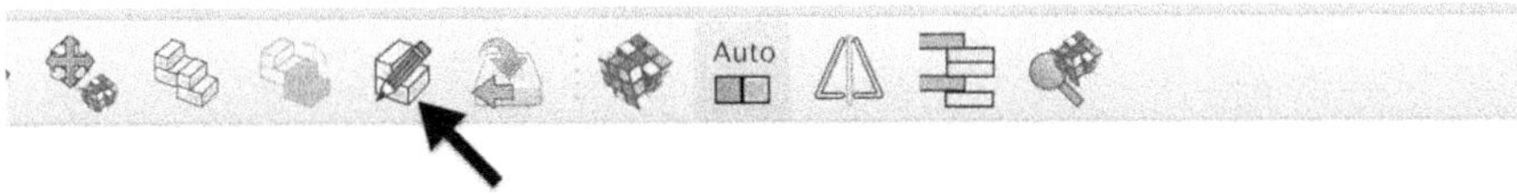
Auto

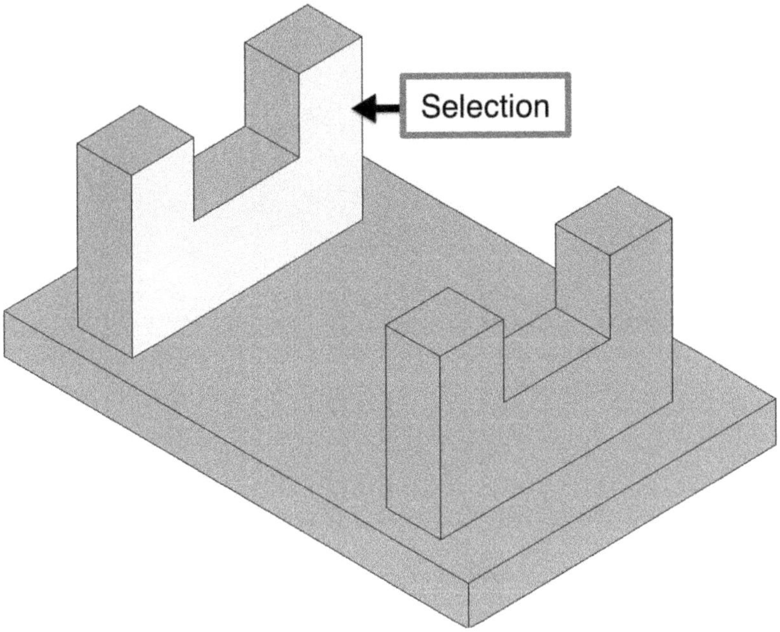
Selection

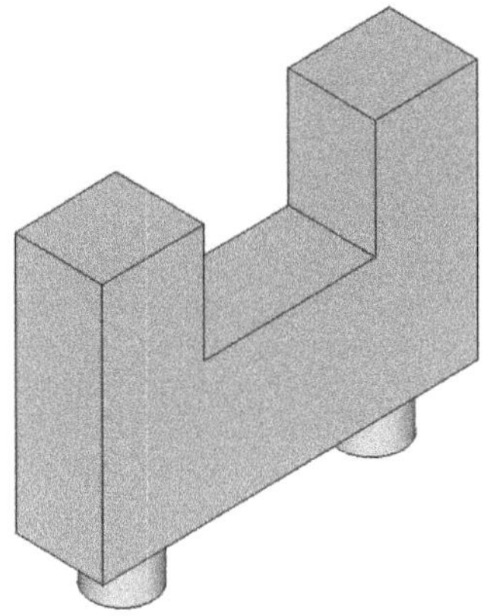

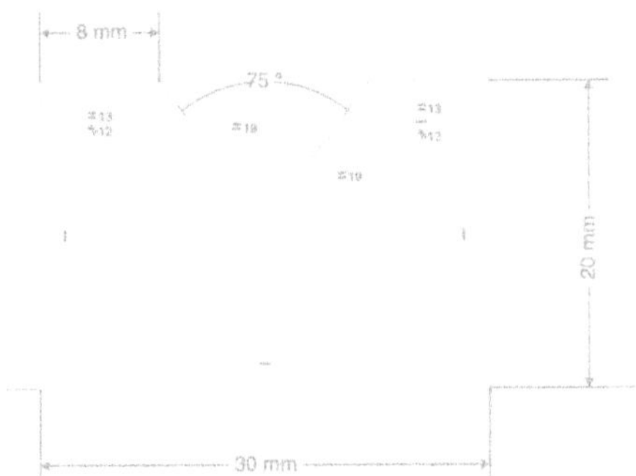
8 mm
75 °
20 mm
30 mm

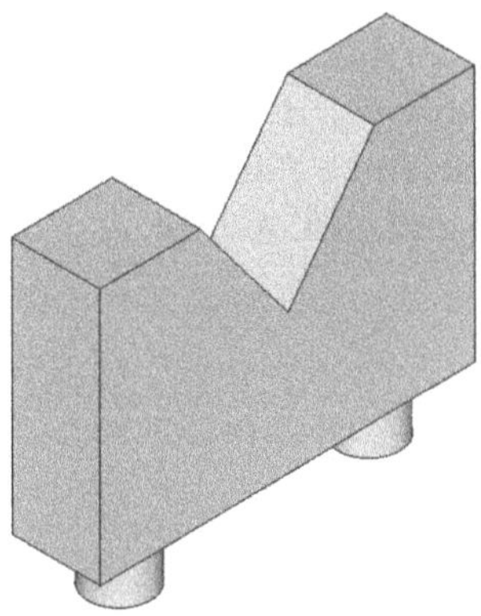

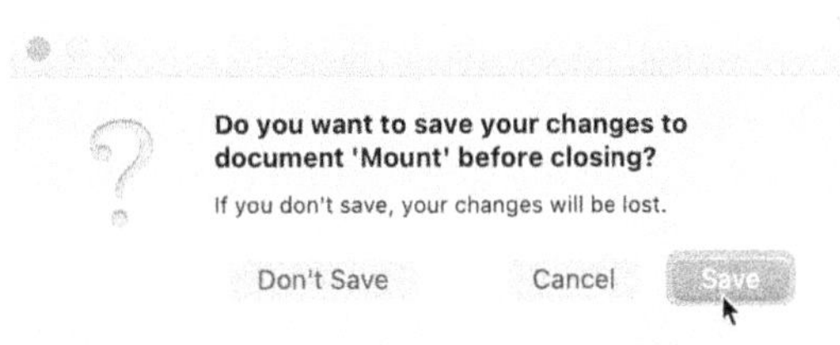
Do you want to save your changes to document 'Mount' before closing?
If you don't save, your changes will be lost.
Don't Save
Cancel
Save

Assembly : 1*
Mount : 1*

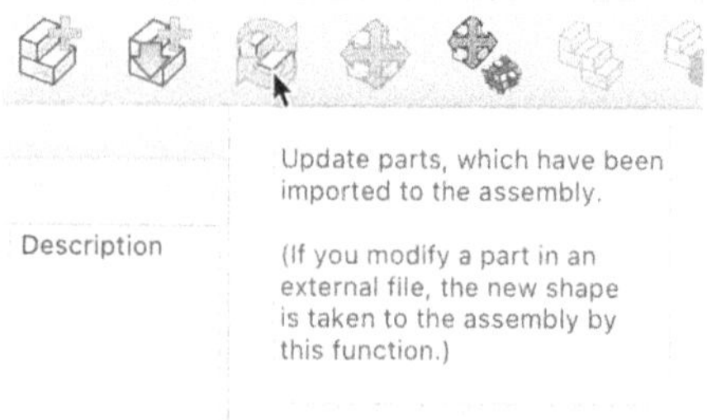

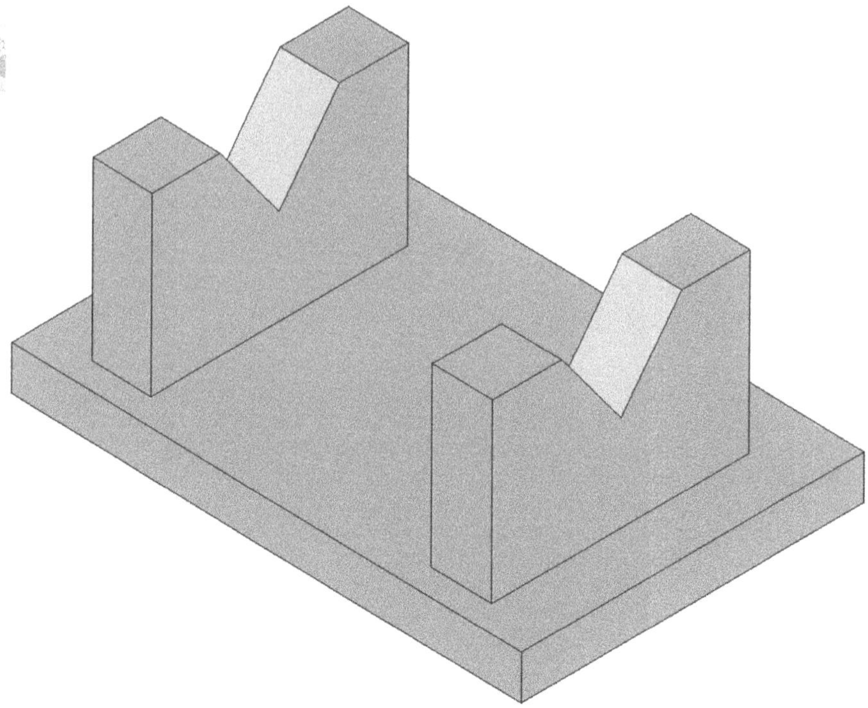

Edit selected constraint

You can also edit constraints in an assembly. To do this, expand the part model in the Model tab of the Combo View panel. Next, select the constraint to be modified and click the Edit selected constraint icon on the **A2p_constraint** toolbar (or) click **A2plus > Constraint > Edit selected constraint** on the menu bar. Next, change the settings on the **Constraint Properties** dialog and click **Accept**.

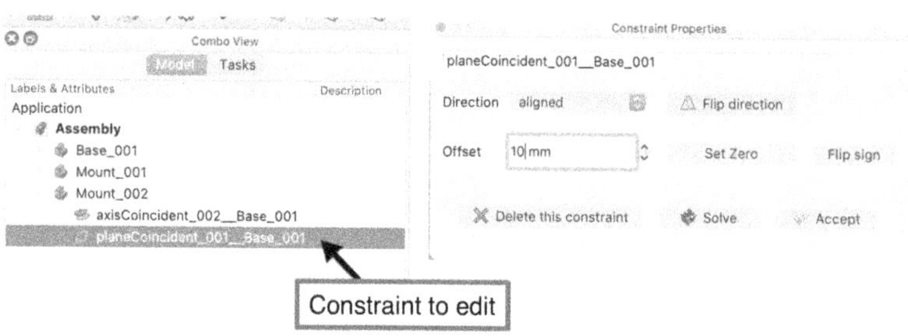

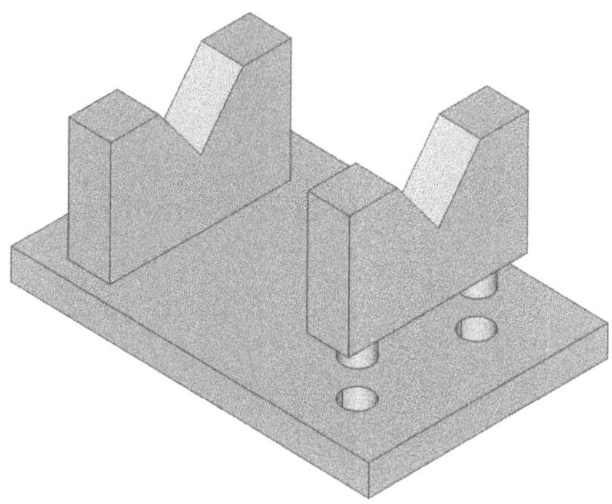

Sub-assemblies

The use of sub-assemblies has many advantages in FreeCAD 0.19. Sub-assemblies make large assemblies easier to manage. They make it easy for multiple users to collaborate on a single large assembly design. They can also affect the way you document a large assembly design in 2D drawings. For these reasons, it is important for you to create sub-assemblies in a variety of ways. The easiest way to create a sub-assembly is to insert an existing assembly into another assembly. You need to simply use the **Add a part from an existing file** command to insert the subassembly into an existing assembly. Next, apply constraints to constraint the assembly. The process of applying constraints is straightforward. You are required to apply constraints between only one part of a sub-assembly and a part of the main assembly. In addition to that, you can easily hide or suppress a group of parts with the help of sub-assemblies.

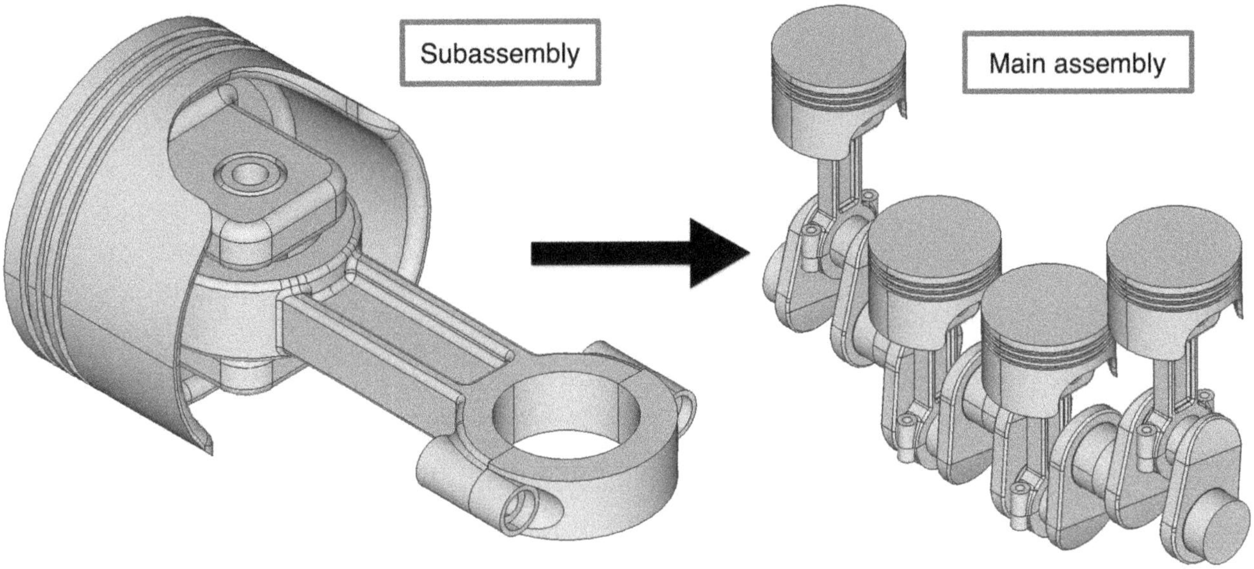

Tutorial 1 (Bottom-Up Assembly)

In this example, you create the assembly shown below.

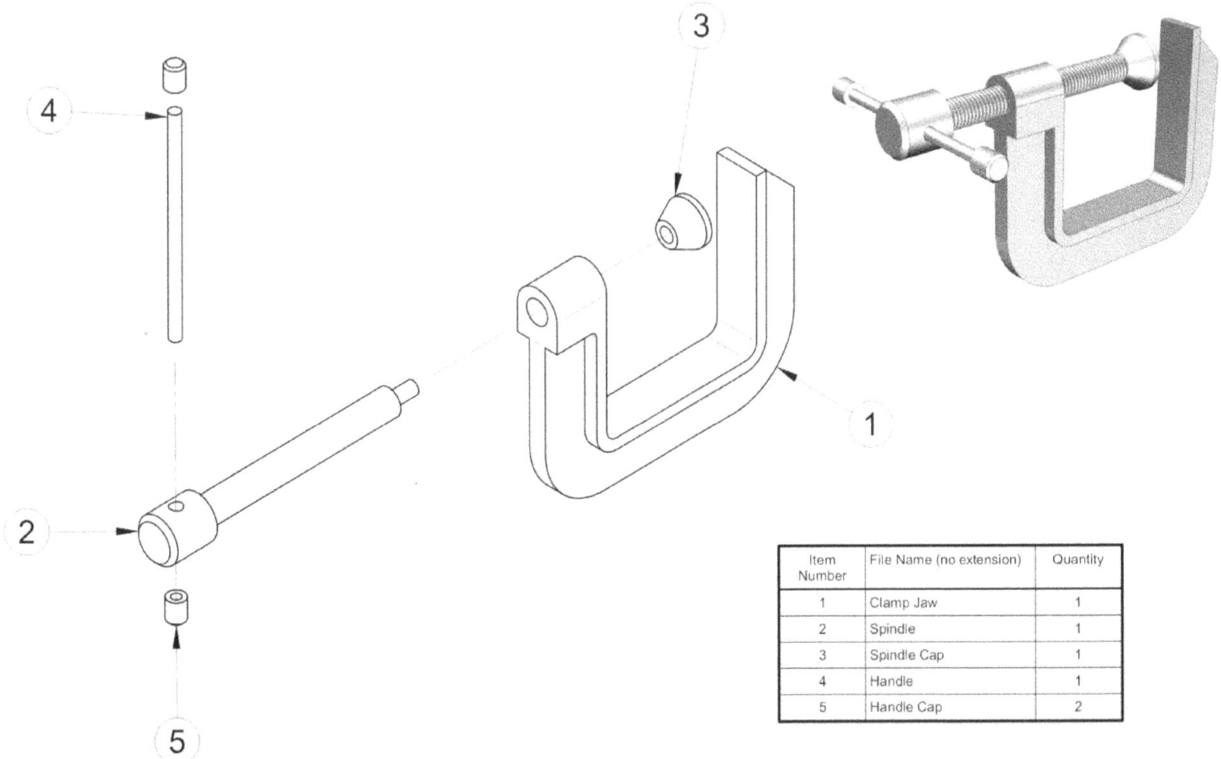

Item Number	File Name (no extension)	Quantity
1	Clamp Jaw	1
2	Spindle	1
3	Spindle Cap	1
4	Handle	1
5	Handle Cap	2

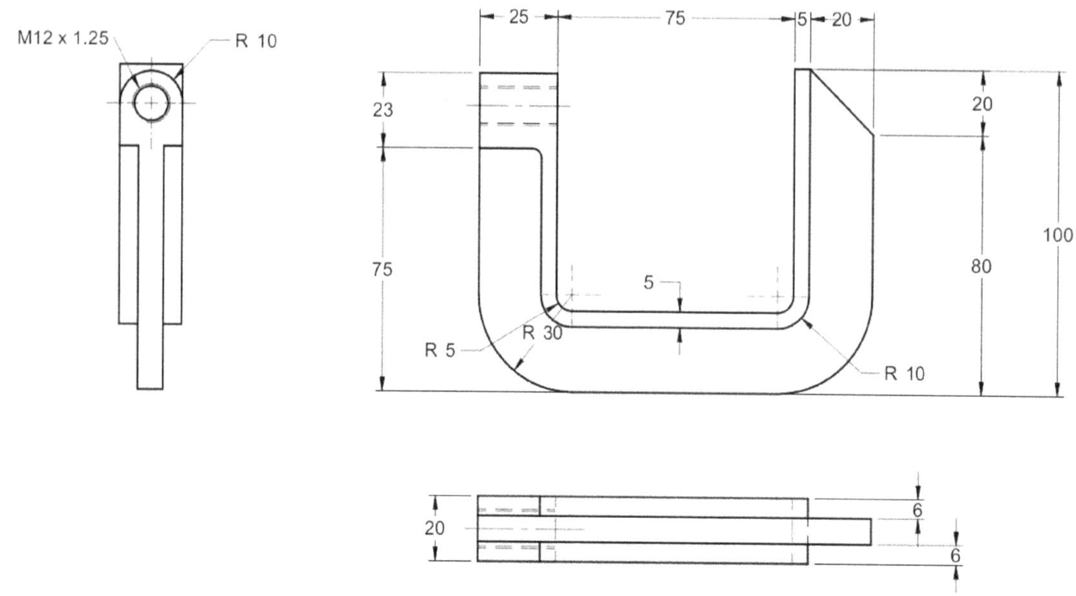

CLAMP JAW

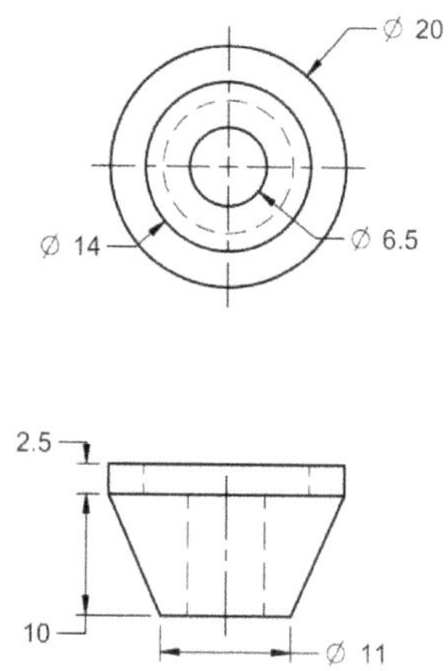

SPINDLE CAP

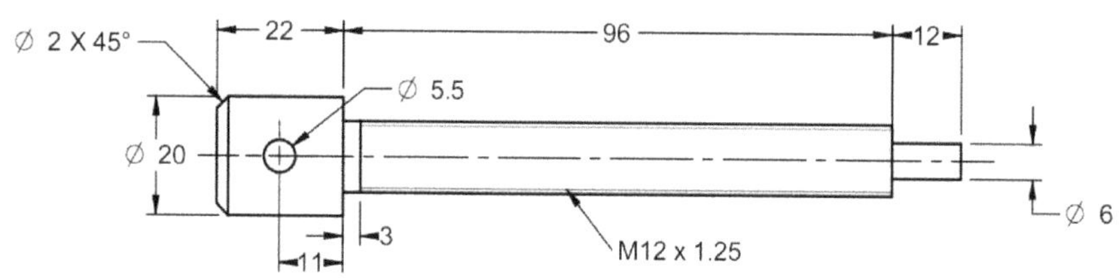

SPINDLE

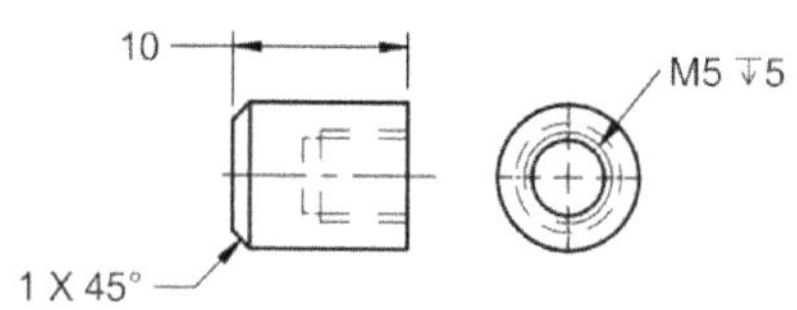

HANDLE CAP

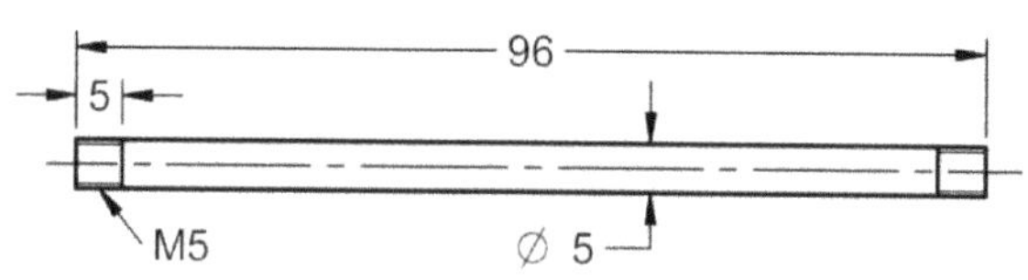

HANDLE

Creating the Assembly file

1. Create the parts of the assembly. You can also get the part files by sending us an email to *freecadtuts@gmail.com*
2. Save all the part files in a single folder.

3. Click the **FreeCAD 0.19** icon on your Desktop to start FreeCAD.
4. On the menu bar, click **File > New**; it creates a new document.
5. On the Menu bar, click **Tools > Addon manager**; the **FreeCAD** message box appears showing that the addons are not part of FreeCAD. Click **OK** to display the **Addon Manager** dialog.
6. Click the **Workbenches** tab and select the **A2plus** addon. Next, click **Install/update**.
7. Click **Close**; the **Please restart FreeCAD for changes to take effect** message appears. Click **OK**.
8. Close the **FreeCAD** application window. Next, click the **FreeCAD 0.19** icon on the Desktop.
9. On the **Workbench** toolbar, select **Workbench** drop-down > **A2plus**.
10. Click **Edit > Preferences** on the **Menu** bar; the **Preferences** dialog appears on the screen.
11. Click **Units** tab and select **Unit System > Standard(mm/kg/s/degree)**.
12. Select **Number of decimals > 2** and click **OK** on the **Preferences** dialog.
13. Click **File > Save** on the Menu bar. Next, go to the location of the folder in which all the part files of the assembly are located. Enter **Tutorial 1** in the **File name** box and click **Save**.

Inserting part files in the Assemblies

1. Click the **Add a part from an external file**  icon on the **A2p_Part** toolbar (or) click **A2plus > Add a part from an external file** on the menu bar; the **Select FreeCAD document to import part from** dialog appears.
2. On the **Select FreeCAD document to import part from** dialog, go to the folder in which all the part files of the assembly are located.
3. Select the **Clamp Jaw.FCStd** part file from the list, and then click **Open**; the clamp jaw is placed in the graphics window.
4. Click the **Add a part from an external file**  icon on the **A2p_Part** toolbar (or) click **A2plus > Add a part from an external file** on the menu bar.
5. On the **Select FreeCAD document to import part from** dialog, click **Spindle.FCStd**, and then click **Open**. Click in the graphics window to place the component.

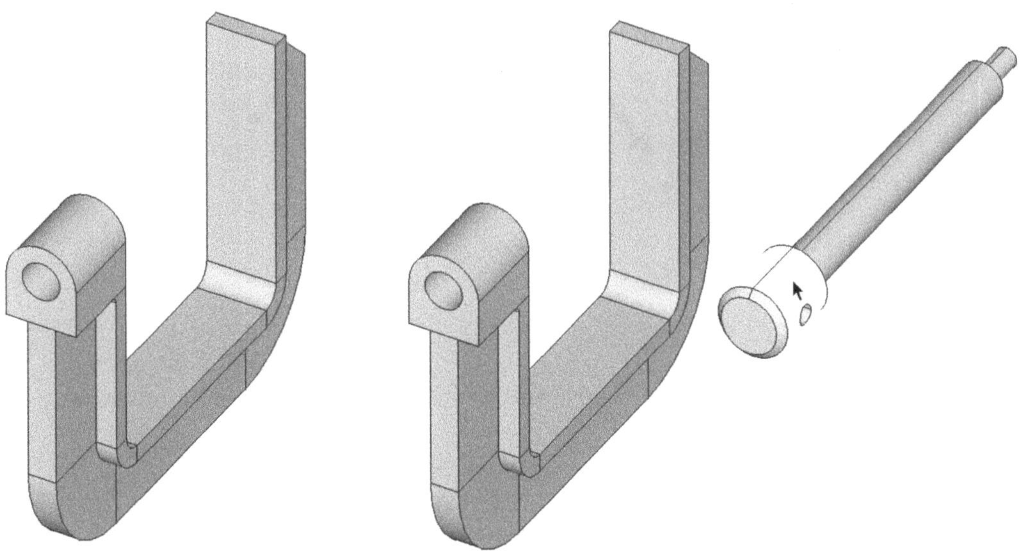

6. Likewise, insert the *Spindle Cap, Handle,* and two instances of *Handle Caps.* Next, click the green check on the **Insert parts and assemblies** dialog.

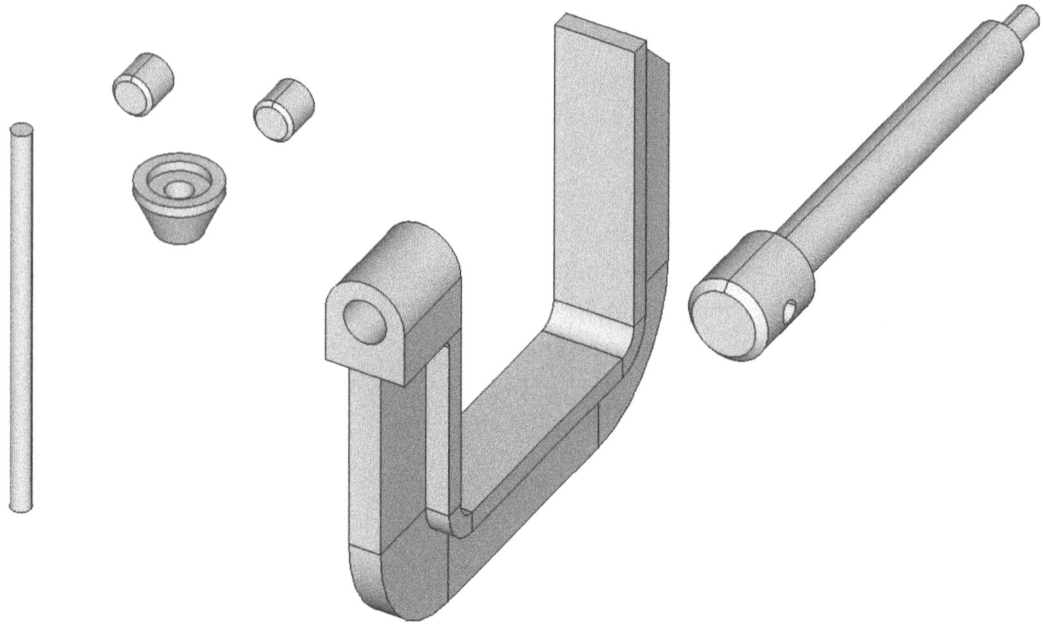

Defining Constraints

1. Press and hold the Ctrl key and select the flat face of the *Clamp Jaw*, as shown in the figure.
2. Rotate the model view and select the flat face of the model, as shown.
3. Click the **Add planeCoincident constraint**  icon on the **A2p_Constraints** toolbar (or) click **A2plus > Constraint > Add planeCoincident constraint** on the menu bar.
4. On the **Constraint Properties** dialog, select **Direction > opposed**. Next, type 31 in the **Offset** box, and then click **Flip sign**. Click **Accept** to create the planeCoincident constraint.

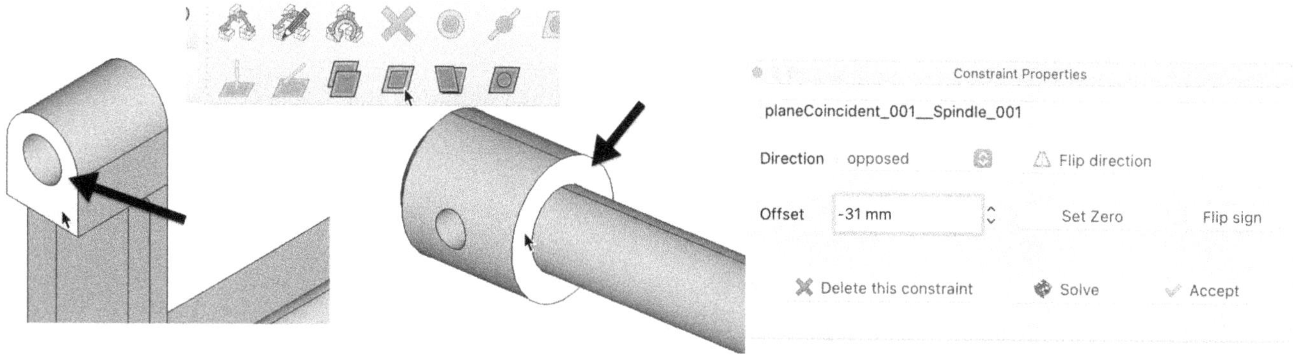

5. Click the **Define constraints**  icon on the **A2p_Constraint** toolbar (or) click **A2plus > Constraint > Define constraints** on the menu bar.
6. Press and hold the Ctrl key and select the cylindrical face of the *Spindle* and hole of the *Clamp Jaw*.
7. Click the **Add axis Coincident constraint**  icon on the **Constraint Tools** dialog.

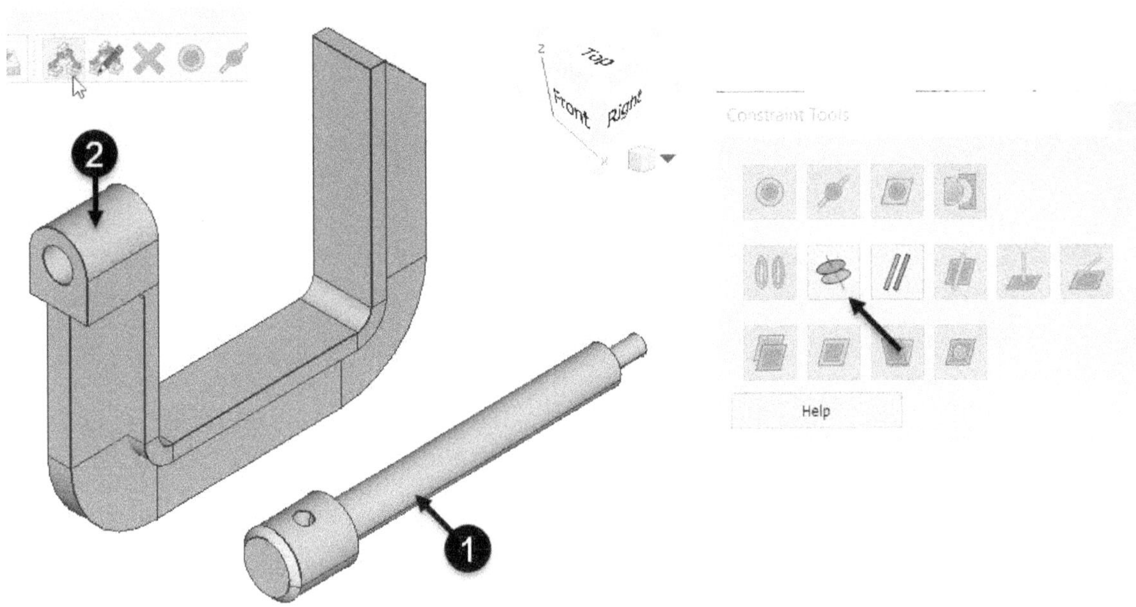

8. On the **Constraint Properties** dialog, select **Lock Rotation > True**. Click **Accept**.

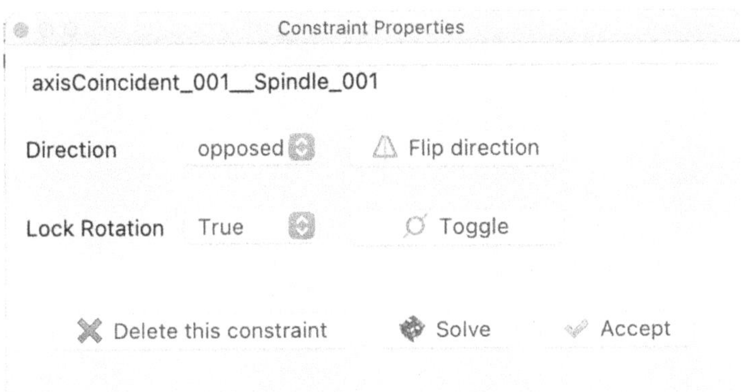

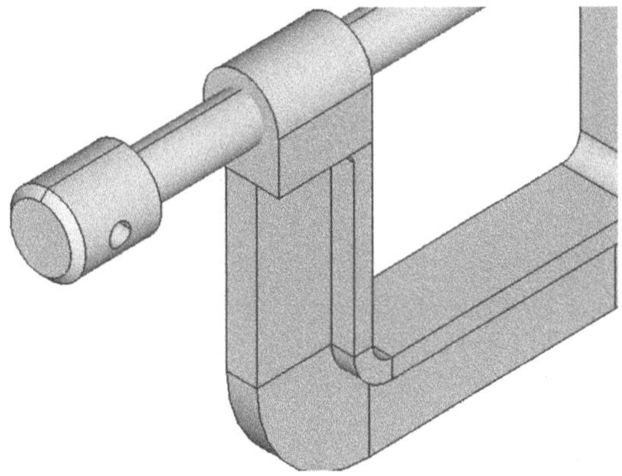

9. Click on the circular edge of the *Spindle Cap* hole, as shown.
10. Press and hold the Ctrl key and click on the circular edge of the *Spindle*, as shown.
11. Click the **Add circularEdge constraint**  icon on the **Constraint Tools** dialog.
12. Select **Direction > opposed** on the **Constraint Properties** dialog; the *Spindle* and *Spindle Cap* are axially aligned and positioned opposite to each other.

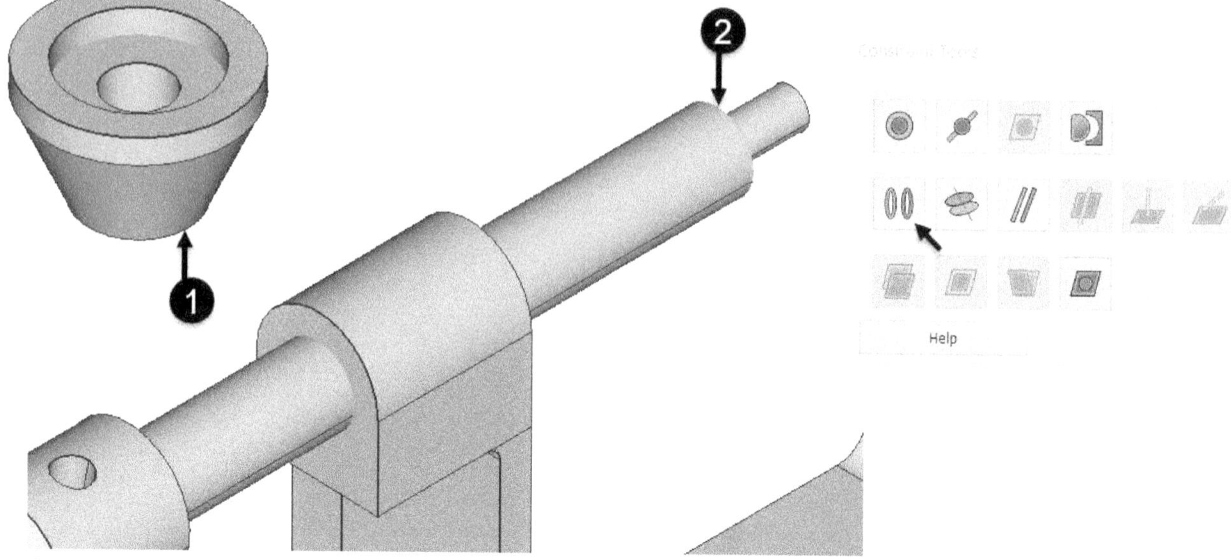

13. Select **Lock Rotation** > **True**, and then click **Accept**.

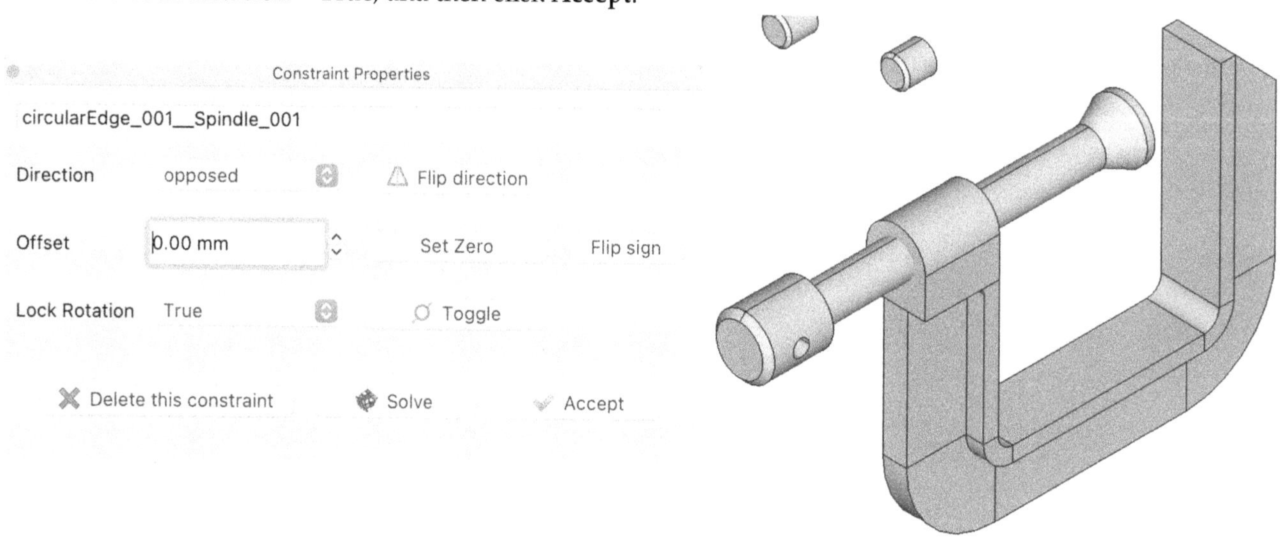

14. Press and hold the Ctrl key and select the cylindrical face of the *Handle* and the hole located on the *Spindle*.

15. Click the **Add axis Coincident constraint**  icon on the **Constraint Tools** dialog.
16. Select **Lock Rotation** > **True**, and then click **Accept**.

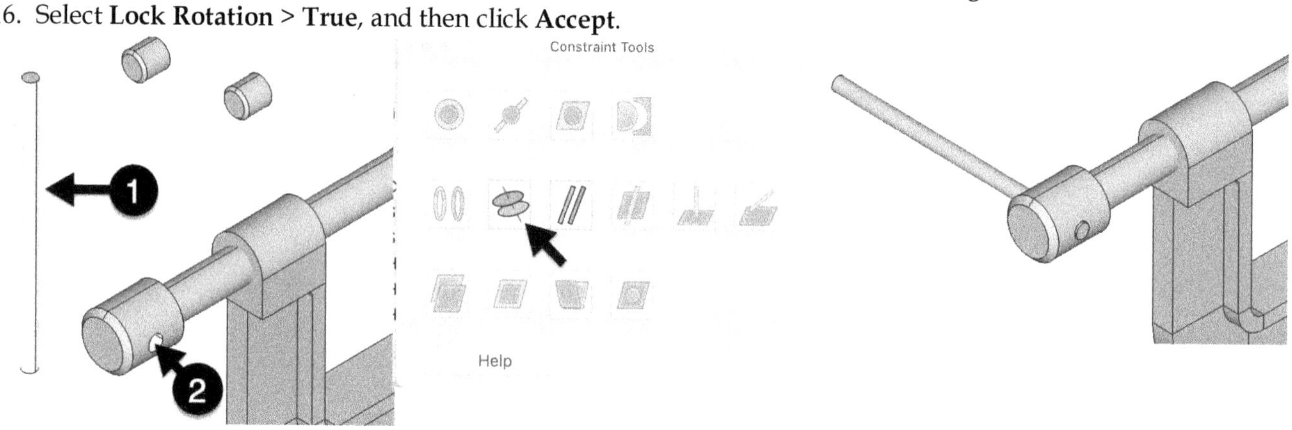

17. Close the **Constraint Tools** dialog.

18. Click the **Move the selected part under constraints** icon on the **A2p_Part** toolbar.
19. Select the handle, move the pointer toward right and click; the handle is moved along the constrained axis.

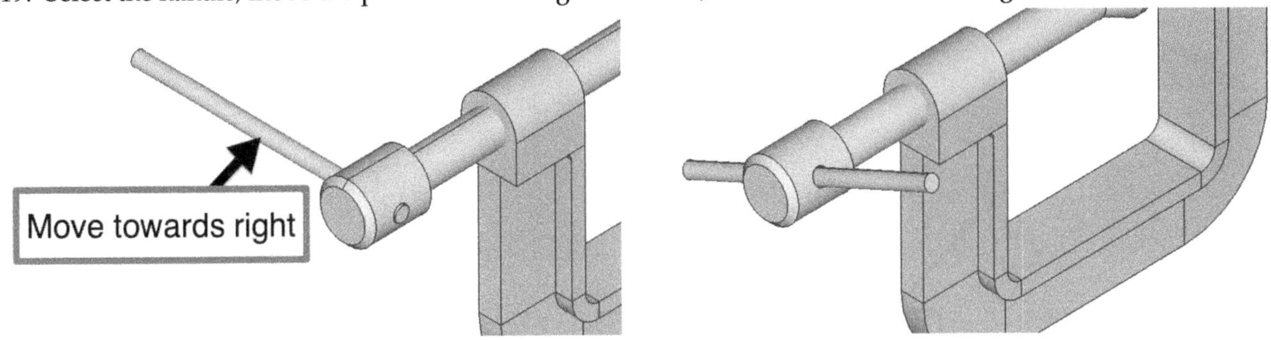

20. Press and hold the Ctrl key and click on the circular edge of the *Handle*, as shown.
21. Rotate the model and click on the inner circular edge of the *Handle cap*, as shown.

22. Click the **Add circularEdge constraint** icon on the **Constraint Tools** dialog.
23. Select **Direction > aligned** on the **Constraint Properties** dialog.
24. Select **Lock Rotation** > **True**, and then click **Accept**.

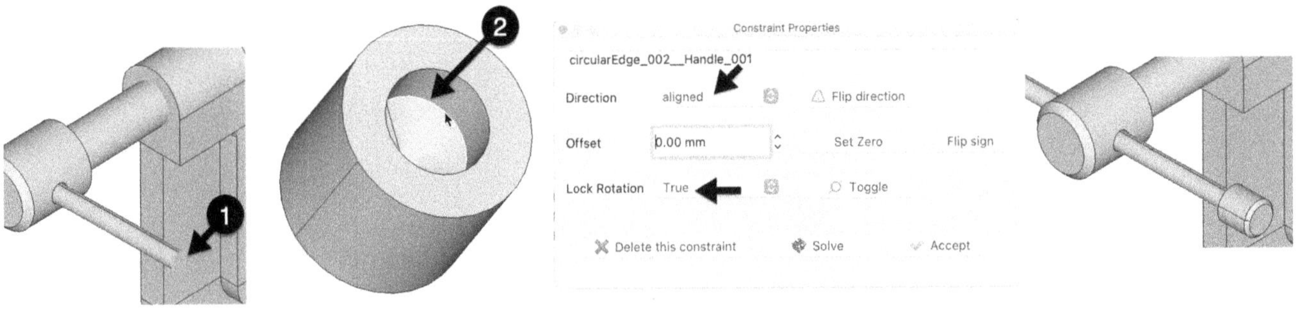

25. Press and hold the Ctrl key and click on the left circular edge of the *Handle* hole, as shown.
26. Rotate the model and click on the inner circular edge of the second *Handle cap*, as shown.

27. Click the **Add circularEdge constraint** icon on the **Constraint Tools** dialog.
28. Select **Direction > opposed** on the **Constraint Properties** dialog.
29. Select **Lock Rotation** > **True**, and then click **Accept**.

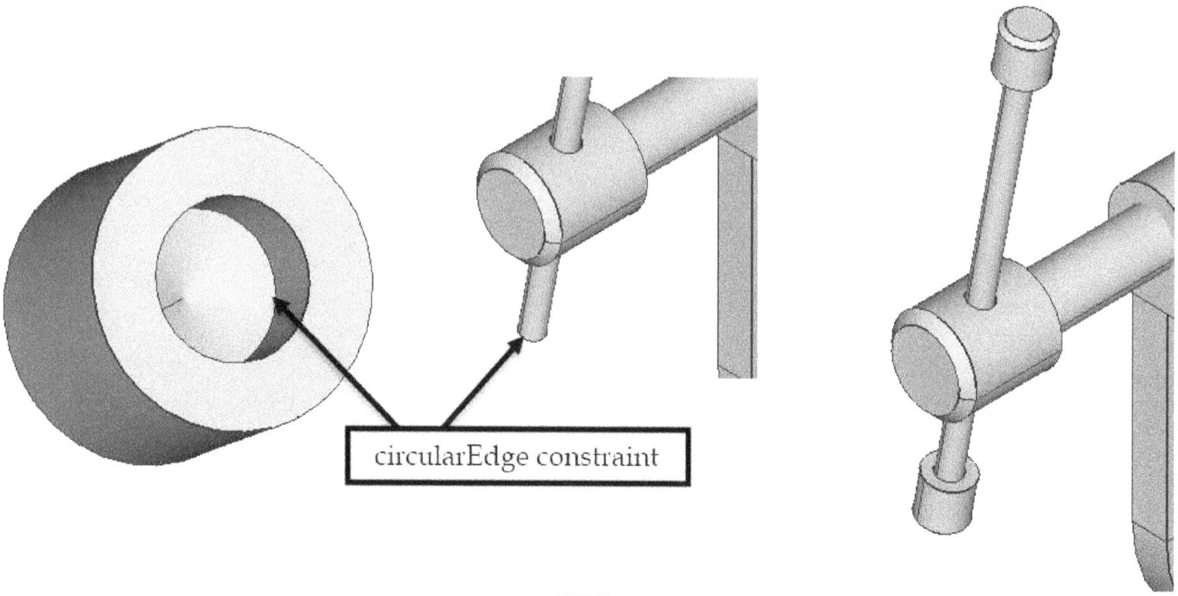

30. Click the **Print detailed DOF information** icon on the **A2p_View** toolbar (or) **click A2plus > View > Print detailed DOF information** on the menu bar; the degrees of freedom of all the parts are displayed.

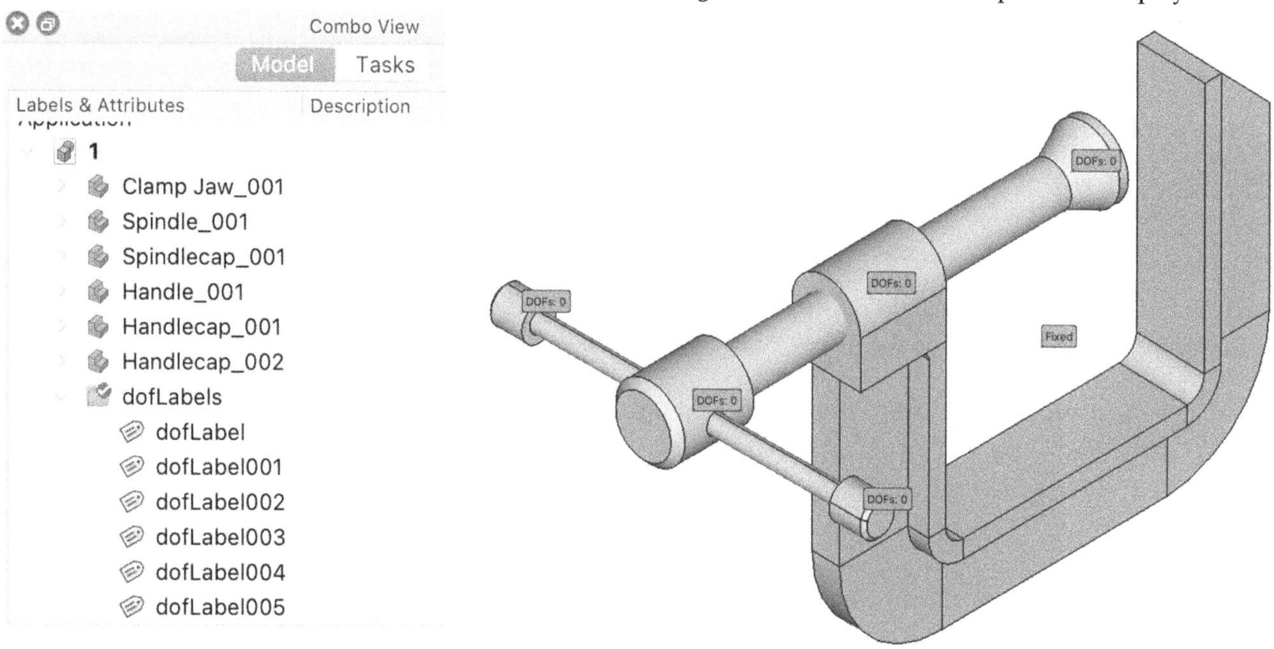

31. Click the **Print detailed DOF information** icon on the **A2p_View** toolbar to hide the DOF labels.
32. Save and close the assembly file.

Questions

1. How do you start an assembly?
2. What is the use of the **Define Constraints** command?

3. What is the use of the **Angle Planes** command?
4. How to create a sub-assembly?

Exercises

Exercise 1

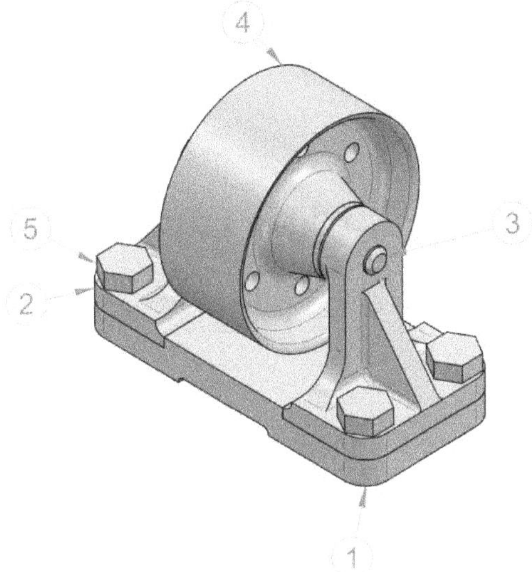

Item Number	File Name (no extension)	Quantity
1	Base	1
2	Bracket	2
3	Spindle	1
4	Roller-Bush assembly	1
5	Bolt	4

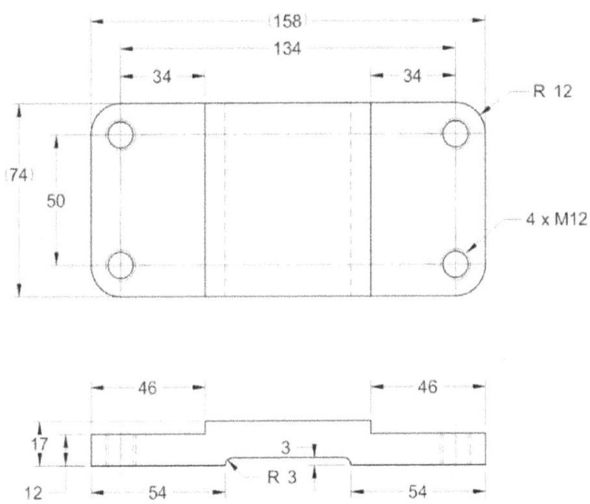

Base

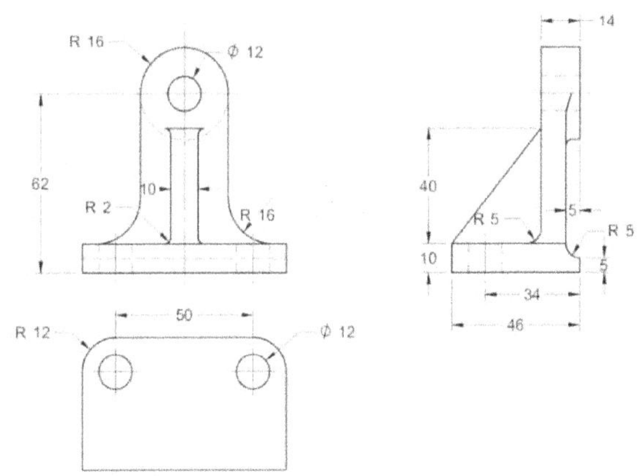

Bracket

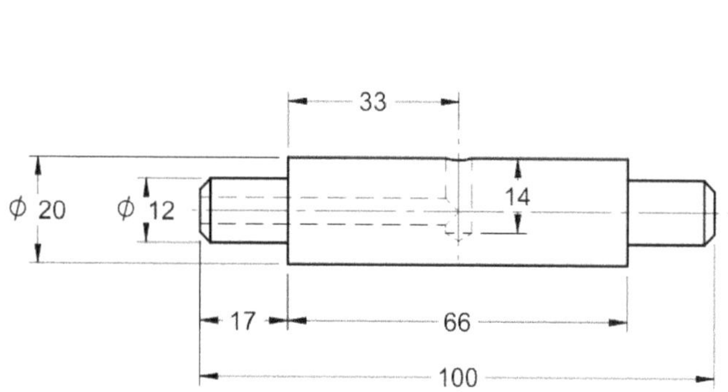

SPINDLE

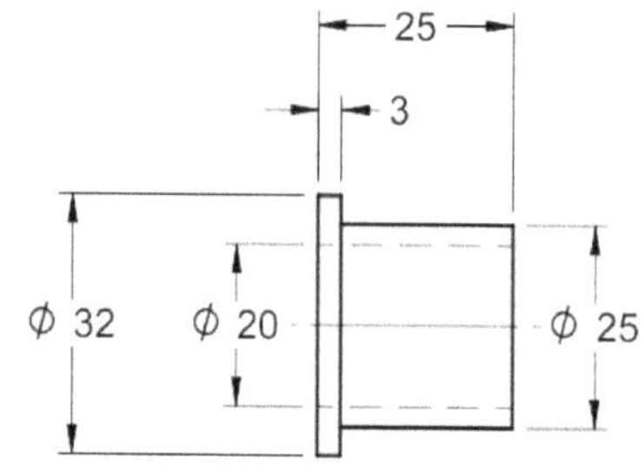

BUSH

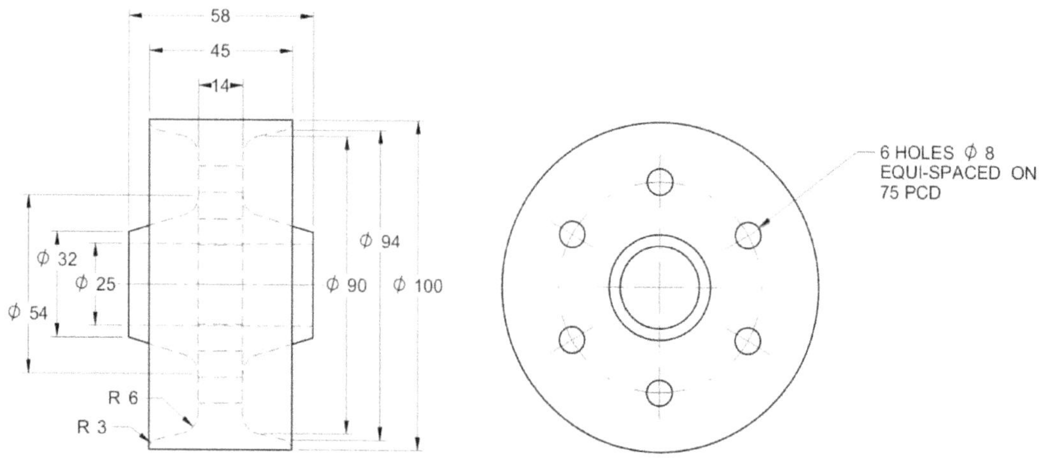

Roller

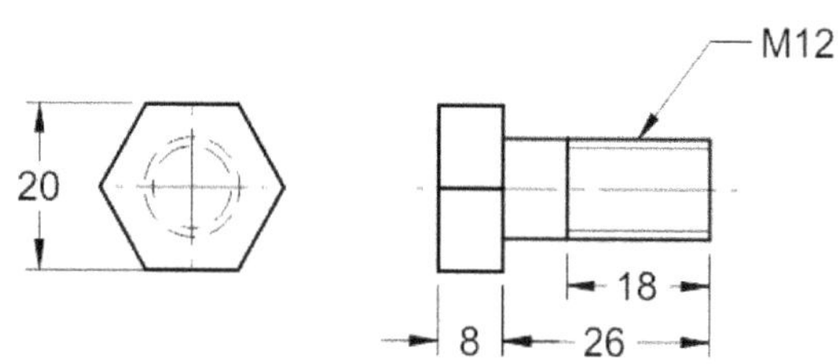

Bolt

Chapter 10: Drawings

Drawings are used to document your 3D models in the traditional 2D format, including dimensions and other instructions useful for manufacturing purposes. In FreeCAD 0.19, you first create 3D models and assemblies and then use them to generate drawings. There is a direct association between the 3D model and the drawing. When changes are made to the model, every view in the drawing will be updated. This relationship between the 3D model and the drawing makes the drawing process fast and accurate. Because the 2D drawings are widely used in the mechanical industry, drawings are one of the three main files you can create in FreeCAD 0.19.

The topics covered in this chapter are:

- *Insert Views*
- *Projected Group*
- *Section views*
- *Detail views*
- *Centerlines*
- *Dimensions*
- *Text*

Starting a Drawing

To start a new drawing, first, open the part file of which you want to create the drawing. Next, select the **TechDraw** option from the **Workbench** drop-down (or) click **View > Workbench > TechDraw** on the menu bar. Click **Edit > Preferences** on the menu bar. Next, click the **TechDraw** option on the left side of the **Preferences** dialog. Click the **General** tab on the **Preferences** tab and select **Projection Group Angle > Third** from the **Conventions** section.

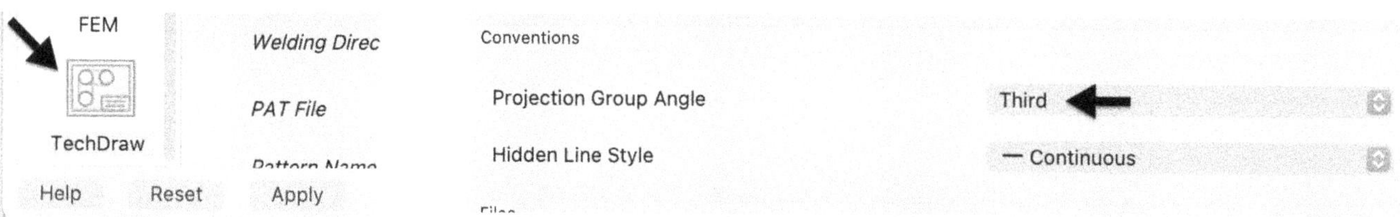

Click the **Browse** button next to the **Default Template** drop-down in the **File** section. Select the required template from the **Templates** folder and click **Open**. Click **OK** on the **Preferences** dialog.

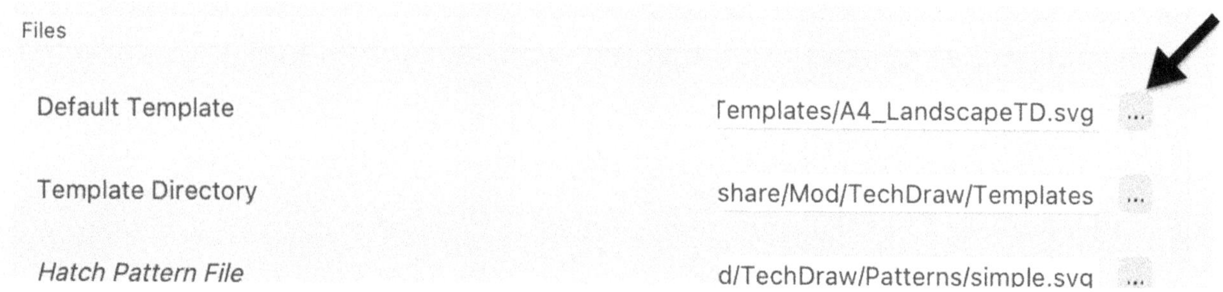

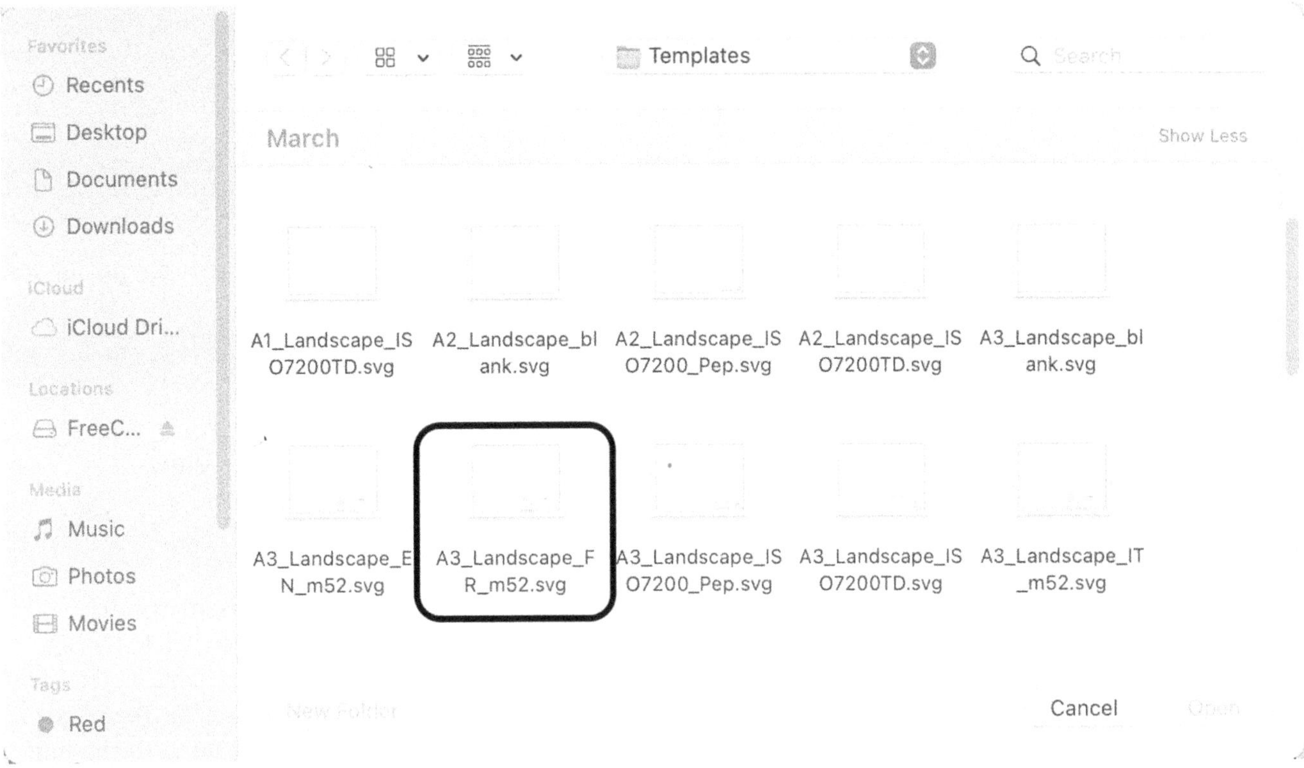

Inserting the Drawing page

Click **Insert Default Page** icon on the **TechDraw Pages** toolbar (or) click **TechDraw > Insert Default Page** on the menu bar. The drawing page with the default template is inserted into the graphics window. Now, you can create drawing views and add annotations and dimensions.

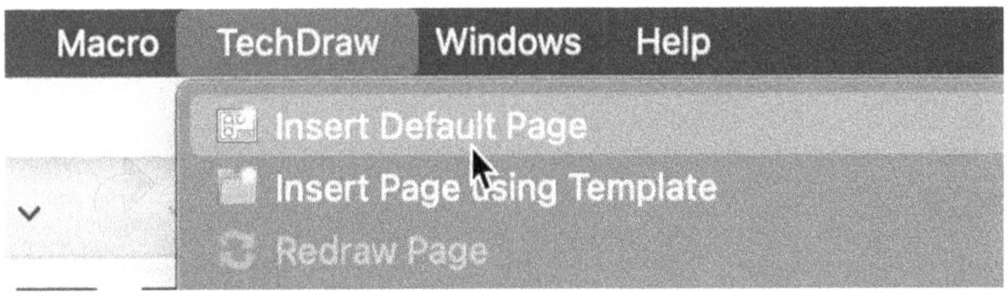

Insert View

After inserting the drawing page, you need to insert the drawing views on it. To do this, first switch to the 3D graphics window and specify the orientation of the model using the **Standard view** options available on the View toolbar. Next, select a face or edge from the model and click the **Insert view** icon on the **TechDraw Views** toolbar (or) click **TechDraw > Insert View** on the menu bar. Next, click the **Page** tab at the bottom of the graphics window; the model view will be displayed on the page.

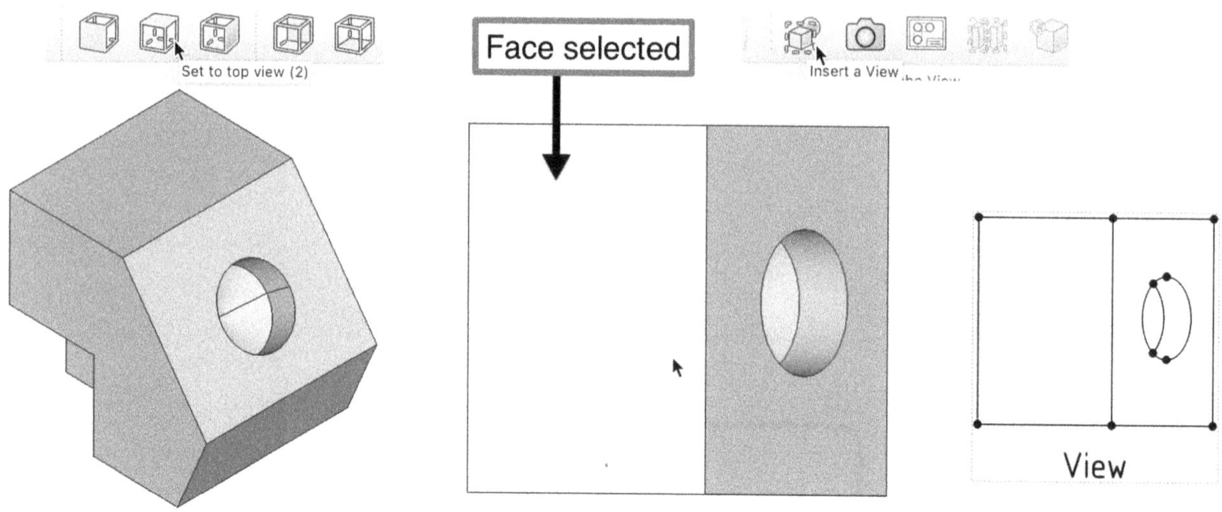

Insert Projection Group

There are different standard views available in a 3D component, such as front, right, top, and isometric. In FreeCAD 0.19, you can create these views using the **Insert Projection Group** command. First, switch to the 3D graphics window and specify the orientation of the model using the **Standard view** options available on the **View** toolbar. Next, select a face or edge from the model and click the **Insert Projection Group** icon on the **TechDraw Views** toolbar (or) click **TechDraw > Insert Projection Group** on the menu bar. Click the **Page** tab at the bottom of the graphics window; model view will be displayed on the page.

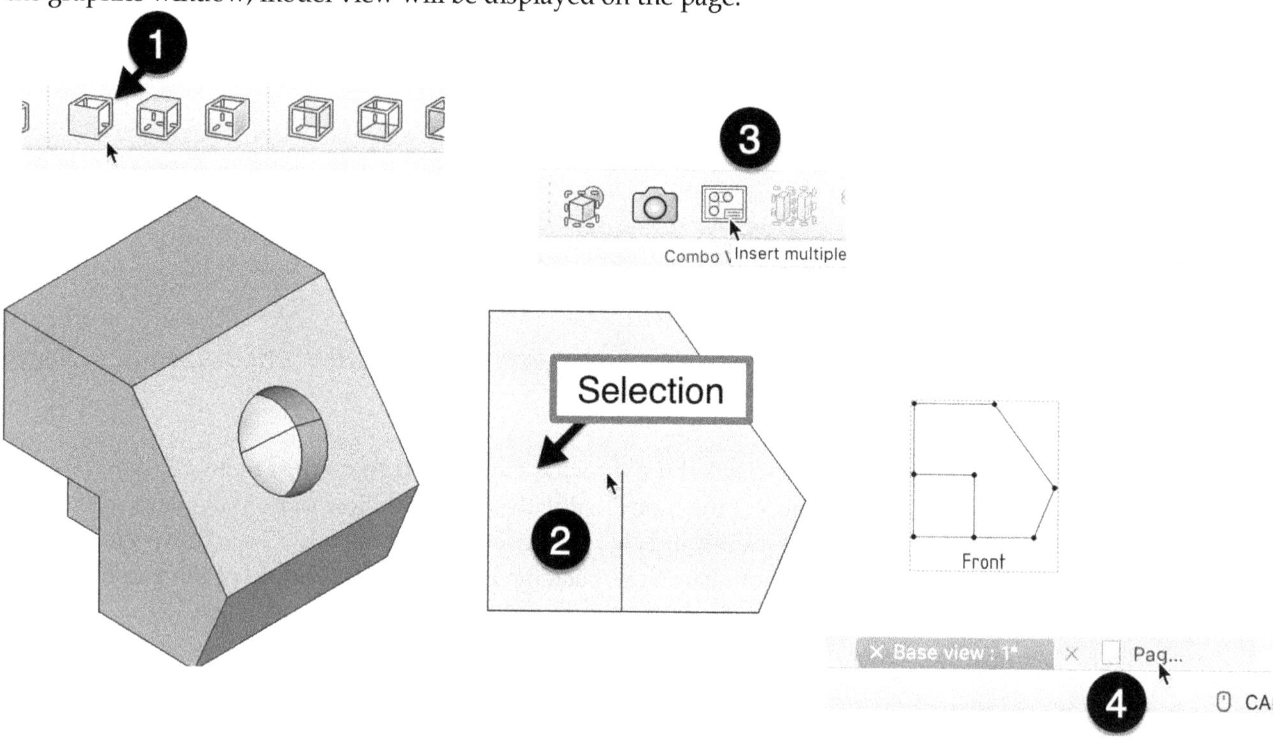

Specify the angle of projection using the **Projection** drop-down.

Change the scale factor in the **Scale** drop-down to adjust the size of the view to sheet size.

Use the arrow buttons in the **Adjust Primary Direction** section **Projection Group** panel to change the orientation of the view.

Select the checkboxes in the **Secondary Projections** section. For example, select the **Top**, **Right**, and **RightFrontTop** checkboxes to create the top, right, and Isometric views.

Adjust the spacing between the views by entering values in the **X Spacing** and **Y Spacing** boxes. Next, click **OK** to create the projection group.

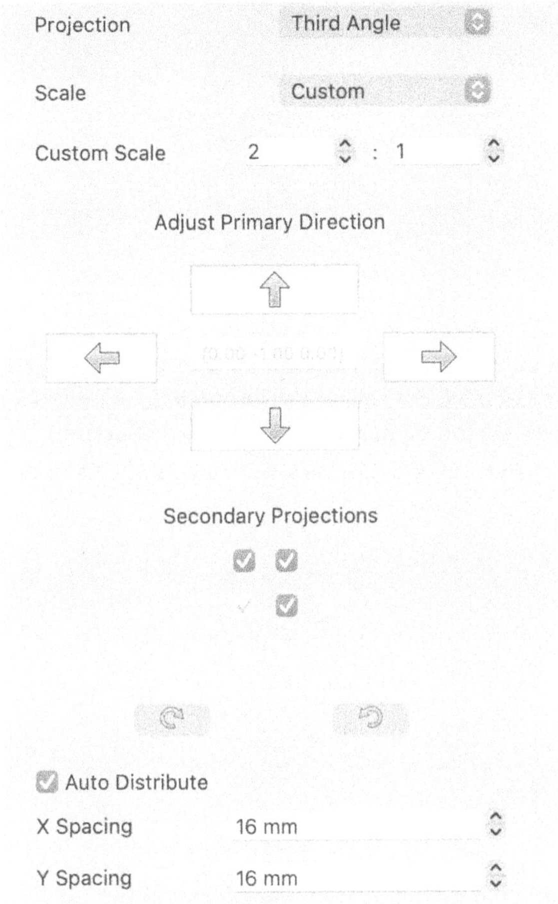

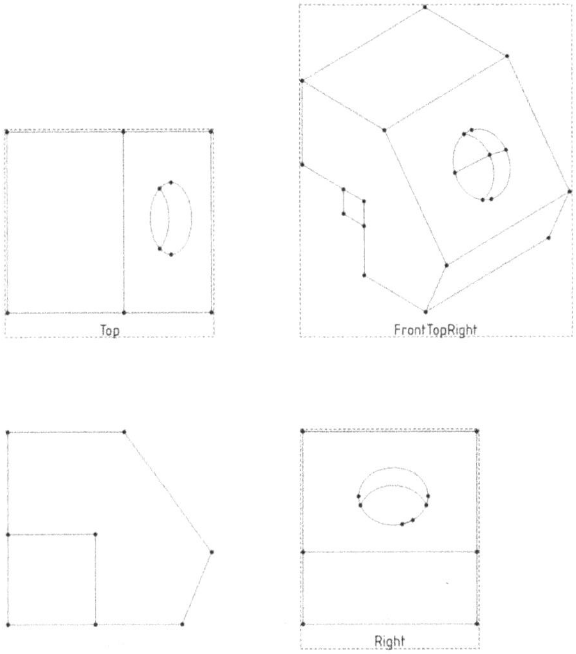

Insert Active View

This command inserts the snapshot of view of the model inside the 3D graphics window on to the drawing page. To do this, switch to the 3D graphics window and select a face or edge from the model. Click the **Insert Active View** icon on the **TechDraw Views** toolbar (or) click **TechDraw > Insert Active View (3D View)** on the menu bar. Next, specify the **Line Width** and **Render Mode** on the **Active view to TD view** panel, and then click **OK.**

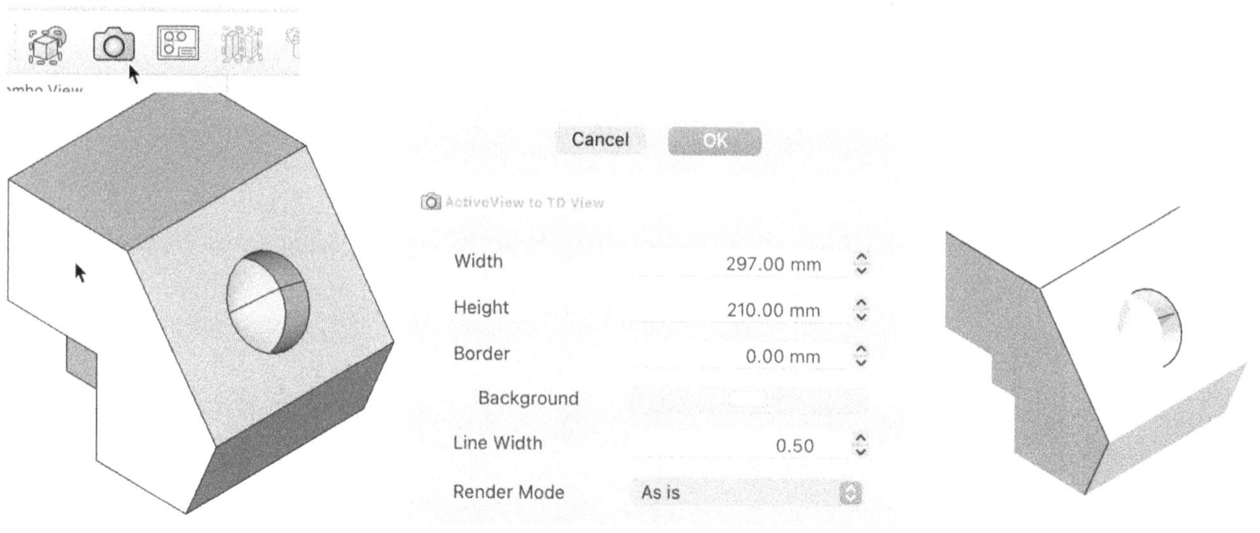

Section View

One of the most common views used in 2D drawings is the section view. Creating a section view in FreeCAD 0.19 is very simple. First, select the base view by clicking on its frame. Next, click the **Insert section view** icon on the **TechDraw Views** toolbar (or) click **TechDraw > Insert Section View** on the menu bar. On the **Create Section View** panel, select an option from the **Section Orientation** section. Click and drag the section view to the desired location.

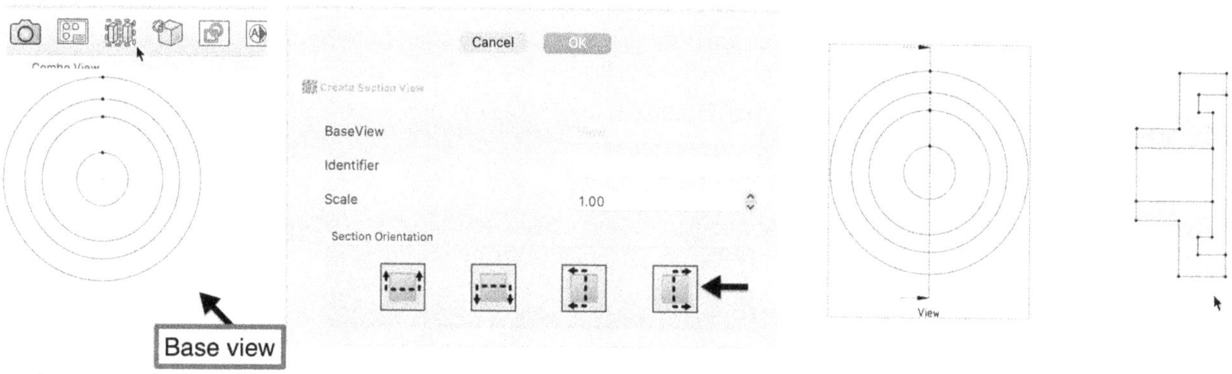

Enter a letter in the **Identifier** box to name the section view. You can change the section plane location by changing the X, Y or Z values. Next, click **OK** to create the section view.

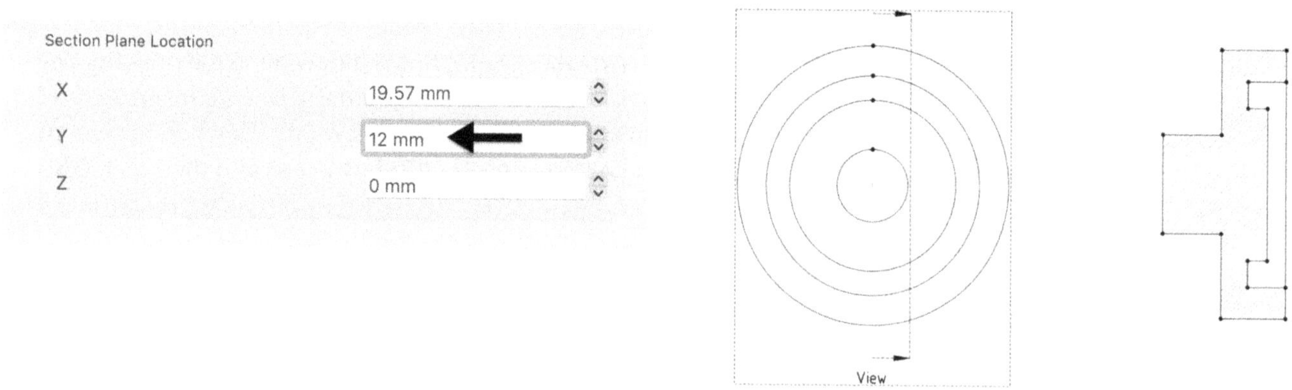

Detail View

If a drawing view contains small features that are difficult to see, a detailed view can be used to zoom in and make things clear. To create a detailed view, first select the base view. Next, click the **Insert Detail View** icon on the **TechDraw Views** toolbar (or) click **TechDraw > Insert Detail View** on the menu bar. Click the **Drag Highlight** button on the **New Detail View** panel.

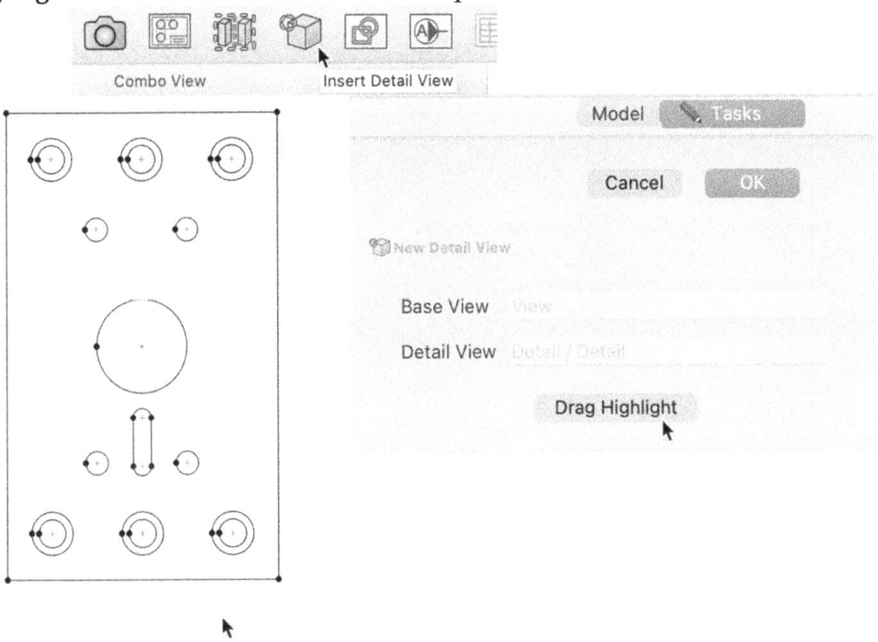

Click and drag the highlighter to the desired location. Next, change the radius of the highlighter by entering a value in the **Radius** box. Also, specify the magnification value by entering a value in the **Scale factor** box. Click and drag the detail view to the desired location.

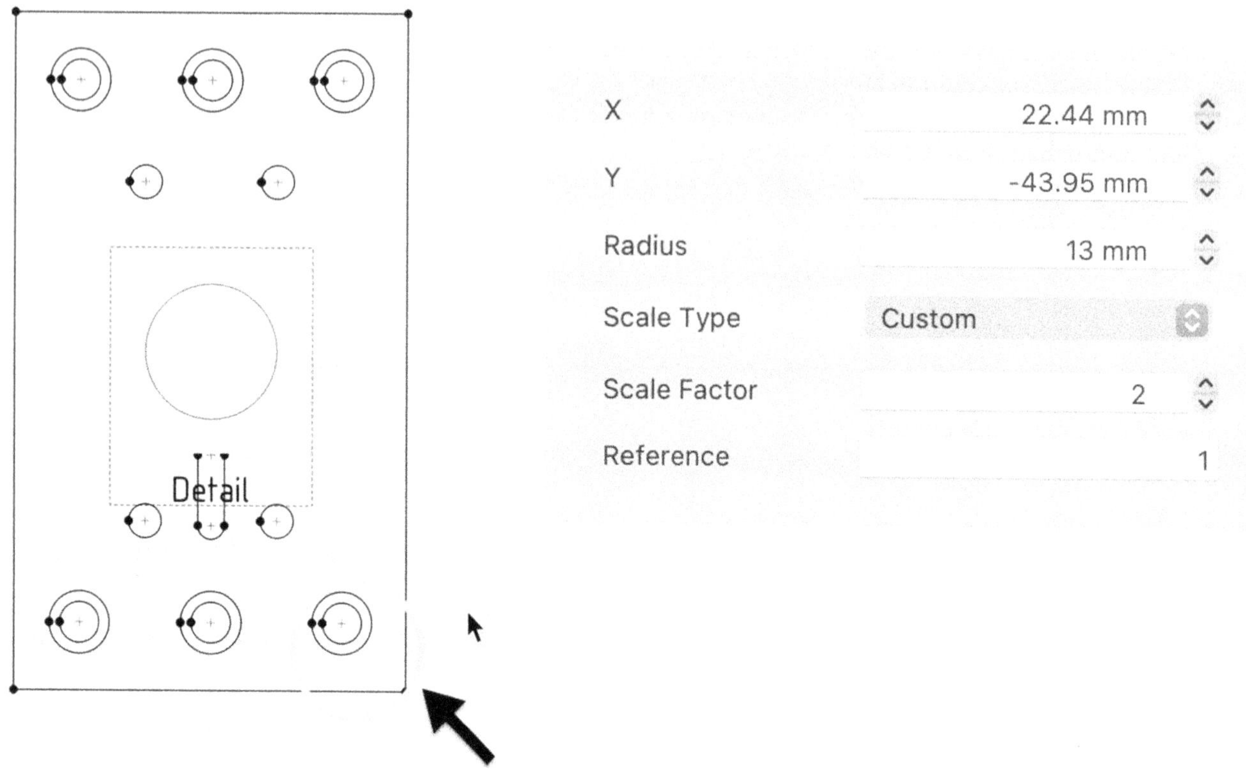

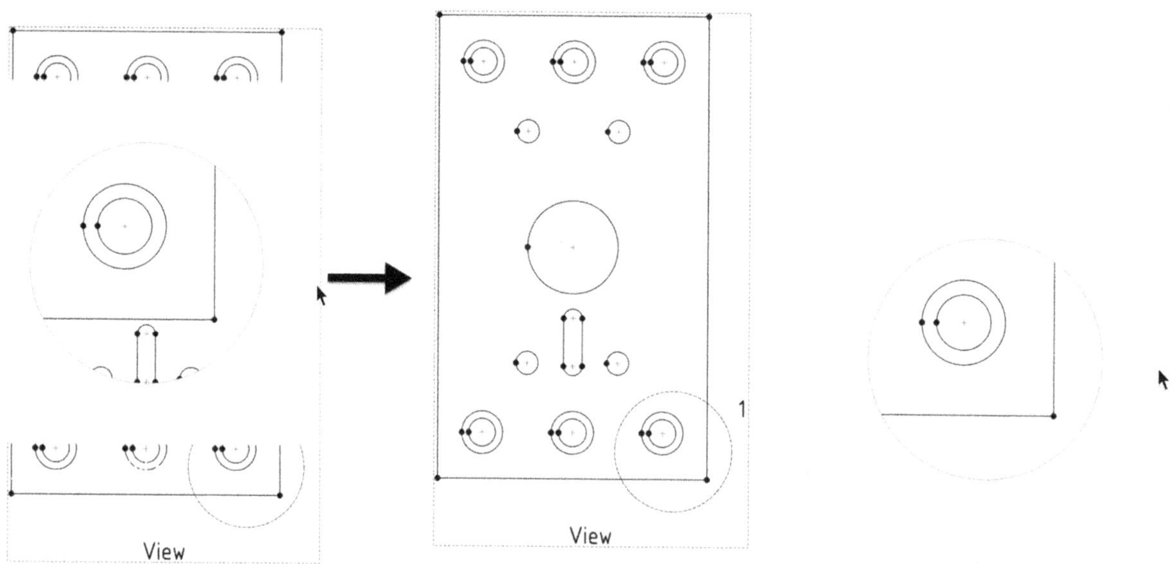

Next, click **OK** to complete the detail view.

Add Centerlines to Faces

To add centerlines, first select a cylindrical face from the drawing view. Next, click the **Add Lines** drop-down > **Add Centerlines to Faces** on the **TechDraw Annotations** toolbar (or) click **TechDraw > Add Lines > Add Centerlines to Faces** on the menu bar. Specify the **Weight** and **Style** settings on the **Create Center Line** panel, and then click **OK**.

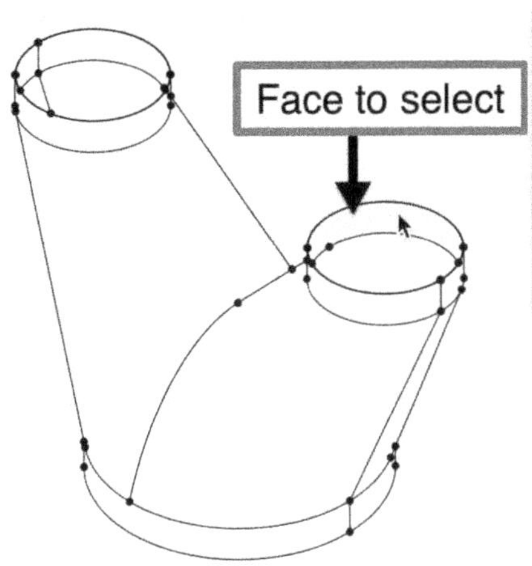

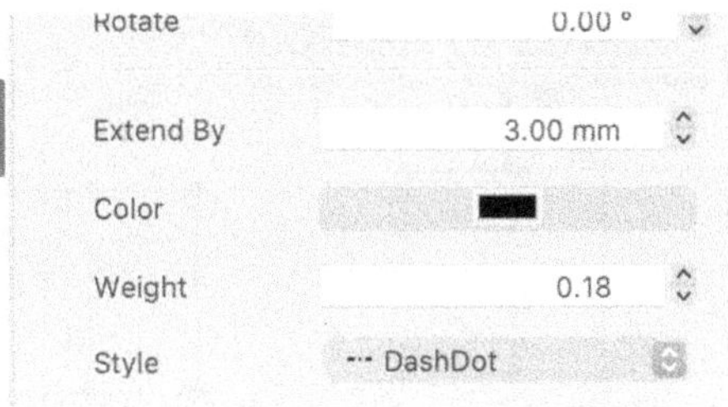

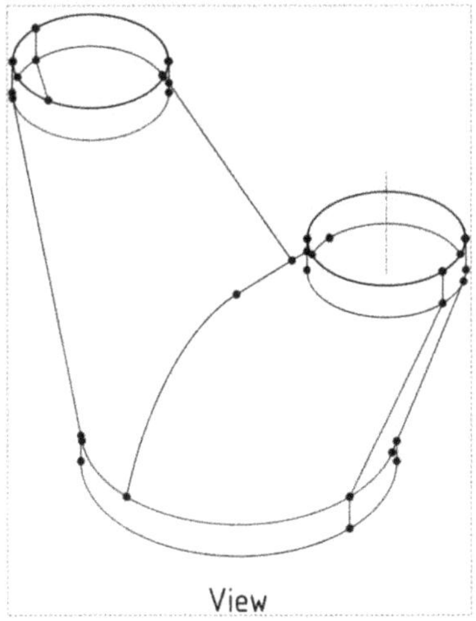

Add Centerline between 2 Lines

The **Add Centerline between 2 Lines** command is used to create a centerline bisecting two lines. This command is beneficial while creating a centerline on the section view or projected views. Press and hold the CTRL key (COMMAND key for Mac users) and then click on two edges of the drawing view. Next, click the **Add Lines** drop-down > **Add Centerlines between 2 Lines** on the **TechDraw Annotations** toolbar (or) click **TechDraw > Add Lines > Add Centerlines between 2 Lines** on the menu bar. Specify the **Weight** and **Style** settings on the **Create Center Line** panel, and then click **OK**; a centerline will be created between the two lines.

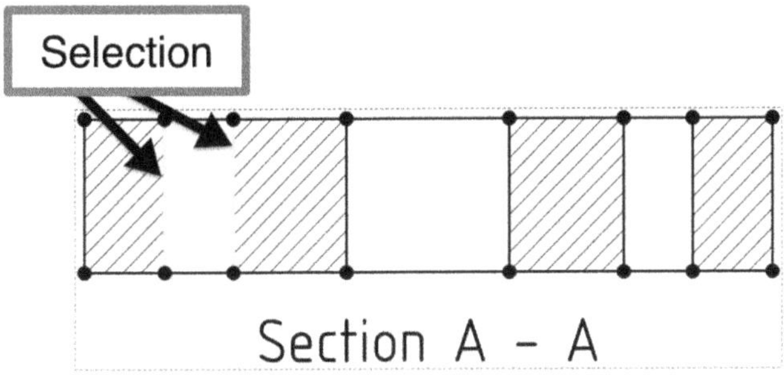

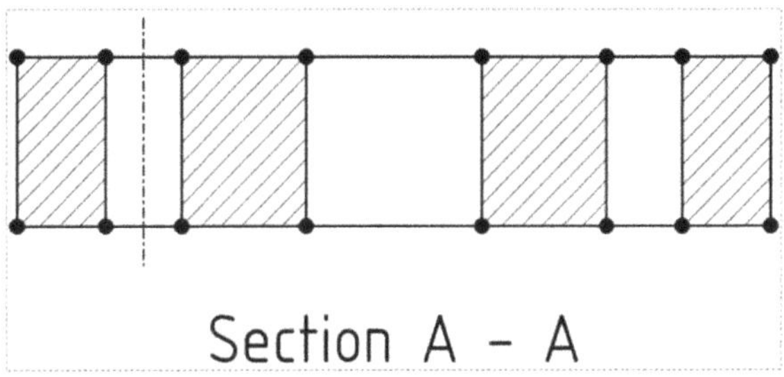

Add Centerline between 2 Points

The **Add Centerline between 2 Points** command is used to create a centerline between two points. Press and hold the CTRL key (COMMAND key for Mac users) and then click on two vertices of the drawing view. Next, click the **Add Lines** drop-down > **Add Centerlines between 2 Points** on the **TechDraw Annotations** toolbar (or) click **TechDraw > Add Lines > Add Centerlines between 2 Points** on the menu bar. Specify the **Weight** and **Style** settings on the **Create Center Line** panel, and then click **OK**; a centerline will be created between the two Points.

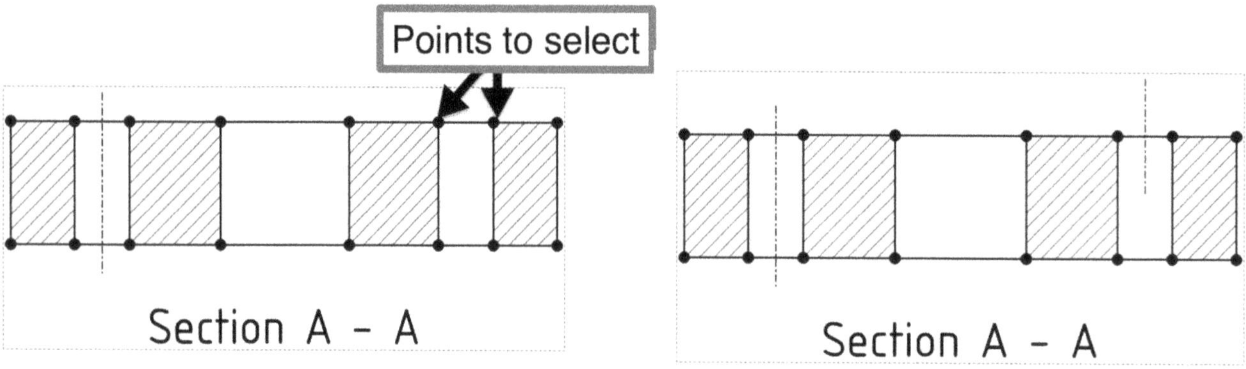

Dimensions

FreeCAD provides you with different ways to add dimensions to the drawing. The commands to add dimensions to the drawing are available on the **TechDraw Dimensions** toolbar.

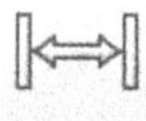

Insert Horizontal Dimension

This command adds a horizontal dimension to a selected line or between two selected points. Select a line (or) press and hold the CTRL key and select two points. Next, click the **Insert Horizontal Dimension** icon on the **TechDraw Dimensions** toolbar (or) click **TechDraw > Dimensions > Insert Horizontal Dimension** on the menu bar. Click and drag the horizontal dimension to the desired location.

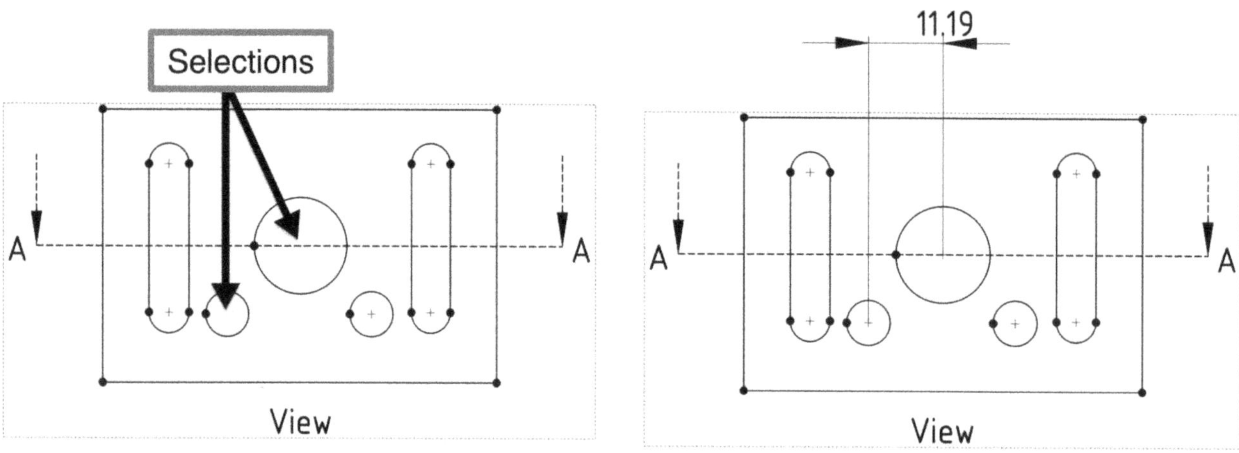

Insert Vertical Dimension

This command adds a vertical dimension to a selected line or between two selected points. Select a line (or) press and hold the CTRL key and select two points. Next, click the **Insert Vertical Dimension** icon on the **TechDraw Dimensions** toolbar (or) click **TechDraw > Dimensions > Insert Vertical Dimension** on the menu bar. Click and drag the vertical dimension to the desired location.

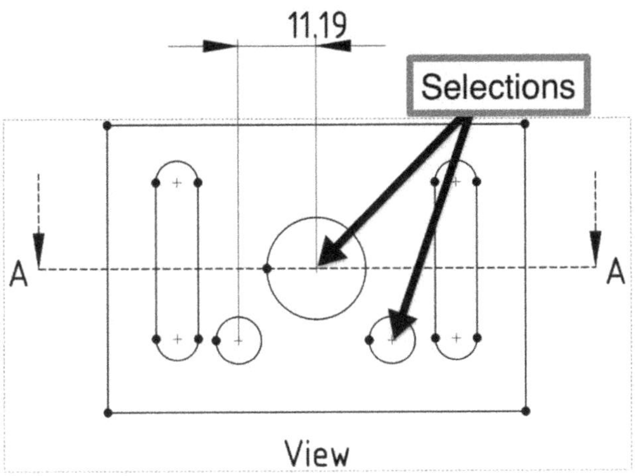

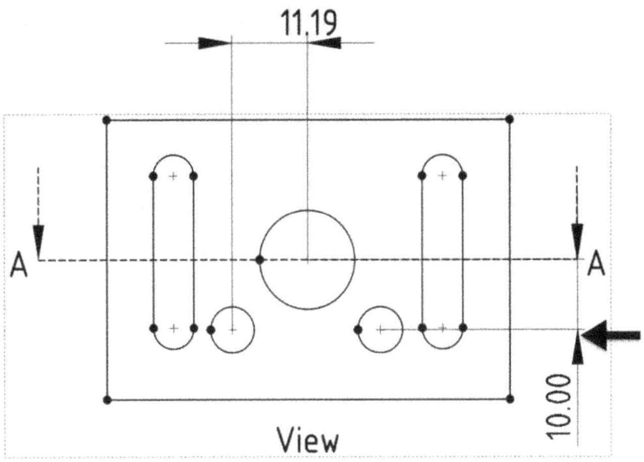

Insert Radius Dimension

This command adds a radius dimension to an arc or circle. To do this, select a circular edge and click the **Insert Radius Dimension** icon on the **TechDraw Dimensions** toolbar (or) click **TechDraw > Dimensions > Insert Radius Dimension** on the menu bar.

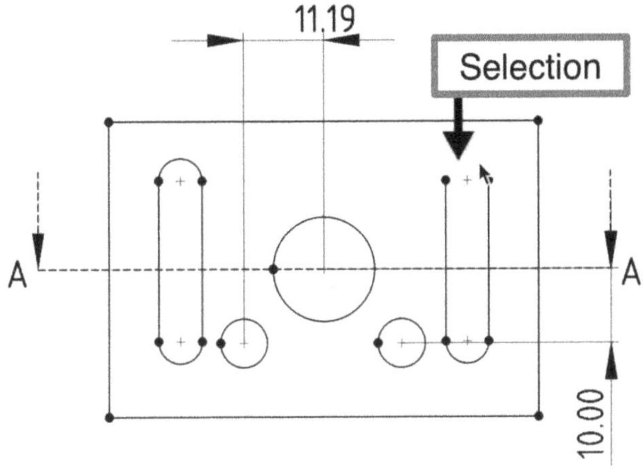

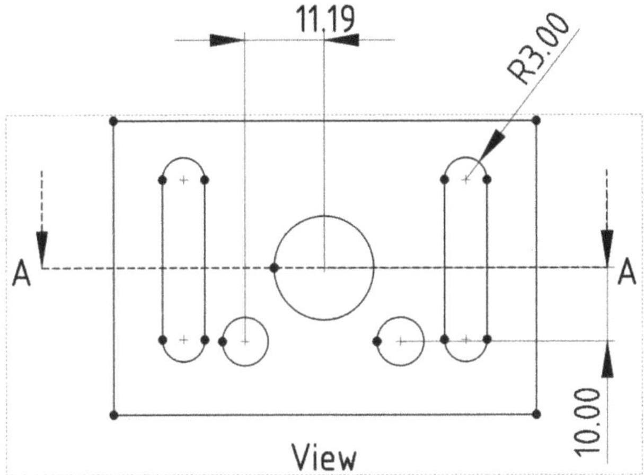

Insert Diameter Dimension

This command adds a diameter dimension to an arc or circle. To do this, select a circular edge and click the **Insert Diameter Dimension** icon on the **TechDraw Dimensions** toolbar (or) click **TechDraw > Dimensions > Insert Diameter Dimension** on the menu bar.

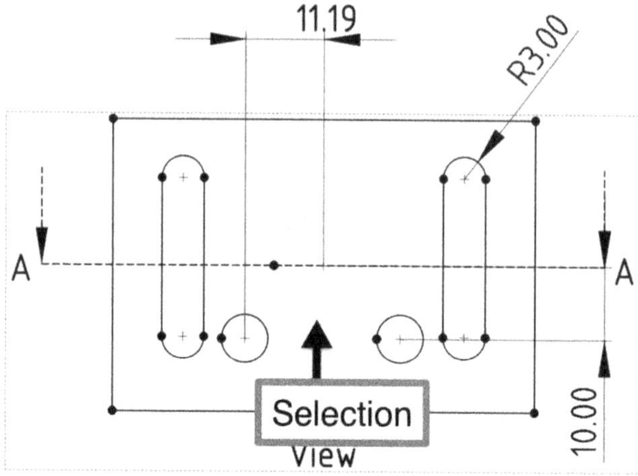

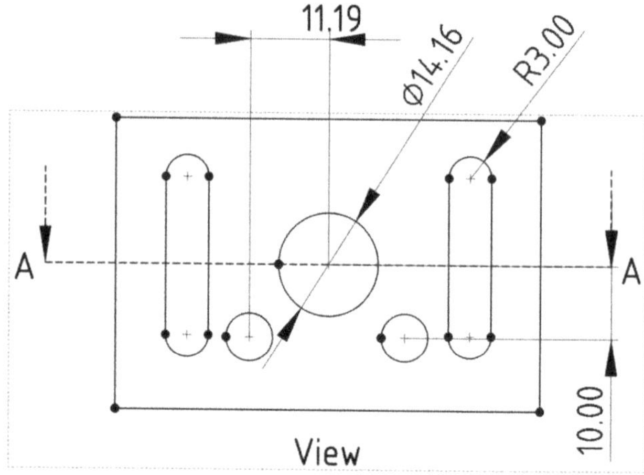

Insert Length Dimension

This command adds a linear dimension between two selected points. To do this, select linear edge (or) press and hold the CTRL key and select two points. Next, click the **Insert Length Dimension** icon on the **TechDraw Dimensions** toolbar (or) click **TechDraw > Dimensions > Insert Length Dimension** on the menu bar.

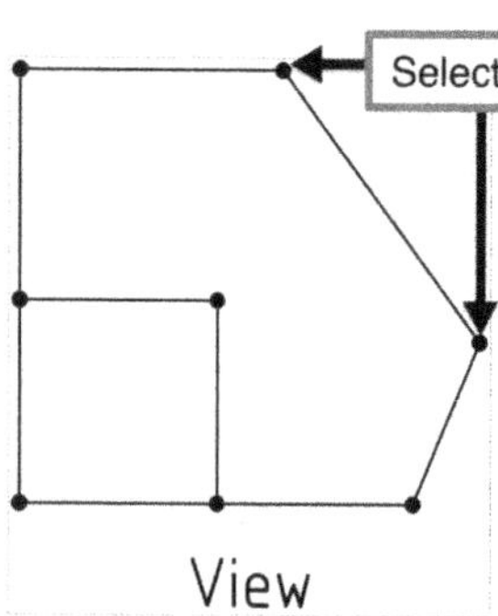

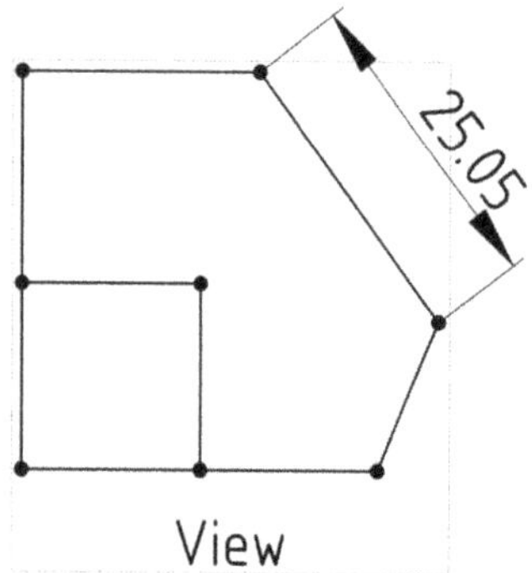

Insert Angle Dimension

This command adds an angular dimension between two selected lines. To do this, press and hold the CTRL key and select two lines. Next, click the **Insert Angle Dimension** icon on the **TechDraw Dimensions** toolbar (or) click **TechDraw > Dimensions > Insert Angle Dimension** on the menu bar.

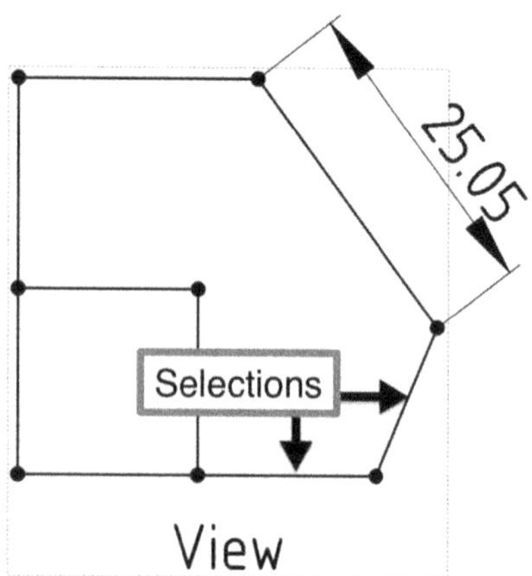

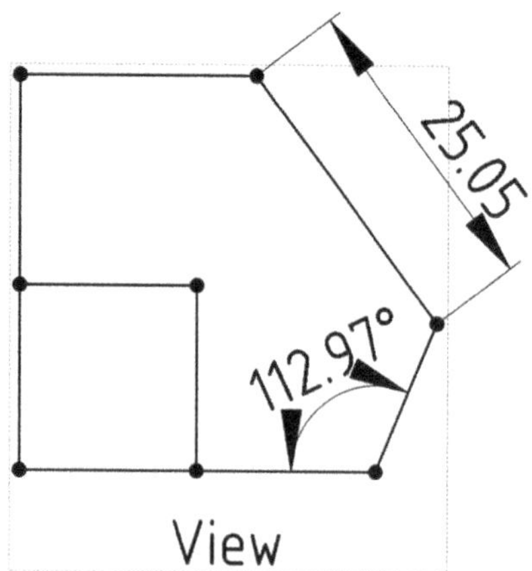

 Insert 3-Point Angle Dimension

This command adds an angular dimension between two points. To do this, press and hold the CTRL key and select first point, centerpoint, and the second point. Next, click the **Insert 3-Point Angle Dimension** icon on the **TechDraw Dimensions** toolbar (or) click **TechDraw > Dimensions > Insert 3-Point Angle Dimension** on the menu bar.

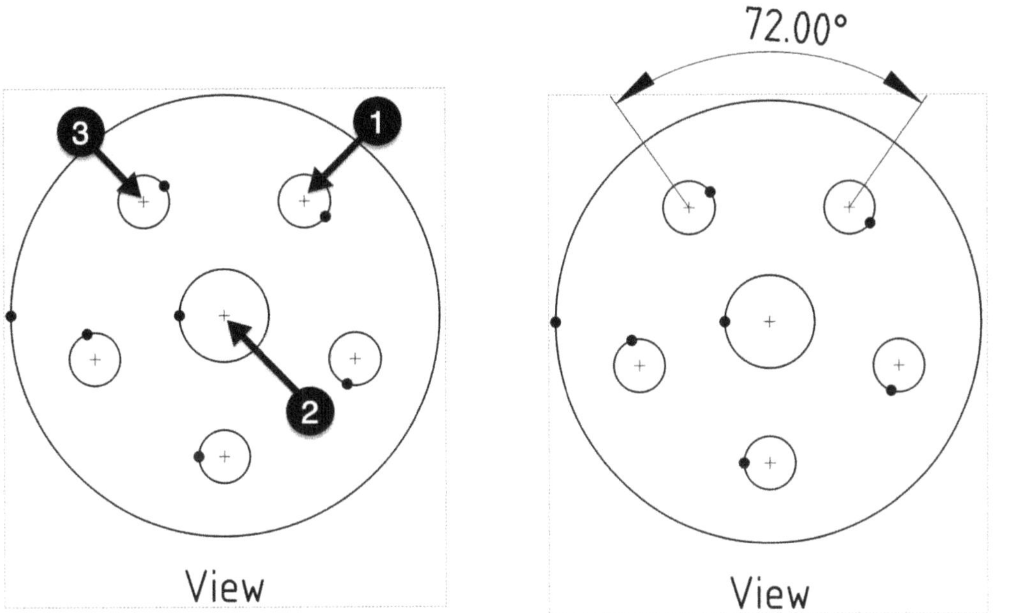

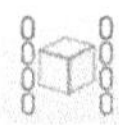 Insert Horizontal Extent Dimension

This command adds a horizontal extent dimension to the view. To do this, select the view and click the **Extent Dimensions** drop-down > **Horizontal Extent** on the **TechDraw Dimensions** toolbar (or) click **TechDraw > Dimensions > Insert Horizontal Extent Dimension** on the menu bar.

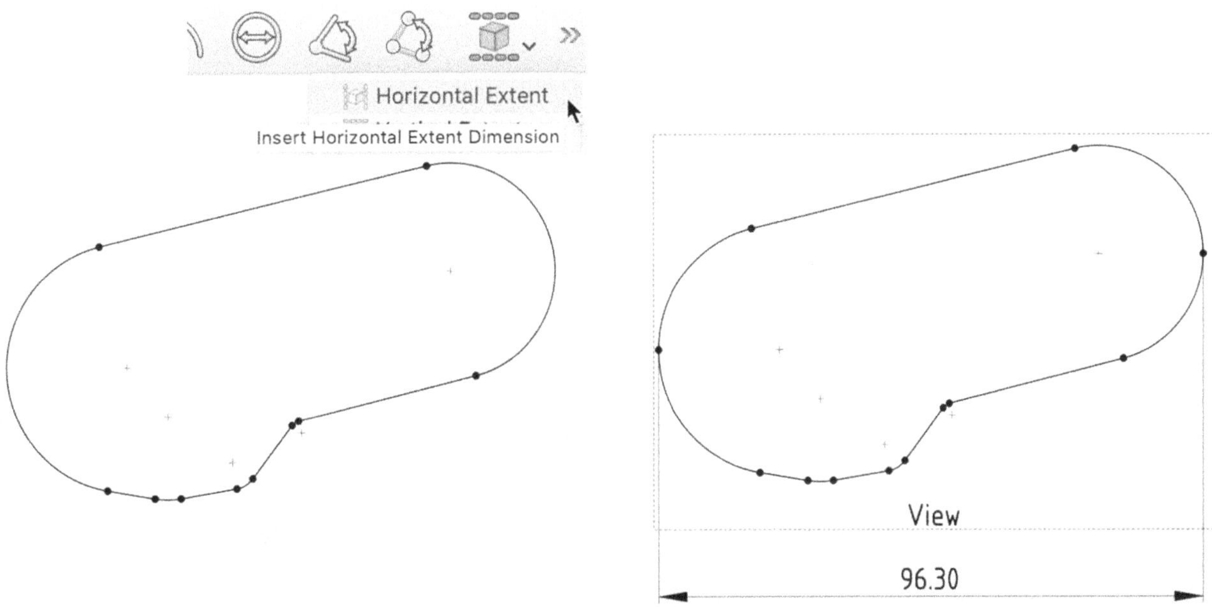

Insert Vertical Extent Dimension

This command adds a vertical extent dimension to the view. To do this, select the view and click the **Extent Dimensions** drop-down > **Vertical Extent** on the **TechDraw Dimensions** toolbar (or) click **TechDraw > Dimensions > Insert Vertical Extent Dimension** on the menu bar.

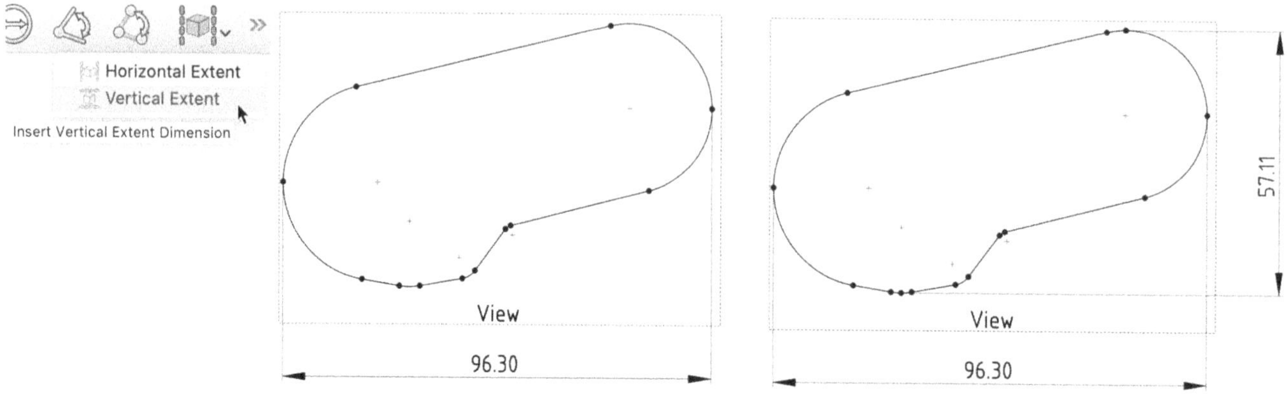

Insert Annotation

Annotation is an important part of a drawing. You add annotation to provide additional details, which cannot be done using dimensions and annotations. To add an annotation, click the **Insert Annotation** icon on the **TechDraw Annotation** toolbar (or) click **TechDraw > Annotations > Insert Annotation** on the menu bar. Next, click and drag the annotation frame to the desired location. Place the pointer in the **Text** box of the **Properties** panel. Next, click the **Annotation text** icon displayed next to it. Type the annotation text in the **List** dialog and click **OK**.

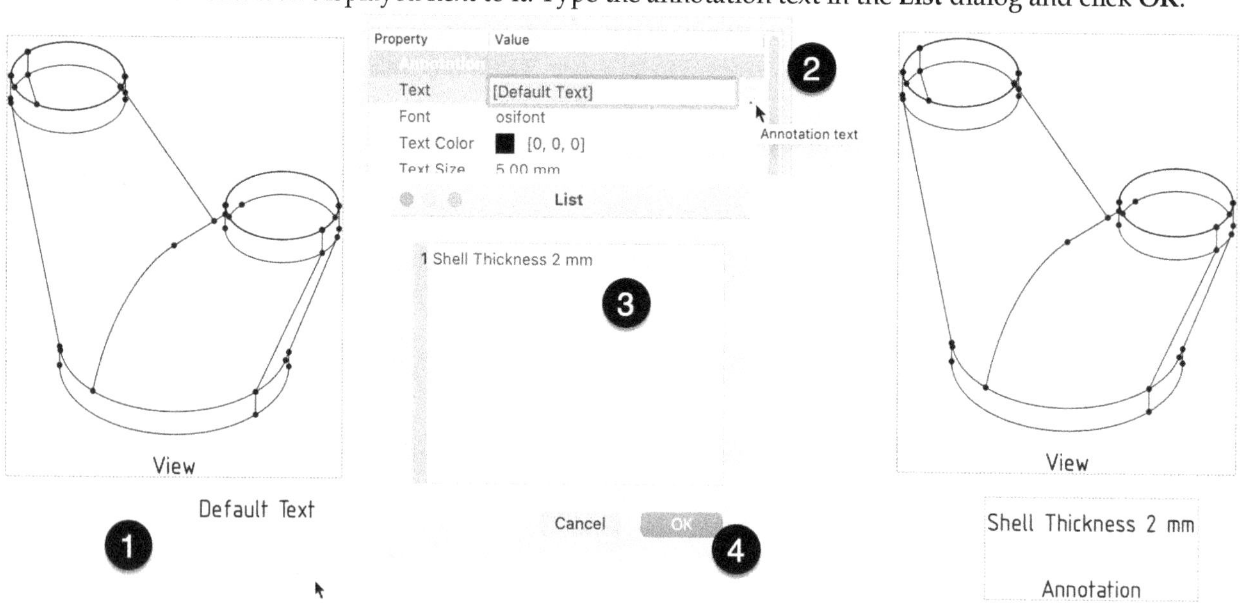

Tutorial 1

In this example, you create the 2D drawing of the part shown below.

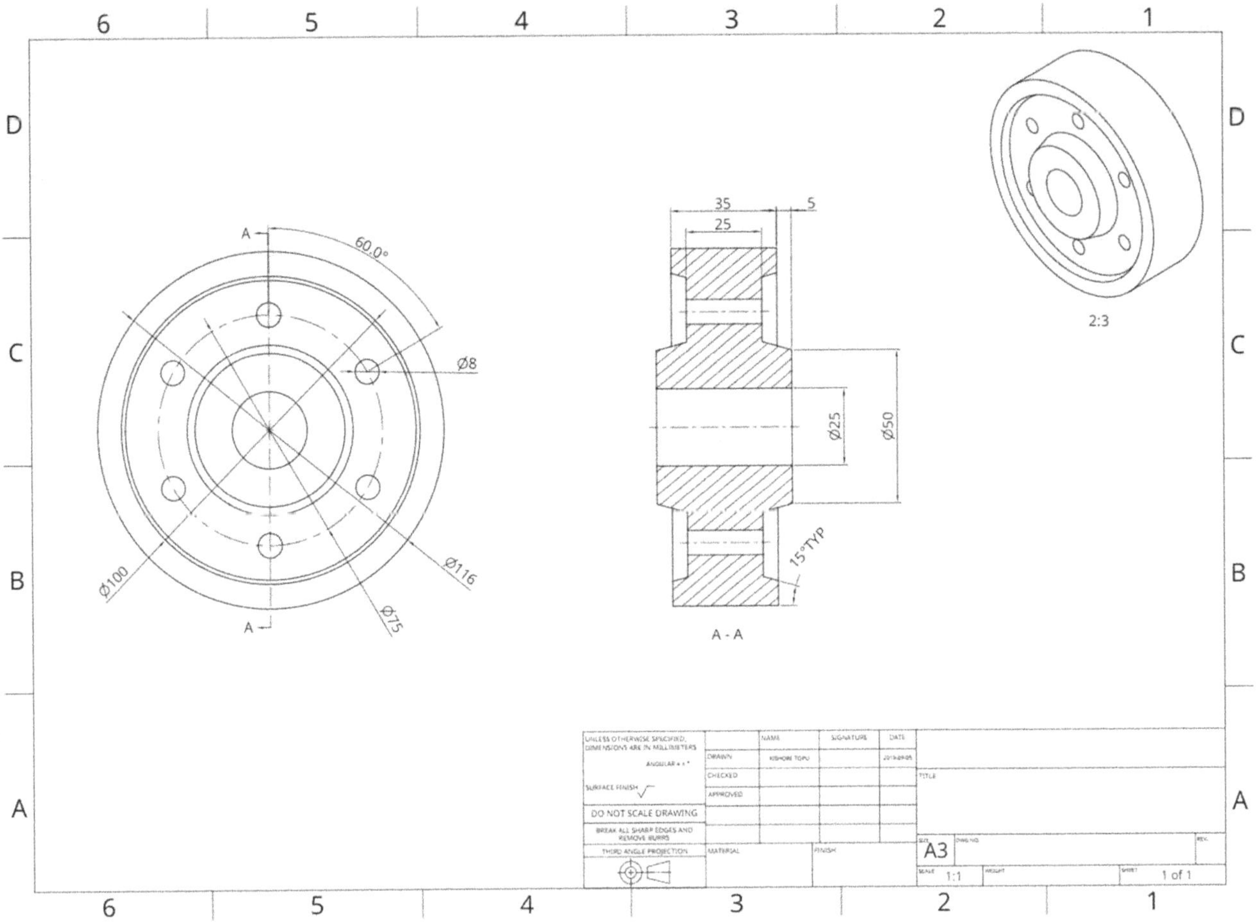

Creating a New Drawing

1. Download the **Chapter 10** part files from the companion website. Next, extract the zip file.
2. Click the FreeCAD icon on the Desktop.
3. Click **File > Open** on the Menu bar. Next, go to the location of the download files and double-click on the Tutorial1 file.
4. Select the **TechDraw** option from the **Workbench** drop-down (or) click **View > Workbench > TechDraw** on the menu bar.
5. Click **Edit > Preferences** on the menu bar. Next, select the **TechDraw** option from the left side of the **Preferences** dialog.
6. Select **Projection Group Angle > Third.**
7. Click the **Annotation** tab, and then select **Center Line Style > DashDot**. Next, click **OK**.

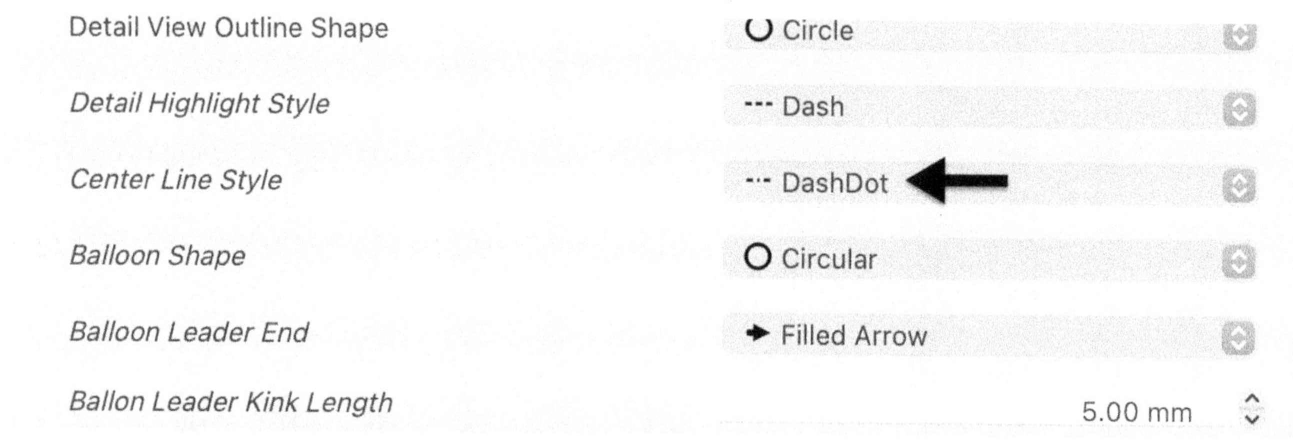

8. Click the **Insert Page using Template** icon on the **TechDraw Pages** toolbar (or) click **TechDraw > Insert Page using Template** on the menu bar.
9. Go to the Templates folder and select the **A3_LandscapeTD** template. Next, click **Open**.

Inserting the Base View

1. Click the **Tutorial 1** tab on the bottom of the window. Next, select the front face of the model.

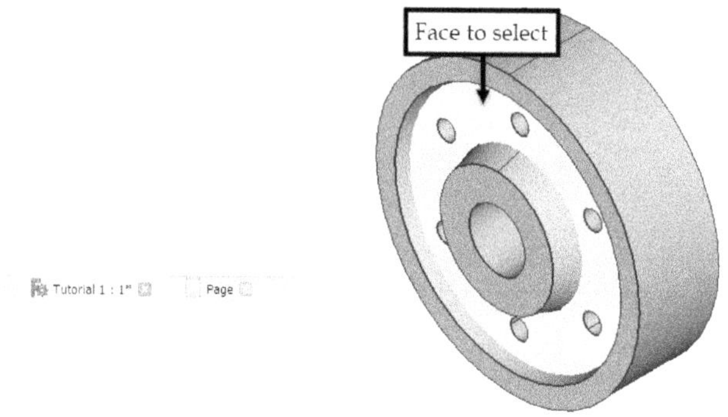

2. Click the **Page** tab on the bottom of the window.
3. To generate the base view, click the **Insert View** icon on the **TechDraw Views** toolbar.
4. Click on the dotted borderline of the view. Next, press and hold the left mouse and drag the view to left.

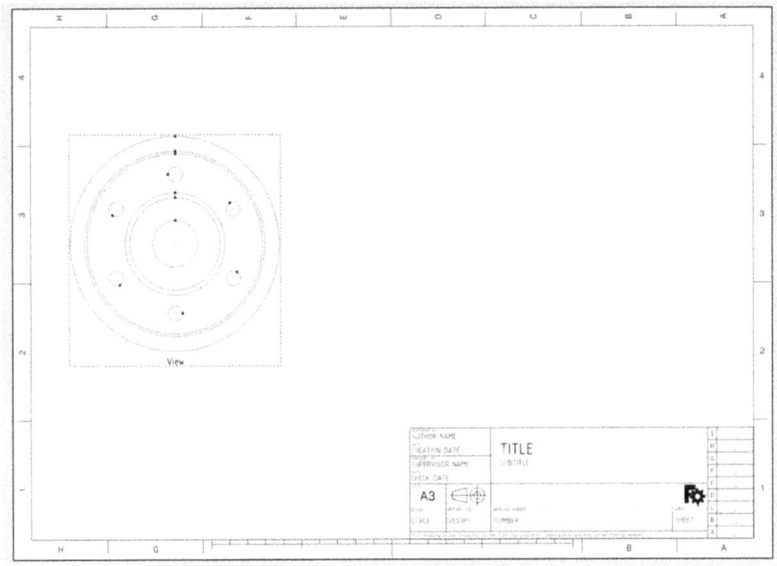

Generating the Section View

1. Select the base view.
2. Click the **Insert section view** icon on the **TechDraw Views** toolbar.
3. Click the **Looking left** icon on the **Section Orientation** section.
4. Type **0** in the **X**, **Y**, and **Z** boxes, respectively. These values define the location of the section plane.
5. Type **A** in the **Identifier** box and click **OK** to create the section view.
6. Drag the section view and position it correctly. Make sure that it is horizontally in-line with the base view.

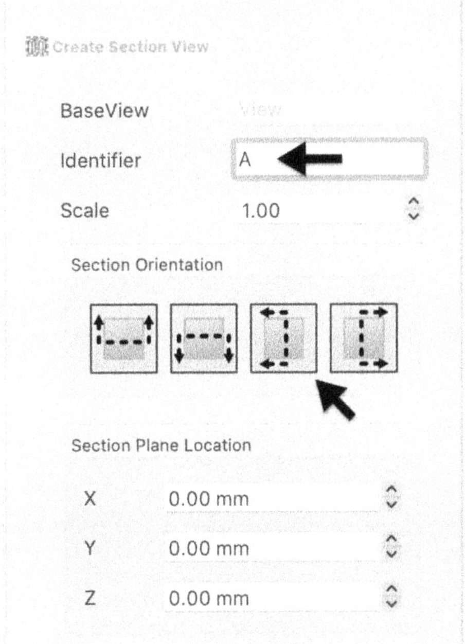

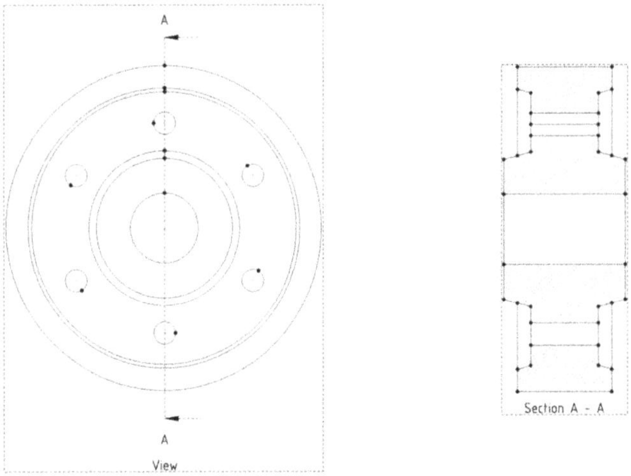

Inserting the Isometric View

1. In the **Combo View** panel, select the **Body** from the **Model** tab.

2. Click the **Insert View** icon on the **TechDraw Views** toolbar; the isometric view of the model is displayed on the drawing page.
3. Select the isometric view and click the **Data** tab on the **Property** panel.
4. Scroll to the **Base** section and change the **Scale** value to 0.5.
5. Click and drag the isometric view to the top-right corner.

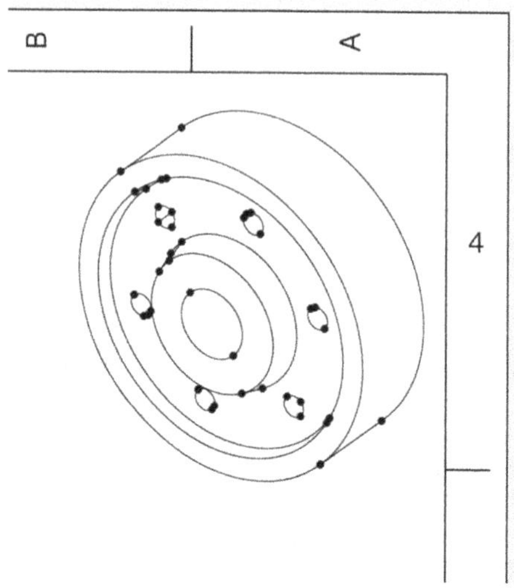

Adding the center lines to the drawing view

1. Select the base view from the drawing page. Next, click the **View** tab on the **Property** panel.
2. Scroll to the **Decoration** section and select **Horiz Center Line > true**.

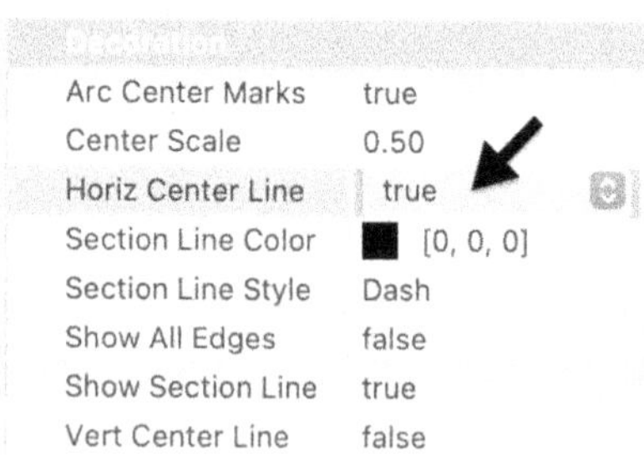

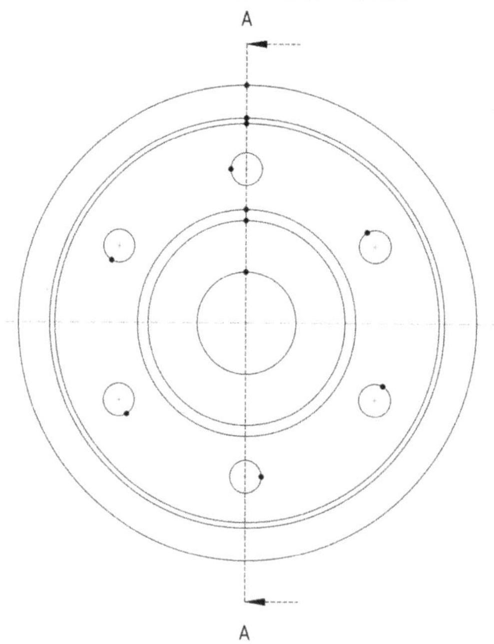

3. Select the section view from the drawing page. Next, click the **View** tab on the **Property** panel.
4. Scroll to the **Decoration** section and select **Horiz Center Line > true**.

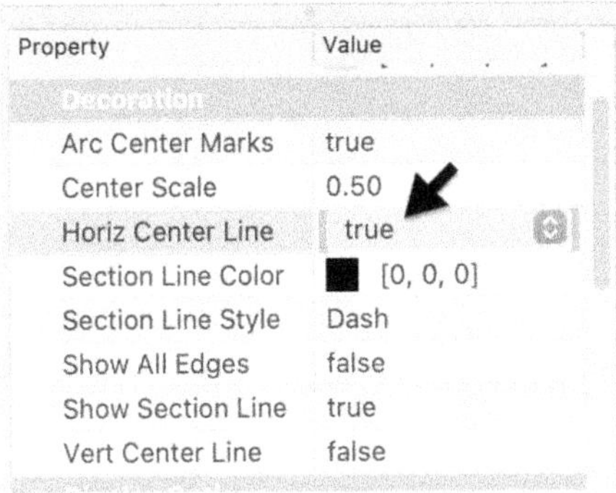

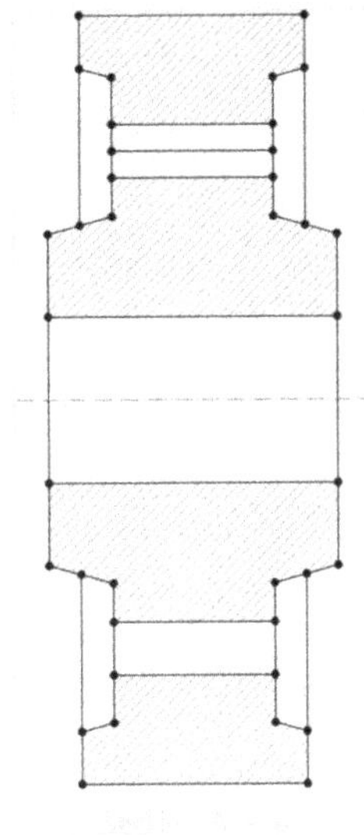

Adding Dimensions

Now, you add dimensions to the drawing.

1. Select the outer circular edge of the base view. Next, click the **Insert Diameter Dimension** icon on the **TechDraw Dimensions** toolbar (or) click **TechDraw > Dimensions > Insert Diameter Dimension** on the menu bar.
2. Click and drag the dimension outside the view.

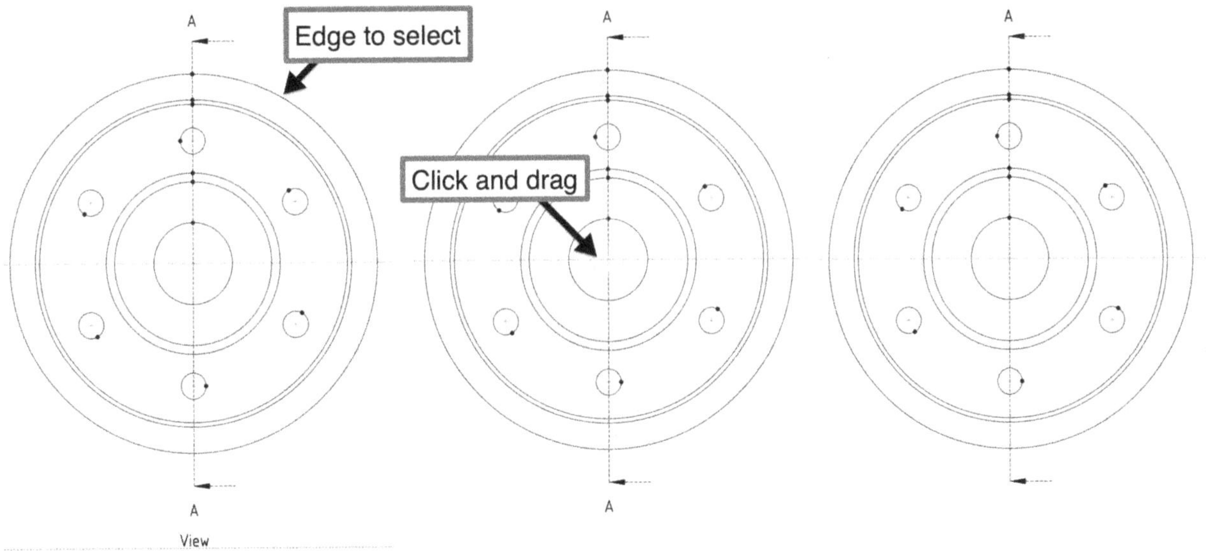

3. Likewise, add remaining dimensions to the base view, as shown.

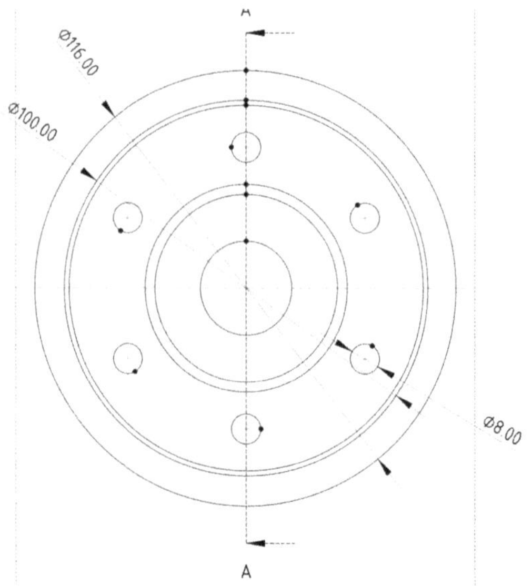

4. Press and hold the CTRL key (COMMAND key for Mac users) and select the three diameter dimensions.
5. Click the **View** tab on the **Property** panel. Next, select **Standard and Style > ASME Referencing**.
6. Select **Rendering Extent > Minimal**.

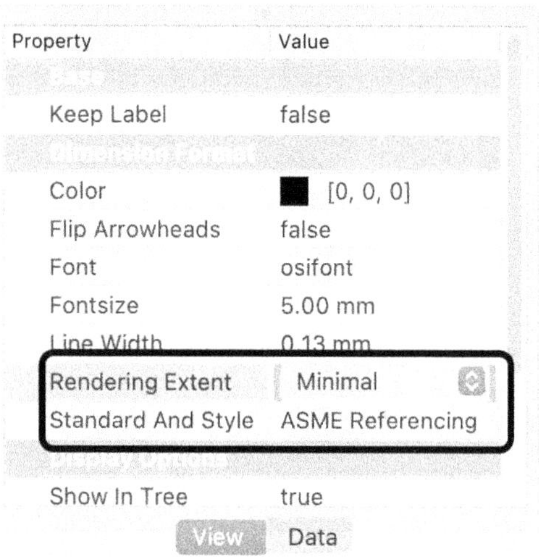

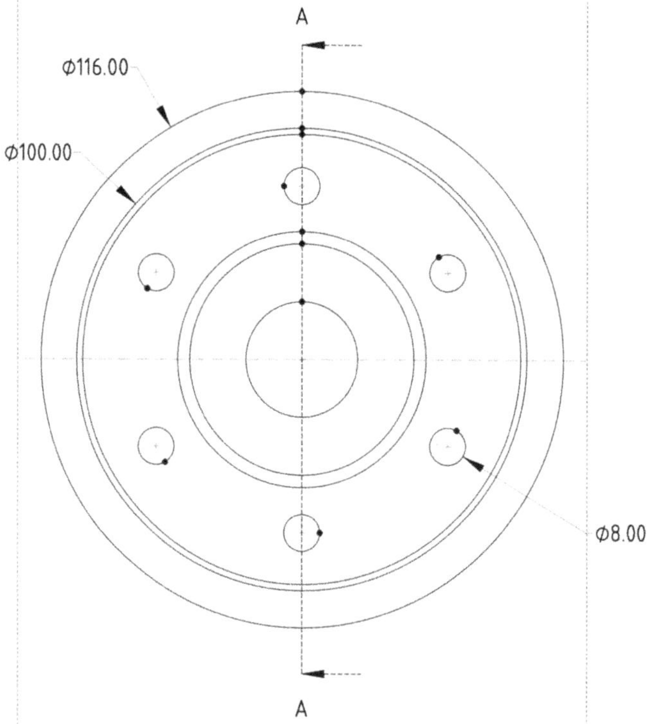

7. Select the diameter dimension of the small hole. Next, click the **Data** tab on the **Property** panel.
8. Click in the **Format Spec** box, as shown. Next, type **6 Holes** and press the SPACEBAR.
9. In the **Format Spec** box, click next to the diameter value and press the SPACEBAR.
10. Type **Equi-spaced on 75 PCD**. Next, click and drag the dimension.

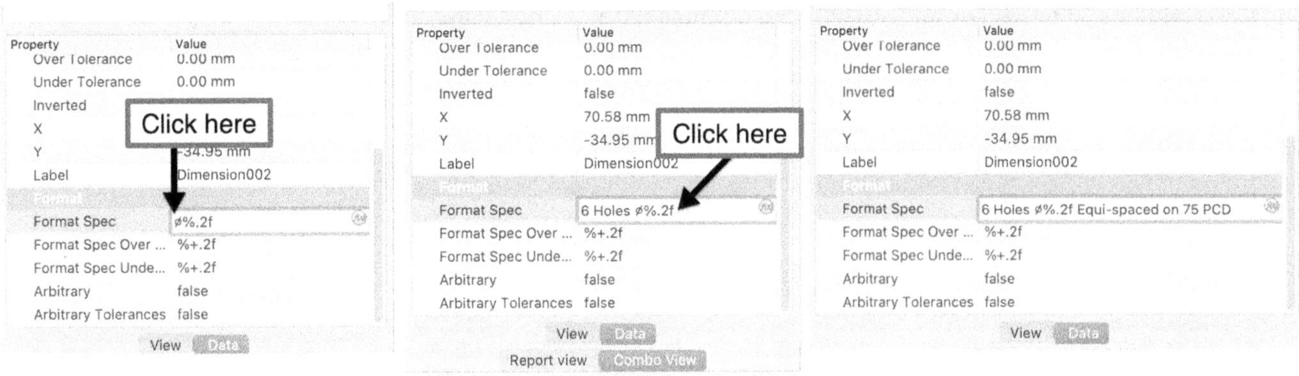

11. Press and hold the Ctrl key and select the vertices of the section view, as shown.
12. Click the **Insert Horizontal Dimension** 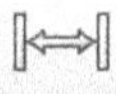 icon on the **TechDraw Dimensions** toolbar.
13. Click and drag the dimension upward.

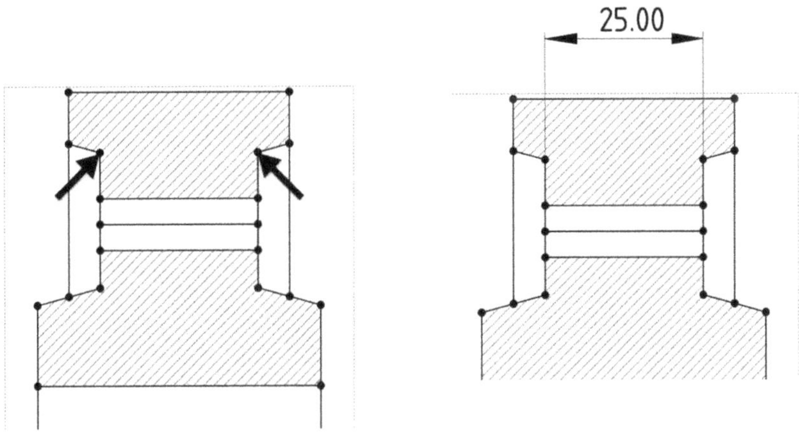

14. Press and hold the Ctrl key and select the vertices of the section view, as shown.
15. Click the **Insert Horizontal Dimension** 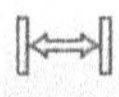 icon on the **TechDraw Dimensions** toolbar.
16. Likewise, create another horizontal dimension.
17. Drag the dimensions upward, as shown.

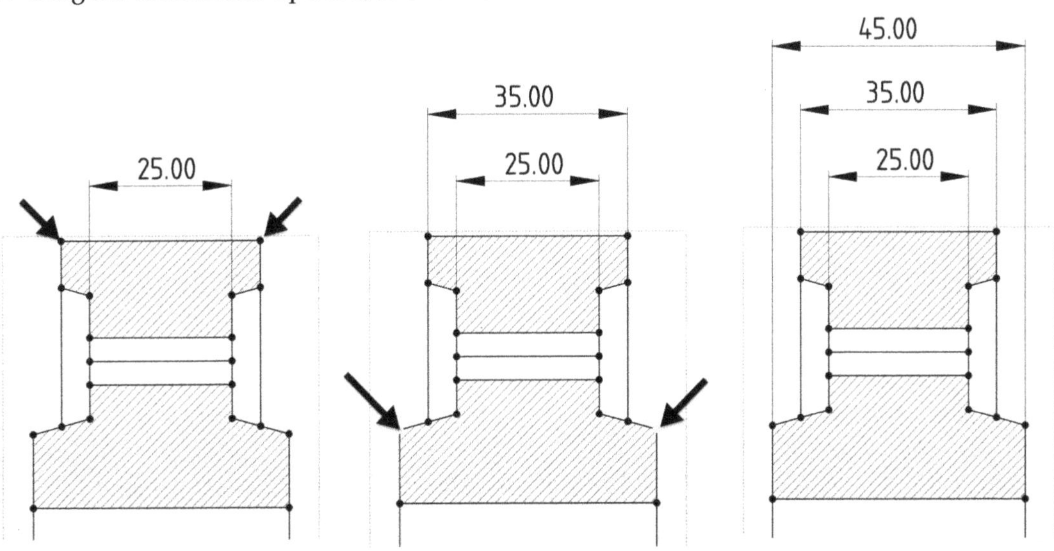

18. Press and hold the Ctrl key and select the two points of the section view, as shown.
19. Click the **Insert Vertical Dimension** icon on the **TechDraw Dimensions** toolbar.
20. Likewise, create another horizontal dimension.
21. Drag the dimensions toward the right, as shown.

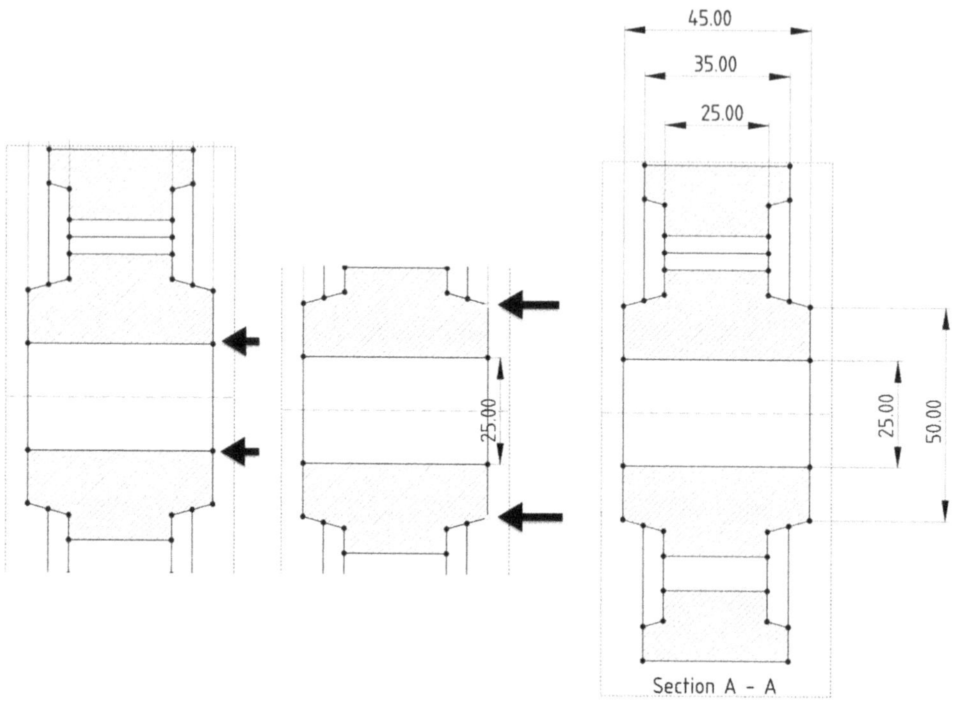

22. Press and hold the Ctrl key and select the inclined and horizontal edges of the section view, as shown.
23. Click the **Angle Dimension** icon on the **TechDraw Dimensions** toolbar.
24. Select the angular dimension value. Press and hold the left mouse button and drag the pointer toward left. Position the dimension, as shown.

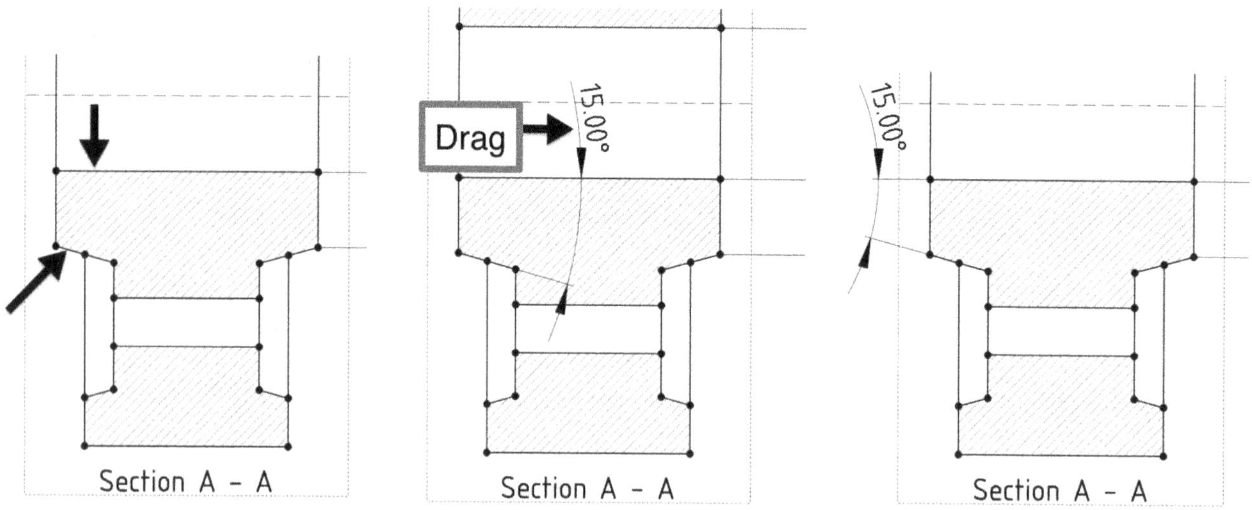

25. Select the angular dimension from the section view. Next, click the **Data** tab on the **Property** panel.
26. In the **Format Spec** box, click next to the dimension value and press the SPACEBAR. Next, type **TYP**.

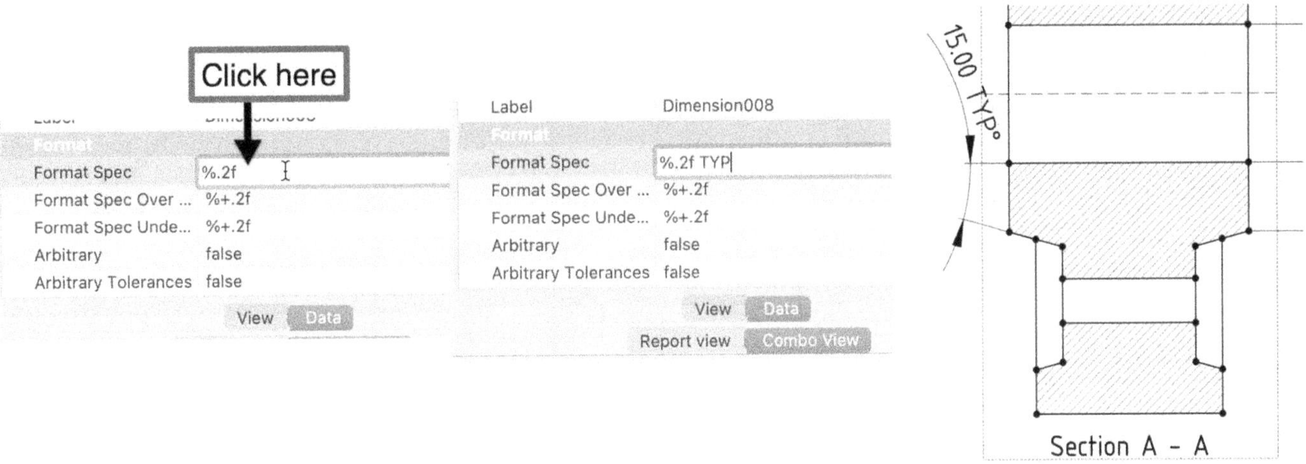

27. Select any one of the diameter dimension from the base view. Next, click the **Data** tab on the **Property** panel
28. Select the diameter symbol from the **Format Spec** box. Next, right click and select **Copy**.

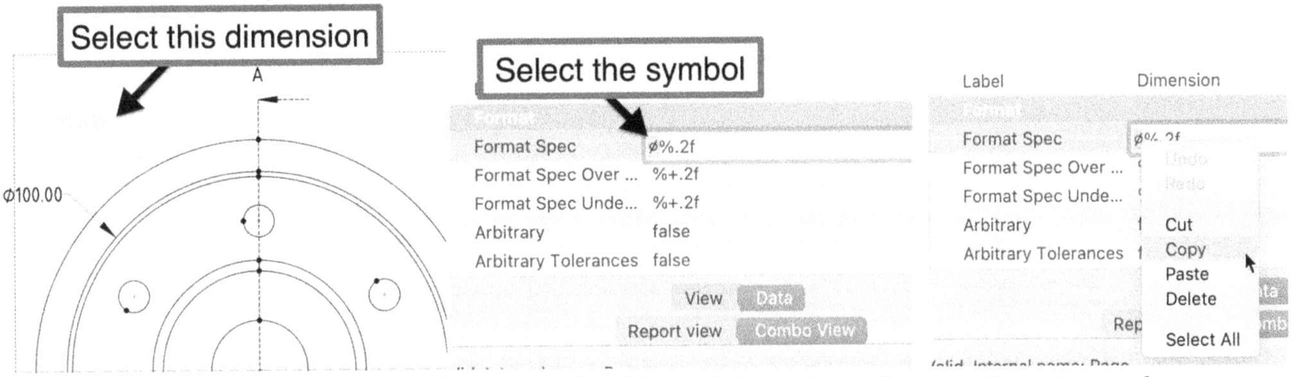

29. Select the vertical dimension from the section view. Next, click the **Data** tab on the **Property** panel
30. In the **Format Spec** box, click before the dimension value. Next, right click and select **Paste**; the diameter symbol is pasted.
31. Click on the drawing sheet to update the dimension.

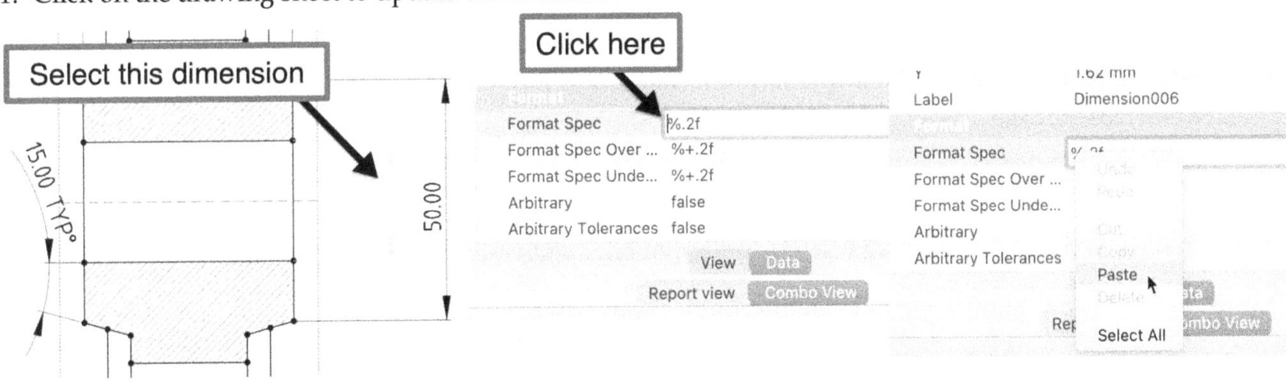

32. Likewise, add the diameter symbol to the vertical dimension with the value 50.

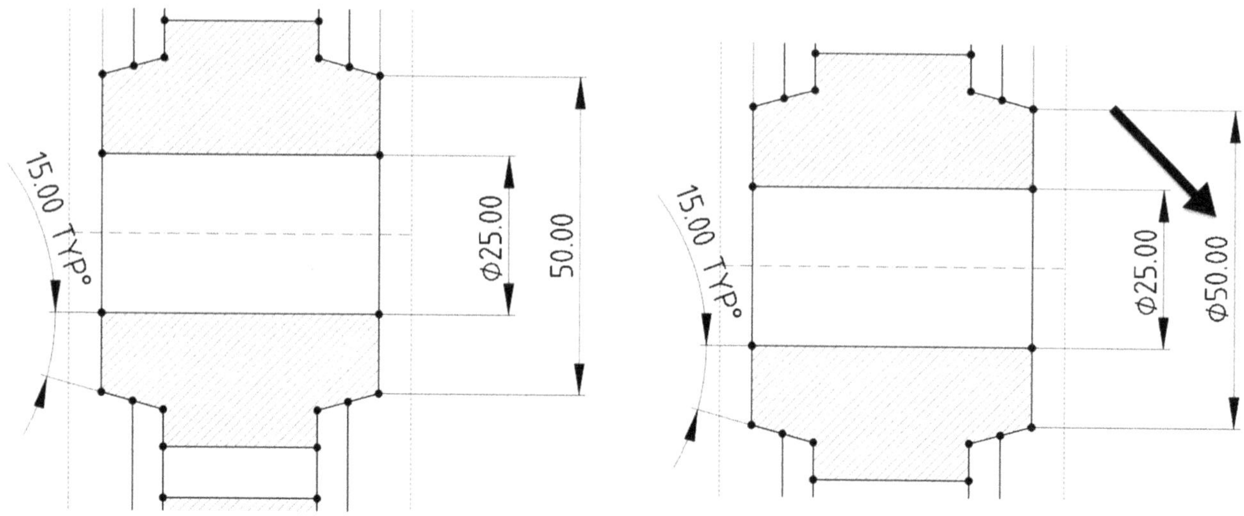

Populating the Title Block

1. Zoom in to the title block area. Next, double-click on the green square displayed on TITLE.
2. Type **Ch10_Tutorial_1** in the **Value** box, and then click **OK**.

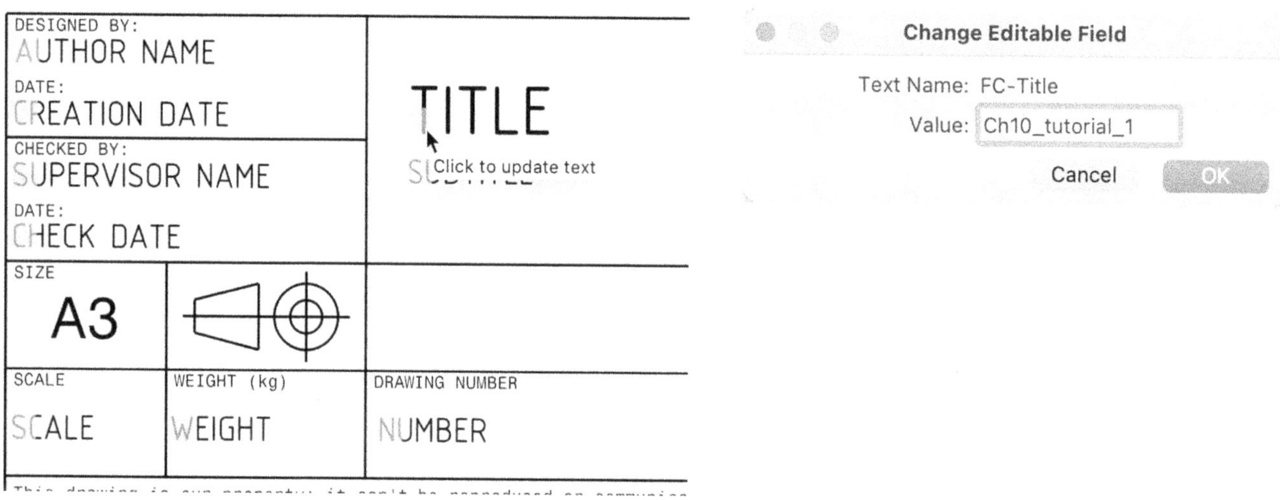

3. Likewise, add data to the remaining fields in the Title block.
4. Save and close the file.

Questions

1. How to create drawing views from an existing part or assembly file?
2. List the commands used to create centerlines and center marks.
3. How to add symbols and texts to a dimension?
4. How to create section views?

Exercises

Exercise 1

Create orthographic views of the part model shown below. Add dimensions and annotations to the drawing.

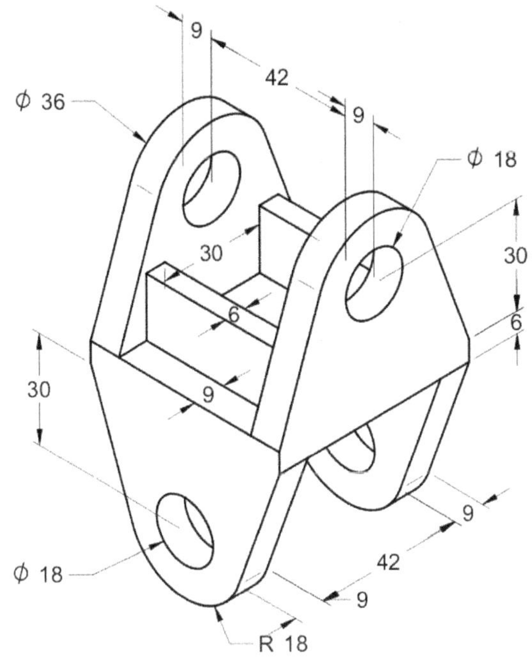

Exercise 2

Create orthographic views and an auxiliary view of the part model shown below. Add dimensions and annotations to the drawing.

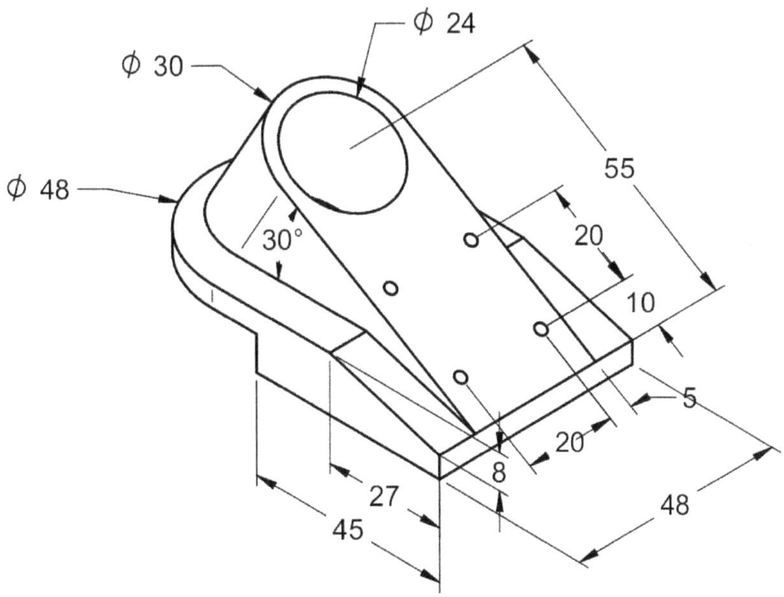

Lightning Source UK Ltd.
Milton Keynes UK
UKHW032117230222
399149UK00003B/152

9 798201 306939